Earthquake Hazard and Risk

Advances in Natural and Technological Hazards Research

VOLUME 6

The titles published in this series are listed at the end of this volume.

Earthquake Hazard and Risk

Edited by

VLADIMÍR SCHENK

Institute of Rock Structure and Mechanics,
Academy of Sciences,
Prague, Czech Republic

KLUWER ACADEMIC PUBLISHERS

DORDRECHT / BOSTON / LONDON

A C.I.P. Catalogue record for this book is available from the Library of Congress

ISBN 0-7923-4008-6

Published by Kluwer Academic Publishers,
P.O. Box 17, 3300 AA Dordrecht, The Netherlands.

Kluwer Academic Publishers incorporates
the publishing programmes of
D. Reidel, Martinus Nijhoff, Dr W. Junk and MTP Press.

Sold and distributed in the U.S.A. and Canada
by Kluwer Academic Publishers,
101 Philip Drive, Norwell, MA 02061, U.S.A.

In all other countries, sold and distributed
by Kluwer Academic Publishers Group,
P.O. Box 322, 3300 AH Dordrecht, The Netherlands.

Printed on acid-free paper

TABLE OF CONTENTS

Preface vii

Part 1: Seismicity, Stress Fields and Focal Mechanism

G. ALESSIO, A. GORINI, G. VILARDO, G. IANNACCONE / Low Energy Sequences in Areas With High Seismic Potential: Benevento (Southern Apennines), April 1990 3

V. PEÇI / The Probability of the Time Occurrence of Strong Earthquake for some Zones in Albania 17

B. OSHER / Statistical Estimation of the Maximum Magnitude and Its Uncertainty from a Catalogue Including Magnitude Errors 25

J. LAPAJNE, B. ŠKET MOTNIKAR / Estimation of Upper Bound Magnitude in Earthquake Hazard Assessment in Slovenia 39

A. V. SOLONENKO, N. V. SOLONENKO, V. I. MELNIKOVA, E. A. SHTEIMAN / The Analysis of the Spatial-Temporal Structure of Seismicity in the Baikal Rift Zone 49

YU. P. SKOROVODKIN, T. V. GUESEVA, L. S. BEZUGLAYA / Changes of the State of Stress and Deformation in Epicentral Zones During the Aftershock Period 63

S. G. TYULENEVA, S. L. YUNGA / Dispersion and Probability Density Function for Focal Mechanism Tensors 77

Part 2: Earthquake Hazard and Seismic Risk

D. SLEJKO (Coord. CNR-GNDT Working Group on Hazard Assessment) / Preliminary Seismic Hazard Assessment for the Italian Seismic Code 87

V. SCHENK, Z. SCHENKOVÁ, P. KOTTNAUER / Earthquake Hazard Assessment for the Czech Republic and Adjacent Area 125

M. MUCCIARELLI, D. ALBARELLO, M. STUCCHI / Sensitivity of Seismic Hazard Estimates to the Use of Historical Site Data 141

D. GIARDINI, E. BOSCHI / Recent Trends in Regional and Global Seismic Hazard Assessment 153

K. ATAKAN, M. R. KARPUZ, S.O. DAHL / Importance of Geological Data in Probabilistic Seismic Hazard Assessments: A Case Study from Etne, Western Norway 169

Z. YUEQING, A. VOGEL, H. PING / Expert System for Earthquake Hazard Assessment: ESEHA 199

C. YONG, CH. XINGLIAN, F. ZHENGXIANG, Y. ZHIQIAN, Y. MANDONG / Estimating Losses from Future Eartquakes in China 211

G. A. PAPADOPOULOS, A. ARVANITIDES / Earthquake Risk Assesment in Greece 221

G. SOBOLEV, YU. TYUPKIN, I. FROLOV, N. SERGEYEVA, S. AREFEV, A. GOVOROV, E. ROGOZHIN, R. TATEVASYAN, M. ZHIZHIN, YU. SCHUKING, P. LOCKRIDGE, L. WHITESIDE, R.E. HABERMANN, H. MEYERS, A. MALITZKY / The Optical Disk "Spitak Earthquake of 1988" 231

Part 3: Strong Ground Motion and Microzonation

K. MAKROPOULOS, D. DIAGOURTAS, N. VOULGARIS, J. DRAKOPOULOS / A Study of Site Effects on Strong Motion Records Obtained from Local Network Around Gulf of Corinth (Central Greece) 241

A. MENDEZ, F. PACOR, R. BERNARDI, C. PETRUNGARO / Strong Motion Modelling at a Near-Source Site in Avezzano, Italy 253

J. AKAMATSU, H. YASUI, K. NISHIMURA, M. KOMAZAWA / Application of Microseisms to Microzoning of an Urban Area in a Sedimentary Basin 267

A. ROMAN, V. ALCAZ / Peculiarities of Ground Seismic Reaction in Kishinev, Moldova 281

PREFACE

Earthquakes still constitute a major problem to many countries because of the natural disasters they produce, which frequently result in casualties and destructions of buildings, bridges, plants and other vitally important structures. In recent years, the growth of population and the scatter of settlements over the hazardous areas have aggravated the impact of earthquake disasters. Within the International Association for Seismology and Physics of Earth's Interior (IASPEI), in its Sub-Commission of Earthquake Hazard, a unique opportunity of international scientific discussions has been provided for a long time concerning the topics mentioned. The open discussions help critically evaluate new methodological procedures for improving the earthquake hazard evaluation and seismic risk assessment. Such international contacts are also an efficient tool for a transfer of technologies to developing countries.

The Symposium "Earthquake Hazard and Risk" was held during the 27th IASPEI General Assembly in Wellington (New Zealand), on January 11 and 12, 1994. The coconveners of the Symposium were G.A.Sobolev (Russia), A.Shapira (Israel), and Li-Li Xie (China). Thirty-six contributions were accepted for oral presentations and thirty-two for posters. Since within the IASPEI General Assembly no special floors were devoted to the problems of seismicity, seismotectonics, and other related problems of the main topics of the Symposium, such as for example strong ground motions, microzonations and applications in engineering seismology, the scope of the Symposium had to be extended for such contributions. Authors who agreed to publishing their contributions sent the papers to the Editor of these Proceedings who organized for each paper a reviewing by two reviewers. Then according to their recommendations and additional comments authors were responsible for the preparation of camera-ready papers which are included in the Proceedings. Altogether nineteen papers passed successfully the reviewing and were divided with respect to the solved problems into three parts:

- Seismicity, Stress field and Focal Mechanism,
- Earthquake Hazard and Seismic Risk, and
- Strong Ground Motions and Microzonation.

Two papers of the first part deal with a standard investigation of the regional seismicity pattern: *Alessio et al.* describe the seismic sequence of the April 1990 Benevento province and compare the results with other 160 earthquake events which were located within the area of study. *Peçi* investigates independent strong earthquakes ($M \geq 5.9$) having occurred in the main seismogenic zones of Albania in the period 1901-1988. The other two papers of the first part, delivered by *Osher* and *Lapajne & Motnikar*, deal with the same problem: evaluation of the maximum possible earthquake. *Solonenko et al.* analyze stress field round the Baikal Rift Zone and its adjacent areas (Mongolia, South Yakutia) on the basis of focal mechanisms of more than 3000 earthquakes occurred during 1950-1990 period. *Skorovodkin et al.* demonstrate an importance of aftershock instrumental observations succeeded strong earthquakes (examples from Uzbekistan, Georgia, Iran) for a study of seismotectonic processes. *Tyuleneva et al.* describe a procedure of statistical analysis of a set of stochastic second symmetric focal mechanism tensors, their dispersion and probability density function.

The second part of the Proceedings contains eight contributions directed to the main topics of the Symposium. *Slejko* (Coordinator of the Italian WG on Hazard Assessment) presents the recent stage in investigations of the earthquake hazard assessment in Italy. *Schenk et al.* deliver a review of the same problem for the Czech Republic; the paper includes zoning maps and the last versions of the hazard calculations in which the effects of regional sedimentary covers are involved. *Mucciarelli et al.* describe comparative studies in which earthquake hazard estimates are assessed by the use of historical data for three Italian sites. *Giardini & Basham* inform on the progress in the Global Seismic Hazard Assessment Program (GSHAP), proposed by the International Lithospheric Program (ILP) under the sponsorship of the International Council of Scientific Unions (ICSU) in 1992. *Atakan et al.* apply available neotectonic studies and related tectonic data known from the Sunnhordland District (Norway) for seismic hazard estimates in intraplate areas. *Yueqing et al.* describe an expert system for earthquake hazard assessment (ESEHA) based on an application of artificial intelligence techniques. The paper written by *Chen Yong et al.* describes an estimation of losses from earthquakes expected for the future 50 years in China. *Papadopoulos & Arvanitides* attempts to establish an "organized antiseismic strategy" in Greece; examples for highly populated regions of Athens and Thessaloniki are presented. *Sobolev et al.* introduce a complete set of accumulated observational data on the Spitak (Armenia) earthquake of 1988 available on a CD-ROM for standard PC.

The latter part of the Proceeding is devoted to strong ground motion studies and microzonations. *Makropoulos et al.* present three-component accelerograms from 130 events recorded by the strong motion network RASMON installed in July 1991 around the western part of Gulf of Corinth (Central Greece). *Mendez et al.* use possible rupturing scenarios of the Serrone fault in an area of the 1915 Avezzano earthquake (M=6.9) and express the earthquake hazard in broad-band synthetic waveforms. *Akamatsu et al.* apply a microzonation method which spectral ratios of microseisms obtained by means of simultaneous observations in the Kyoto basin, south-west Japan. *Roman & Alkaz* use seismic wave recordings of the Carpathian intermediate earthquakes with M=4.5 to 7.2 recorded in Kishinev (Moldavia) and macroseismic investigations for a definition of the Fourier and response spectra of possible seismic vibrations expected for future earthquakes.

I am confident that the papers published in the Proceedings will help in both methodological applications of the earthquake hazard and seismic risk calculations and in setting forth a strategy for future projects aimed at assessment and mitigation of hazards over wide regions.

From the editoral viewpoint I wish to express my acknowledgements to all reviewers for their efforts and collaboration. I thank Michal Jechumtál and Zuzana

Schenková for assistance in reformatting and rewriting of some manuscripts. Great thanks belong to numerous anonymous referees for careful comments on individual papers. My thanks are also due to the whole staff of the Kluwer Scientific Publishers, namely to Petra van Steenbergen for her kind understanding of editorial problems.

Vladimír Schenk

Convener of the Symposium,
Chairman of the IASPEI Sub-Commission
"Earthquake Hazard and Risk"

Institute of Rock Structure and Mechanics
Academy of Sciences of the Czech Republic,
V Holešovičkách 41, 182 09 Praha 8 - Libeň
Czech Republic
fax: +42-2-6880105
e-mail: schenk@lorien.site.cas.cz

Part 1:

SEISMICITY, STRESS FIELDS AND FOCAL MECHANISM

LOW ENERGY SEQUENCES IN AREAS WITH HIGH SEISMIC POTENTIAL: BENEVENTO (SOUTHERN APENNINES), APRIL 1990.

GIULIANA ALESSIO, ANTONELLA GORINI*, GIUSEPPE VILARDO
Osservatorio Vesuviano - Via Manzoni, 249 - 80123 Napoli, Italy

GIOVANNI IANNACCONE
Dipartimento di Geofisica e Vulcanologia - Largo S. Marcellino, 10 - 80138 Napoli, Italy

Abstract. In the southern sector of the Apennines chain (Italy), destructive earthquakes with medium-high magnitude occurred within a narrow belt along the axis of the chain. Low-energy seismic swarms have frequently been observed.

The Sannio area and particularly the Benevento province, was hit by strongly destructive events of I ≥ IX MCS (1456, 1688, 1702, 1732, 1930, 1962), as well as by seismic swarms (1885, 1903, 1905).

A seismic sequence started in April 1990 near the town of Benevento. This sequence was studied in detail through the analysis of data recorded by temporary seismic networks installed in the area and operating for about two months. Over 300 microearthquakes were located in a small area of about 100 km^2, close to the town of Benevento. Cross sections of hypocenters indicate that the depths of the events were concentrated within the first 15 km of the crust.

The maximum seismic energy was released at the beginning of the sequence, with the highest magnitude M_L=3.6. Most events were felt by the population, but they caused no damage.

Fault plane solutions calculated for a selected number of events show that most of the mechanisms have T axes oriented in an anti-apenninic direction (NE-SW).

In this paper a small sequence is analyzed in order to evaluate its relations with the strongest earthquakes and the tectonic lineaments associated with them, for seismic hazard assessment.

Key words. Southern Apennines, seismicity, seismic hazard

1. Introduction

The April 1990 seismic sequence occurred in the Southern Apennines near the town of Benevento, about 50 km to the NE of Naples (Italy).

This paper reports a detailed study of the whole sequence (April-July 1990) through the analysis of data from two local seismic networks.

* now at *Servizio Sismico Nazionale - Via Curtatone, 8 - 00187 Roma, Italy*

V. Schenk (ed.), Earthquake Hazard and Risk, 3–16.

The local stations, together with the regional seismic network of the Osservatorio Vesuviano recorded about 1000 microearthquakes. The seismic energy released by this sequence was very low since the maximum magnitude was 3.6 and only 18 other events had magnitudes greater than 3.0. The cumulative energy of the whole sequence was equivalent to that of one shock with magnitude 4.0.

Due to the proximity of the epicentral area to the town of Benevento, all the strongest events were felt by the population, without causing any relevant damage.

In spite of its very low energy, the 1990 Benevento seismic sequence appears important for understanding the Southern Apennines pattern of seismicity. In fact it occurred in a region hit by highly destructive events in historical times, and at present characterized by lack of high-energy seismicity (Scarpa & Zollo, 1985).

2. Geological Outline

The study area is located in the hinge zone of the southern Apennines chain, between a western sector with huge carbonate massifs (the Taburno-Camposauro mountains and, northwards, the Matese massif), and an eastern sector where terrigenous deposits form a hilly landscape (geological map of Fig. 1).

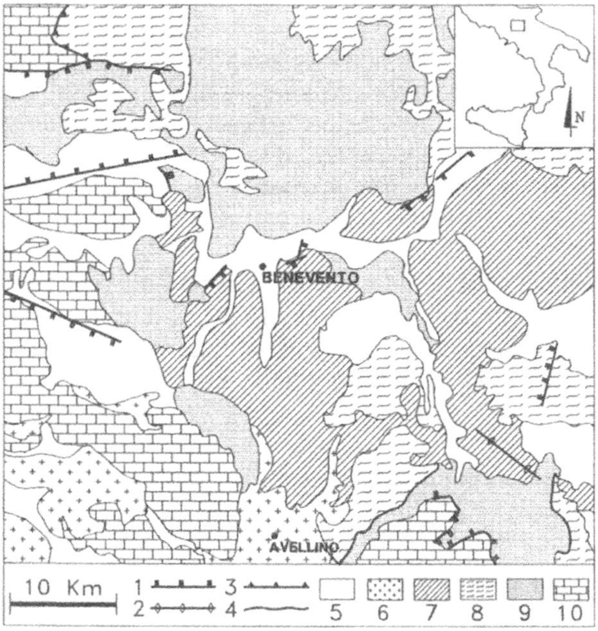

Fig. 1: Simplified geological map of the Benevento province (after C.N.R.-P.F.G., 1992a). Legend : 1= Normal faults; 2 = Axis of anticline; 3 = Contact between groups of different tectonic units; 4 = Undeterminated tectonic contacts; 5 = Alluvial and lacustrine deposits (Holocene-Upper Pleistocene); 6 = Volcanic rocks (mostly Pleistocene); 7 = Terrigenous marine deposits with subordinate limestones, evaporities (Middle Pliocene-Upper Miocene); 8 = Chaothic complex of variegated clays and shales (Eocene-Upper Cretaceous); 9 = Terrigenous resediments including chaotic clays, olistolites of limestones and calcareous turbidites (Tortonian-Langhian), resedimentated quartzarenites (Lower Miocene); 10 = Shallow-water limestones and dolomites, marly and terrigenous deposits (Miocene-Upper Triassic).

In the Benevento area, marine clastic successions and evaporites which were deposited during the late Miocene-middle Pliocene widely outcrop. To the north, siliciclastic turbidites of an even older cycle (Langhian-Tortonian) are found. All the rock successions described above were deposited into intra-Apennines basins and/or on the border of the belt nappes; such basins were formed sequentially during the progressive migration of the orogenic system towards the Foreland (NE). North of Benevento the sinorogenic successions cover basinal nappes made up of variegated clays, shales and flysch deposits (upper Cretaceous-Eocene), which had been deformed during Apenninic orogeny.

Westwards, the sinorogenic successions come into contact with the carbonate platform and slope deposits of the Upper Triassic-Miocene age, which form the highest mountain ranges.

As regards the youngest terrains, Quaternary alluvial and lacustrine deposits outcrop alongside the Sabato and Calore rivers; the town of Benevento is situated at the confluence of these rivers.

Structurally, the Benevento area is part of the Southern Apennines. The Apennines is a fold and thrust belt mostly built up in the Miocene-Pliocene during contraction towards NE (D'Argenio et al., 1973; Patacca & Scandone, 1989). Starting from early Pleistocene, the prevailing deformation regime has been characterized by predominant NE-SW and subordinate NW-SE extension along lineaments with apenninic/antiapenninic (WNW-ESE/NNE-SSW) and meridian/parallel (N-S/E-W) orientation (Cello et al., 1989; Gars & Lippman, 1983).

During the deformative phases of Pliocene-Pleistocene, the Benevento area underwent differential vertical movements to a varying extent entity; particularly Pliocene subsidence followed by Pleistocene generalized uplift has been inferred by some authors (Ciaranfi et al., 1983). Moreover, in recent studies (Pescatore et al., 1994) the existence of some fault segments inside the basin of Benevento has been hypothesized.

As regards the faults mapped in this area in the available geological literature, the main tectonic lineaments are represented by the large border faults of the carbonate massifs; small segments of normal faults cut the Pliocene basinal deposits of the Benevento area (Fig. 1) (C.N.R.-P.F.G., 1992a).

3. Historical Seismicity

In historical times the province of Benevento was hit by strong earthquakes causing severe damages both to the town of Benevento and to a large number of neighbouring villages of the Southern Apennines.

All the macroseismic information on such earthquakes have been obtained from the Atlas of isoseismal maps of Italian earthquakes (Postpischl, 1985); figure 2 reports the isoseismal lines of IX MCS for well-documented earthquakes affecting the province of Benevento.

The strongest earthquakes occurred in December 1456; the damaged area extended from the Abruzzo to Basilicata, an area several times larger than the other events mapped in Fig. 2. The last review on these events published by Figliuolo (1988) points out the large extent of the macroseismic field with a maximum intensity of XI MCS, as well as the complexity of the source, probably involving more than one tectonic structure of the Southern Apennines. The area of IX MCS effects, felt by the population, showed a particular zonal distribution with a large region including several villages to the East and North of the town of Benevento (Fig. 2), and other restricted areas far from the Benevento province.

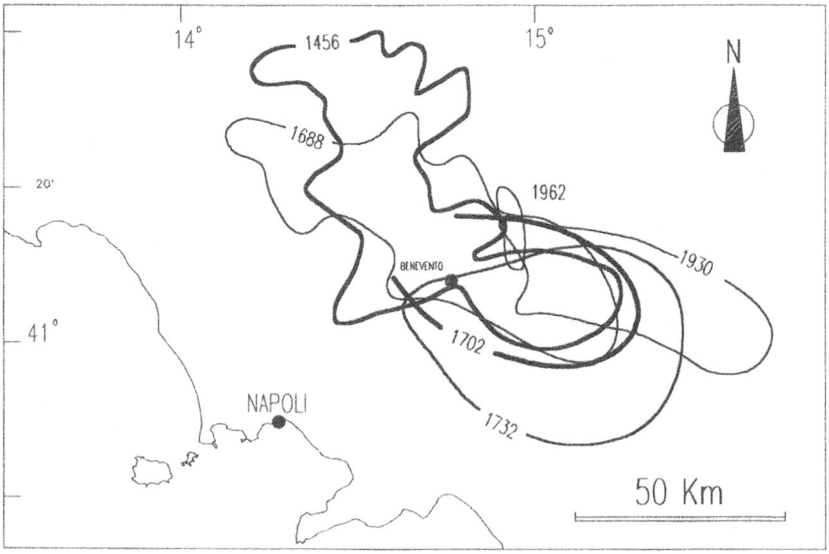

Fig. 2: Isoseismal lines of IX MCS for well documented earthquakes affecting the province of Benevento. The isoseismal line of the 1456 earthquake refers only to the largest area of IX MCS effects; other remote and restricted areas that suffered the same intensity have not been plotted.

The June 1688 earthquake (I = XI MCS) is one of the well-documented seismic events which have occurred in Southern Italy. It caused huge damages all over the Benevento province and particularly in the villages of the south-west slope of the Matese massif. In the epicentral area, effects corresponding to XI MCS were observed, including fractures in the ground, liquefaction phenomena and landslides, as reported in historical documents contemporary to the event (Serva, 1981).

The March 1702 and November 1732 earthquakes struck the Benevento area with effects corresponding to X MCS; the most damaged villages were those to the South-East of Benevento (fig. 2). The isoseismal lines of the 1688 earthquakes, as well as those of 1702 and 1732, show a clear direction of propagation towards the Apennines chain, along a NW-SE direction.

The earthquakes which have occurred in this century, the July 1930 and August 1962 events, hit the eastern part of the Benevento province, causing minor damage to the town of Benevento. The 1930 event mainly affected the Irpinian area, with intensity I = X MCS; the town of Benevento experienced effects of VIII MCS. The 1962 seismic sequence consisted of at least three strong shocks with maximum intensity of IX MCS. The macroseismic field for the 1962 earthquake shows lengthening transverse to the Apennines chain direction, differently from the pattern of the above-described events (Alessio et al., 1993). Source properties and geometry of faulting associated with the 1962 events have been pointed out in Westaway, (1987).

4. Analysis of the 1990 Seismic Sequence

Before the 1990 seismic crisis, seismic activity of the Benevento province had been of low energy and fairly sparse (Alessio et al., 1988; 1993).

A significant increase both in frequency and in energy of the events, was observed on April 21, 1990 and persisted until July; it was characterized by several phases of time concentration of events. The time evolution of the sequence is shown in Fig. 3. It is evident that the sequence is composed of at least three subsequent periods with moderate seismic energy release, timely spaced by events with minimum energy.

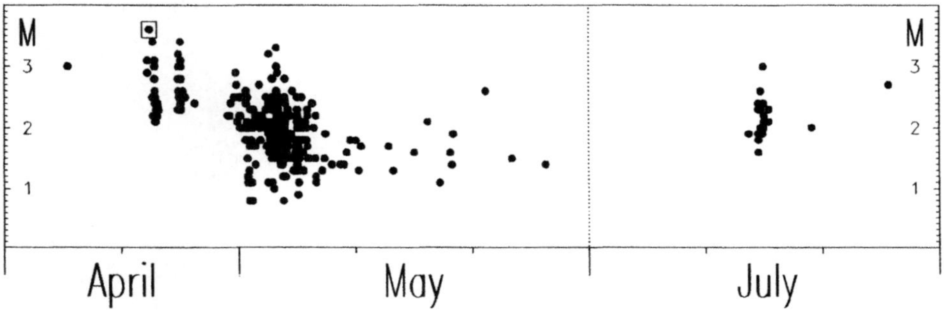

Fig.3: Time distribution of magnitude for the events of the 1990 Benevento seismic sequence. The main shock is marked by the square.

During the first stage of seismicity, occurring between April 21 and 25, several events with the highest magnitude of the whole sequence ($2.0 \leq M \leq 3.6$) took place. Among them, the strongest event ($M_L = 3.6$) occurred on April 22, at 07:45 (GMT). It was located at 16.45 Km of depth with epicentral coordinates 41°12'.30 Lat.N, 14°48'.05 Lon.E.

On the following days, the seismic energy of the events decreased to a minimum (Fig. 3). A second stage of seismicity was observed from April 29; during it the largest number of events were recorded, even if they were of lower energy compared to the previous stage. Finally, after over a month of relative inactivity, a further short stage of seismicity occurred in mid July.

The earthquakes of the Benevento seismic sequence were recorded by several seismic stations operating in the epicentral area (Fig. 4). Particularly, data from 7 stations of the Istituto Nazionale di Geofisica (I.N.G.) national seismic network and 5 stations of the Osservatorio Vesuviano (O.V.) regional seismic network were utilized. Moreover, during the period of strongest seismic activity, 5 digital stations (PCM 5800 Lennartz system) with three component short -period seismometers of the O.V. mobile network and 15 stations of a scientific cooperation program between the University of Napoli and the Institute de Physique du Globe of Paris were installed. The temporary network, operating from April 30, was prevalently equipped with analog stations with vertical components (Sprengnether MEQ-800) recording on smoked paper; only one station (PAD) was equipped with three components digital instrumentation (Iannaccone et al., 1990).

Of the about 1000 events which were recorded by the local networks, 370 earthquakes, with at least 5 P- and 3 S-readings, were localized by means of CHEAP program (Tarantola & Valette, 1982); hypocentres of such events have errors less than 0.5 and 1.5 km for horizontal position and depth, respectively.

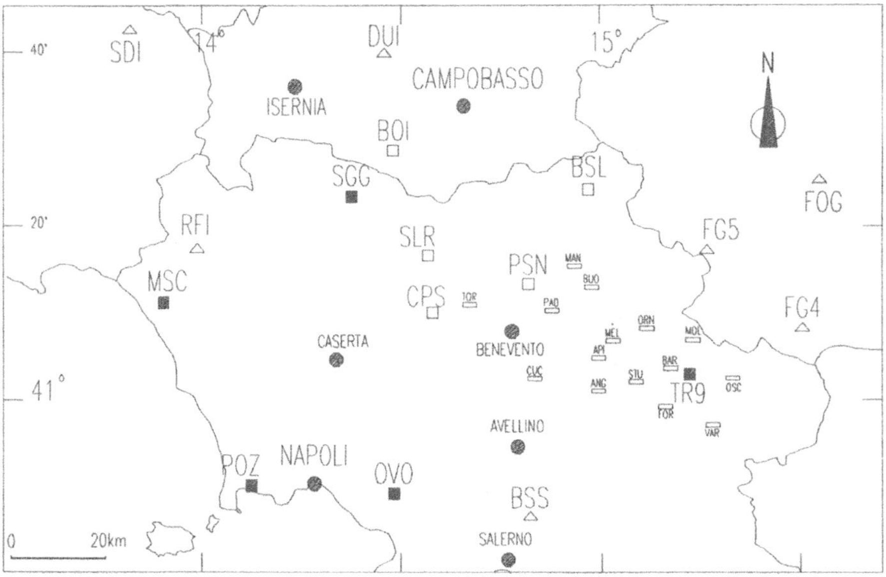

Fig. 4: Location of seismic stations operating in the Benevento area during April-May 1990. The rectangles indicate the 'Benevento Seismic Risk Project' stations; the full and empty squares indicate the O.V. permanent and temporary stations respectively; the triangles indicate the I.N.G. permanent stations.

As regards the velocity model which was used for locating events, data from a seismic refraction profile carried out in October 1992 in the Southern Apennines were used (Biella et al., 1995) profile was in the area previously hit by the 1990 seismic swarm, allowing therefore a more reliable definition of hypocentral depths.

Table 1 shows the crustal model inferred for the Benevento area, with the P-wave velocities relative to the single layers identified. The Vp/Vs ratio was evaluated from Wadati diagrams for the events of this sequence; a value of 1.73 was estimated.

TABLE 1. Velocity model used for earthquake location

P-Wave Velocity (km/s)	Depth (km)	Lithotypes
3.3	0.0	Clays and shales
4.5	2.5	Marls-Silic. rocks
6.0	4.0	Carbonate rocks
6.2	9.0	Cristalline rocks
6.5	20.0	Cristalline rocks

In Fig. 5a and 5b the map of the 370 well located events and a selection of the strongest events with $M \geq 2.5$ are shown. The epicentral area appears to be extremely concentrated to the North of the town of Benevento. In the map of Fig. 5a significant alignments are not evident, while the epicentral distribution of Fig. 5b would suggest a

concentration of epicentres along an anti-apenninic direction (NNE-SSW), between the town of Benevento and Pescosannita.

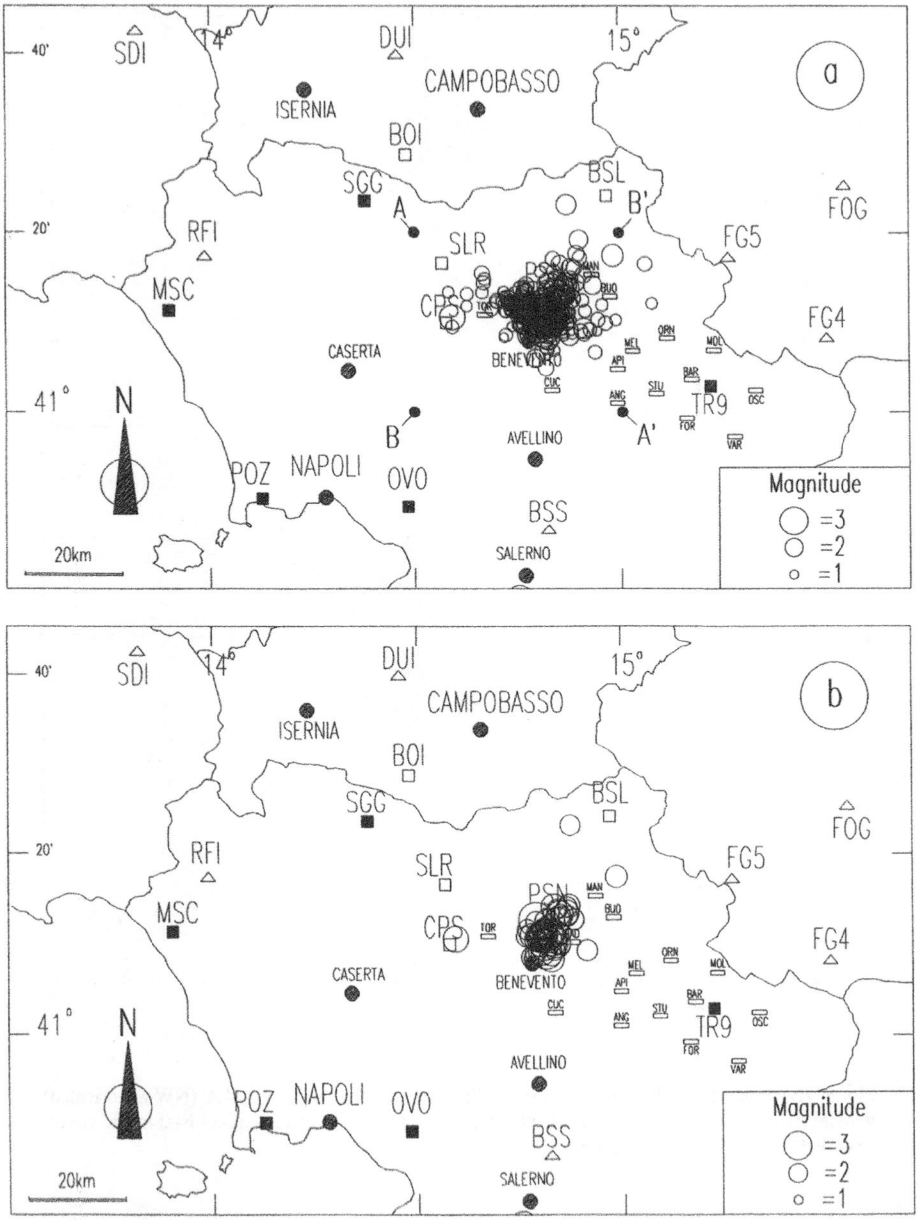

Fig. 5: (a) Epicentral distribution of the 370 best-located events of the 1990 Benevento seismic sequence. Radii of circles are proportional to magnitude of events. AA' and BB' indicate the profiles along which cross sections of hypocenters have been mapped (see Fig. 6a and 6b); (b) Epicentral distribution for earthquakes with $M_L \geq 2.5$.

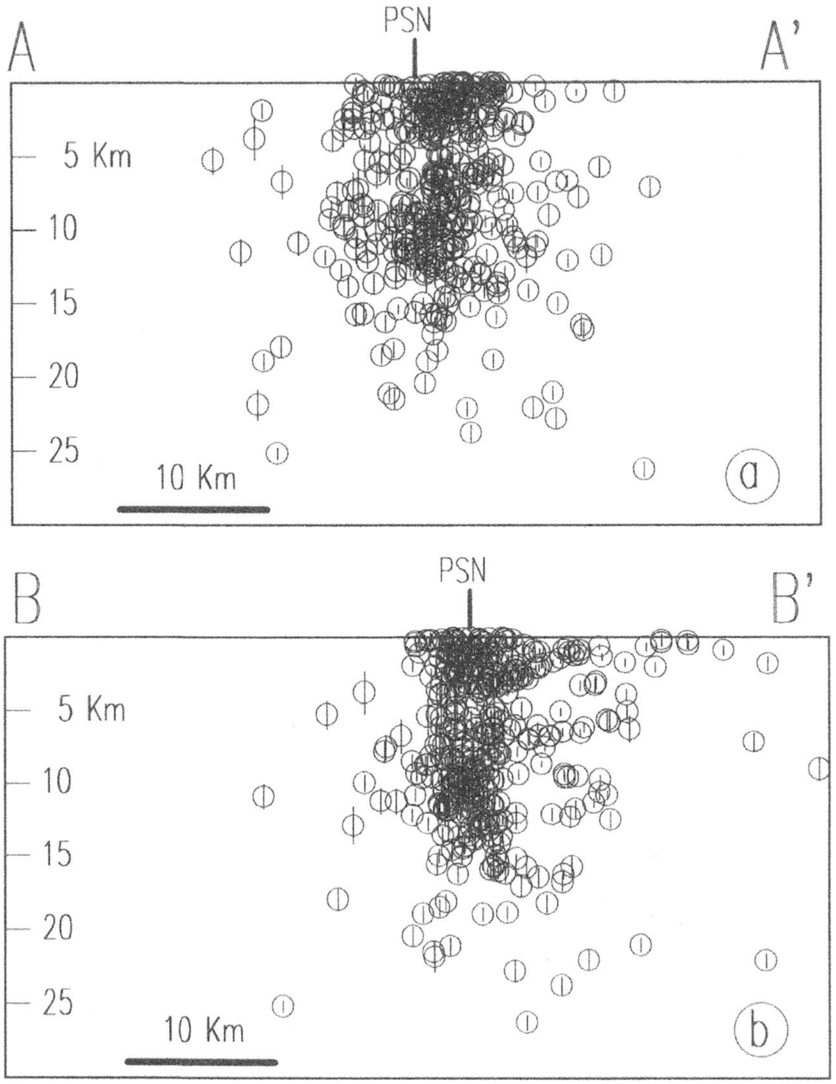

Fig. 6: Hypocenters distribution along the profiles shown in figure 5a. (a) AA' (NW-SE oriented) vertical cross section; (b) BB' (SW-NE oriented) vertical cross section. Vertical bars inside circles represent errors of focal depth in km.

Hypocentral distributions along two orthogonal directions, NW-SE (Fig. 6a) and SW-NE (Fig. 6b), display a quite homogeneous distribution of events from the surface to 15 km of depth or some more, without meaningful alignments of events. As regards the reliability of hypocentral depths, it is to be pointed out that, though the network configuration has changed in time, focal depths of events are fairly well constrained; in fact PSN station (see location in Fig. 6), equipped with digital three-components instrumentation, was located inside the epicentral area and allowed reliable readings of S-P times.

It is interesting to notice that inside each of the above mentioned three phases of seismic activity, most of hypocentral depths are arranged quite homogeneously within the first 15 Km of the crust (Fig. 7), without any dependence on magnitude (Fig. 3).

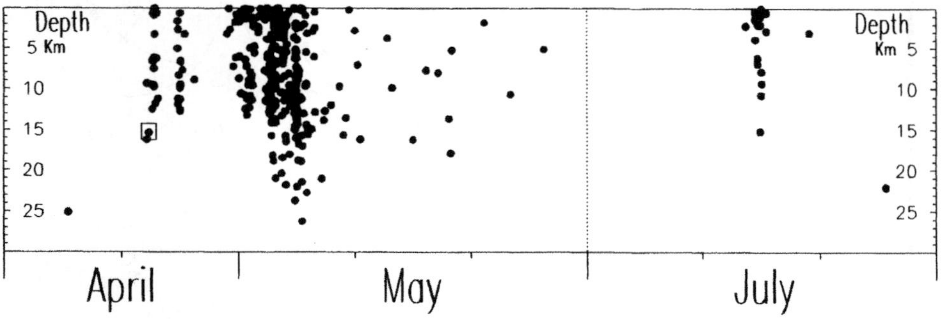

Fig. 7: Time distribution of focal depths for the 370 best located events of the 1990 Benevento seismic sequence. The square marks the main shock.

In order to determine the fault plane solutions relative to this seismic sequence, all the data on the P wave first motion were analyzed. Among the computed solutions with at least 7 reliable P-wave first motion readings, the 19 best constrained focal mechanisms, calculated using the FPFIT program (Reasenberg & Oppenheimer, 1985) are shown in Fig. 8. In table 2 the location parameters of these events are listed.

TABLE 2. Hypocentral parameters of the events whose focal mechanisms are shown in figure 8.

DATE	HOUR	LAT. N	LON. E	DEPTH	M_L
90.04.24	21:39	41-10.83	14-47.67	11.87	3.0
90.04.25	02:22	41-09.58	14-48.48	9.52	3.1
90.05.02	18:34	41-10.59	14-46.28	1.17	2.4
90.05.02	19:46	41-11.20	14-52.29	6.74	2.1
90.05.02	20:39	41-10.79	14-47.17	3.38	2.2
90.05.02	20:49	41-09.29	14-45.74	0.40	2.8
90.05.03	04:57	41-10.58	14-49.26	13.67	3.3
90.05.03	05:14	41-11.22	14-48.58	9.77	3.0
90.05.03	05:20	41-12.86	14-51.01	5.08	2.6
90.05.03	06:34	41-12.09	14-46.66	3.42	2.2
90.05.03	06:49	41-07.79	14-49.04	3.61	2.4
90.05.03	07:23	41-11.22	14-48.88	1.84	2.9
90.05.03	22:13	41-11.55	14-50.06	3.42	2.5
90.05.05	02:16	41-08.47	14-49.87	7.54	1.6
90.05.05	12:01	41-09.26	14-51.37	11.27	2.0
90.05.05	16:51	41-10.17	14-48.53	2.03	2.1
90.05.05	19:21	41-09.63	14-51.12	9.69	2.0
90.05.05	20:44	41-09.76	14-48.46	8.30	2.3
90.05.16	03:45	41-10.92	14-49.24	7.78	2.1

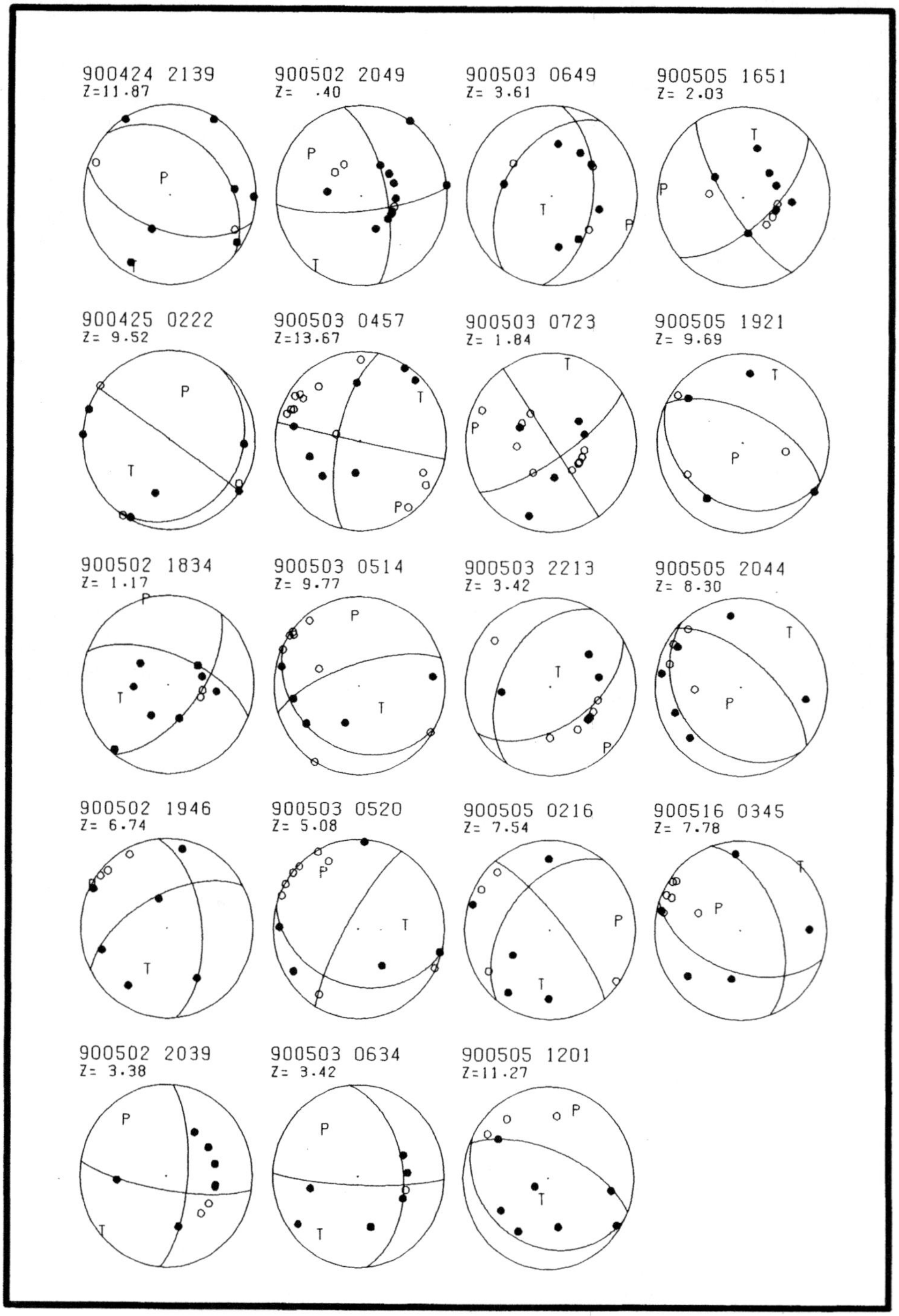

Fig. 8: Fault plane solutions (lower emisphere equal-area projection) of the 19 best constrained mechanisms from the 1990 Benevento sequence. On the top of each mechanism the date, origin time and focal depth (Km) for the relative event are reported.

It is evident that there is no predominant type of faulting motion in the selected mechanisms, which might indicate structural heterogeneity in the medium involved in the stress release process. Moreover, the T and P axes of the 19 selected fault plane solutions (Fig. 8) were plotted on cumulative stereonets for detecting possible significant distributions (Fig. 9). While the P axes appear to be rather scattered, T axes are fairly well aligned along a NE-SW direction, that is orthogonal to the Apenninic chain trend. This result is in good agreement with T axes orientation of most Southern Apennines earthquakes (Gasparini et al., 1980; 1985), and particularly with strain axes orientation obtained for the same area by Federici et al., (1992).

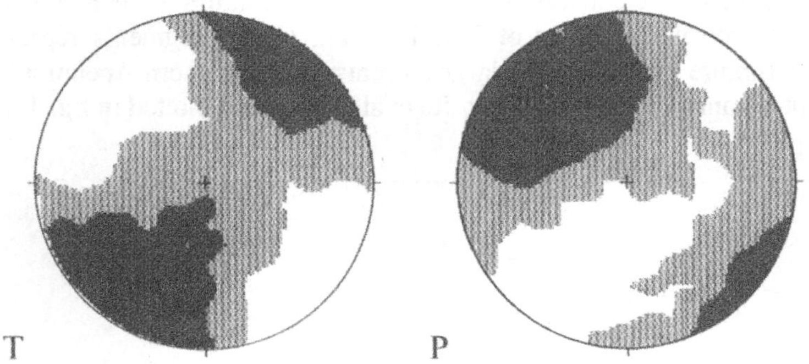

Fig. 9: Lower hemisphere equal-area stereographic projections of contoured T and P axes for the 19 fault plane solutions of fig. 8. Contour interval is equal to 2 sigma.

5. Concluding Remarks

The April 1990 seismic sequence occurred near Benevento, in a region hit by highly destructive events in historical times, and at present characterized by lack of strong seismicity. For this reason, the evolution of seismic activity in this area is worthy being studied in detail, in order to evaluate the seismic potential of this region.

As regards the time and magnitude characteristics of the Benevento seismic sequence, it has been shown that the sequence was composed of three subsequent phases of time concentration of events; the maximum seismic energy was released at the beginning of the sequence, with the highest magnitude $M_L = 3.6$ and only 18 other events had a magnitude greater than 3.0.

As regards earthquake location, about 1000 events were recorded by the local permanent and temporary networks, among which 370 earthquakes were localized in a small area of about 100 km^2, with a distribution extremely concentrated to the North of the town of Benevento. Significant alignments of epicenters are not so evident, as to be interpreted in terms of the local tectonic setting.

Cross sections of hypocenters indicate that depths of events were quite homogeneously distributed within the first 15 km of the crust, without meaningful alignments of events. Similar focal volumes have been reported for numerous seismic sequences in the Central-Southern Apennines (Haessler et al., 1988; Alessio et al., 1988; 1990; 1995).

The best constrained focal mechanisms relative to this seismic sequence show that there are different types of faulting motions associated with each event, since they are probably due to reactivation of local preexisting different structures; therefore such mechanisms might indicate structural heterogeneity in the medium involved in the stress release process. However, the coherent distribution of T axes, orthogonal to the main strike of the Apennines, indicates that kinematics of faulting is consistent with the regional strain field (Cello et al., 1982; Ferranti, 1994).

In conclusion, in order to compare the recent seismicity of the Benevento province with the seismic behaviour of strictly adjacent areas, in fig. 10 the epicentral distribution of the April 1990 seismic sequence has been superimposed on a map with the major structural elements of the Southern Apennines (C.N.R.-P.F.G. 1992b). According to the methodology of Bonilla et al., (1984) segments representing the earthquake rupture lengths for the largest events of the Southern Apennines have been worked out (Branno et al., 1986; Esposito et al., 1992) and plotted in fig. 10.

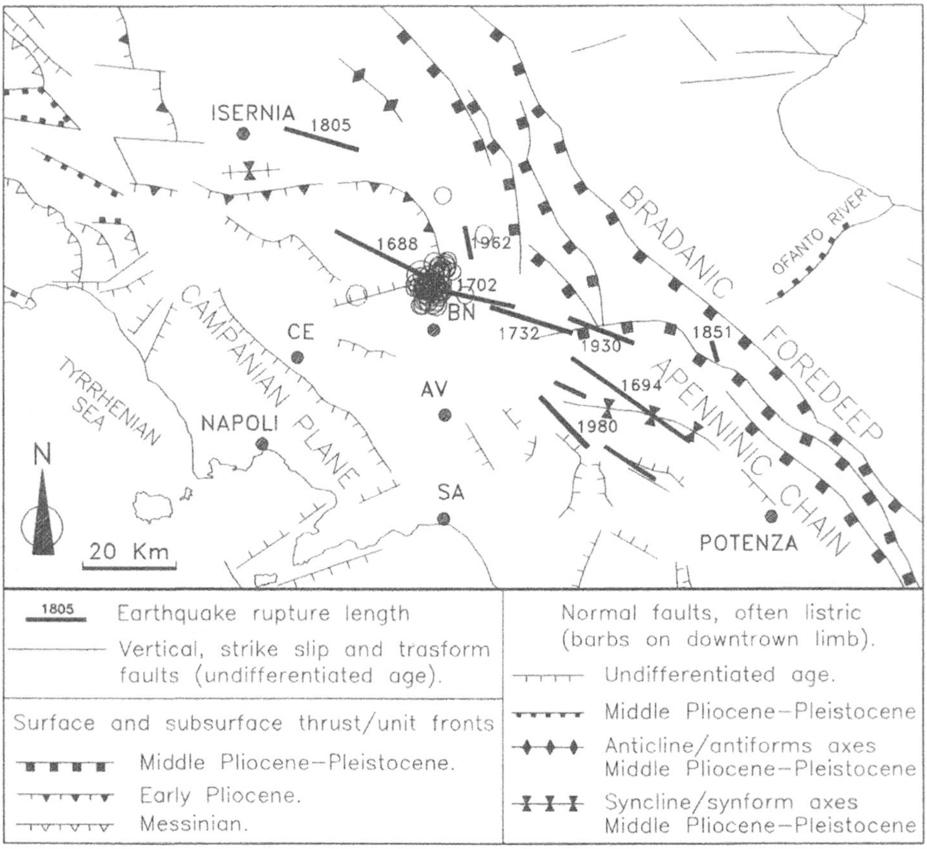

Fig. 10: Epicentral distribution of the best located events with $M_L \geq 2.5$ of the Benevento 1990 seismic sequence. The epicentres have been plotted on a map with the major structural elements of the Southern Apennines (Synthetic Structural-Kinematics Map of Italy, C.N.R.-P.F.G., 1992b). The bold segments represent the earthquake rupture lengths for the largest earthquakes of the Southern Apennines (after Esposito et al., 1992).

It is evident that the April 1990 epicentral area is included between larger systems of seismogenetic structures: to the north-west, the structure responsible for the 1688 earthquake, and, to the south-east, the 1702 and 1732 earthquake faults. For this reason it could be hypothesized that minor structures, which probably interrupt the main longitudinal Apenninic seismogenetic faults, are responsible for minor seismic sequences through which regional stress is released when larger faults are 'silent'. If this is the case, the analysis and interpretation of earthquake parameters relative to small sequences like that of 1990 can provide very useful contributions to the seismic hazard assessment of the whole seismogenetic region.

References

Alessio G., Ferri M., Gorini A., Luongo G. (1988) Sismicità dell'Appennino Meridionale nel periodo 1982 - 1987, *Memorie Soc. Geol. Italiana*, **XLI**, 1129-1137.

Alessio G., Godano C., Gorini A. (1990) A low magnitude seismic sequence near Isernia (Molise, central Italy) in January 1986, *PAGEOPH* **134** (2), 243-260.

Alessio G., Esposito E., Gorini A., Luongo G., Porfido S. (1993) Identification of seismogenic areas in the Southern Apennines, Italy, *Annali di Geofisica* **XXXVI** (1), 227-235.

Alessio G., Esposito E., Gorini A., Porfido S. (1995) Detailed study of the Potentino seismic zone in the Southern Apennines, Italy, *Tectonophysics* (in press).

Biella G., De Franco R., Iannaccone G., Castellano M., Cocco M., Deschamps A., Mirabile L, Romeo R., Zollo A. (1995) A seismic refraction profile across Sannio region (Southern Apennines), Submitted to *Annali di Geofisica*.

Bonilla M.G., Mark R.K., Lienkaemper J.J. (1984) Statistical relations among earthquake magnitude, surface rupture lenght and surface fault displacement, *Bull. Seism. Soc. Am.*, **74** , 2379-2411.

Branno A., Esposito E., Luongo G., Marturano A., Porfido S., Rinaldis V. (1986) The largest earthquakes of the Apennines, Southern Italy, IAEG, AIGI Proc. of the Intern. Symp. on Engineering Geology Problems in Seismic Areas. - Bari, Italy, **IV**, 3-14.

Cello G., Guerra I., Tortorici L., Turco E. and R. Scarpa (1982) Geometry of the neotectonic stress field in Southern Italy: geological and seismological evidence. J. Struct. Geol., **4** (4), 385-396.

Cello G., Tortorici L., Martini N., Paltrinieri W. (1989) Structural styles in the frontal zones of the Southern Apennines, Italy: an example from the Molise district, *Tectonics* **8** (4), 753-768.

Ciaranfi N., Ghisetti F., Guida M., Iaccarino G., Lambiase S., Pieri P., Rapisardi L., Ricchetti G., Torre M., Tortorici L., Vezzani L. (1983) Carta neotettonica dell'Italia Meridionale, *P.F.G. C.N.R.* Roma ,Pubbl. n. 515.

C.N.R.- P.F.G. (1992a) Structural model of Italy and Gravity Map. Scale 1:500.000. Sheet n. 4. Resp.: P. Scandone, **114** (3).

C.N.R. - P.F.G. (1992b) Synthetic Structural-Kinematic Map of Italy in 'Structural Model of Italy' - Scale 1:2.000.000. Sheet n. 5. Resp.: P. Scandone.

D'Argenio B., Pescatore T., Scandone P. (1973) Schema geologico dell'Appennino Meridionale (Campania-Lucania). Proc. Conv. 'Moderne vedute sulla geologia dell'Appennino' *Acc. Naz. Lincei*, Quad. n.**183**.

Esposito E., Luongo G., Petrazzuoli S.M. and S. Porfido (1992) Damage scenarios induced by the major seismic events from XV to XIX century in Naples city with particular references to the seismic responce. Proc. 10th World Conference on Earthquake Engeenering, A.A. Balkema Ed., Rotterdam, 1075-1080.

Federici P., Di Maro R., Marchetti A., Cocco M. (1992) Analisi della sismicità dell'area del Sannio-Matese negli anni 1991-1992, Proc. 11° Natl. Assembly G.N.G.T.S. - Roma, 389-404.

Ferranti L. (1994) Le strutture del bordo meridionale del massiccio del Matese: elementi di tettonica compressiva e distensiva. Boll. Soc. Geol. It., 113, 157-171.

Figliuolo B. (1988) *Il terremoto del 1456*, Edizioni Studi Storici Meridionali.

Gars G., Lippman M. (1983) Nouvelles donées néotectoniques dans l'Apennin campanien (Italie du Sud*), C.R. Acad. Sci., Paris* **298**, 495-500.

Gasparini C., G. Iannaccone G., Scarpa R. (1980) On the focal mechanism of Italian earthquakes, *Rock mechanics* **9**, 85-91.

Gasparini C., Iannaccone G., Scarpa R. (1985) Fault plane solutions and seismicity of the Italian peninsula. *Tectonophysics* **117**, 59-78.

Haessler H., Gaulon R., Rivera L., Console R., Frogneux M., Gasparini C., Martel L., Patau G., Siciliano M., Cisternas A. (1988) Earthquake of 29, April 1984: a microearthquake survey, *Bull. Seism. Soc. Am.* **78** (6), 1948-1964.

Iannaccone G., Romeo, R., Tranfaglia G., Errico L., Lentini E., Bernard P., Deschamps A., Patau G. (1990) Analisi della sequenza sismica di Benevento (aprile-maggio 1990). Proc. 9° Natl. Assembly G.N.G.T.S. Roma, 189-198.

Patacca, E. & Scandone P. (1989) Post-Tortonian mountain building in the Apennines. The role of the passive sinking of a relic lithospheric slab, Acc. Naz. Lincei. In "The Lithosphere in Italy." Roma. *Advances in Earth Science Research*, 157-176.

Pescatore T., Improta L., Romeo R., Iannaccone G. (1994) Geologia della città di Benevento: caratteristiche litostratigrafiche finalizzate alla microzonazione sismica, Submitted to *Boll. Soc. Geol. Ital.*

Postpischl D. editor. (1985) Atlas of isoseismal maps of italian earthquakes - *C.N.R. - P.F.G.* **114** (2A).

Reasenberg P., Oppenheimer D. (1985) FPFIT, FPPLOT and FPPAGE: Fortran computer programs for calculating and displaying earthquake fault plane solutions, *U.S. Geol. Surv. Open File Rep.*, 85-739.

Scarpa R. & Zollo A. (1985) Seismotectonics and seismic quiescence in the Southern Apennines, Italy, *Earthq. Predict. Res.*, **1**, 81-93.

Serva L. (1981), Il terremoto del 1688 nel Sannio. Contributo alla caratterizzazione della sismicità del territorio italiano. *Proc. Natl. Assembly P.F.G. - C.N.R.*, Udine 1981, 209-249.

Tarantola A. & Valette B. (1982) Inverse problems - quest for information, *J. Geophys.* **50**, 159-170.

Westaway R. (1987), The Campania, southern Italy, earthquakes of 1962 August 21. *Geophys. J.R. Astr. Soc.* **88**, 1-24.

THE PROBABILITY OF THE TIME OCCURRENCE OF STRONG EARTHQUAKE FOR SOME ZONES IN ALBANIA

V. PEÇI
Seismological Center, Academy of Sciences of Republic of Albania Tirana, Albania

Abstract

Based on the historical and instrumental data for the period 1901-1988 for earthquakes with magnitude $M \geq 5.9$ ($I_0 = 8$ MSK-64), for some main seismogenic zones and lines of Albania, in this paper we have made use of property that the time interval sequence of earthquakes follows the logarithmic normal distribution, as proved by the test of normality.

Eventually, for the different seismic regions we obtained the different time interval that the next strong earthquake might occur.

The results may help the study of seismic zoning, microzoning, seismic risk and other problems.

1. Introduction

The question that is put forward in seismology, is linked with that, whether the observed instrumental seismicity for a short time gives the datum for the future events, because we know that the seismicity is only a part of elements (phenomena) of the tectonic activity of a longer time.

The zones of the strongest earthquakes are characterized by a seismic activity which change depending on time. For this it is necessary to distinguish the periods with a different activity and particulary, when it is possible, to note the cycles of the seismic regime.

Many scientists think that, except the strong earthquakes which occurr rarely, the seismic regime of big earthquakes may be considered to be stable for a given period of time (Hsu Chung Tsi *et al.*, 1974).

Based on the historical and instrumental data, for earthquakes with magnitude $M \geq 5.9$ ($I_0 = 8$ MSK-64), we have tried to find out the tendency of the development of the seismic activity of the big earthquakes in the main seismogenic zones and lines in our country (Fig. 1.).

V. Schenk (ed.), Earthquake Hazard and Risk, 17–23.

2. Used Method

In order to find out "expected time intervals" of strong earthquakes, all the independent events are used, excluding the foreshocks and aftershocks (Papazachos E.E. *et al.*, 1987), which are dependent on the main seismogenic zone activity and seismogenic line activity in our country and in the whole area of Albania, to find the probability of the occurrence of a high magnitude seismic event in a given period of time.

Based on the historical and instrumental data for the period 1901-1988 for events with M≥5.9 (I_0=8 MSK-64) (Sulstarova E. *et al.*, 1975) we have removed the time series t_1, t_2, t_3, t_4, (Tab. 1, 2, 3, 4) for the zones 1, 2, 3, 4 (See Fig.1).

The extrapolation interval for strong earthquakes is accepted to have the value t_0 equal one month after the last event of each series (Hsu Chung Tsi *et al.*, 1974).

We have done the graphic variation of the t_i or $\log t_i$ (which we call x_i) depending on the number of events of each series (see Fig.2). Based on these series we can find the dispersion function of t:

$$F(t) = \int_0^t f(t')\,dt' \tag{1}$$

but here we must also use t_0 factor, so we can find (Hsu Chung Tsi *et al.*, 1974):

$$\int_{t_0}^{\infty} f(t)\,dt = 1 - F(t_0) \tag{2}$$

The density and dispersion function of the distribution of the occurence time for the event on t_0 + t' (which is the time t of the next event in the future) would be:

$$g(t') = \frac{f(t_0 + t')}{1 - F(t_0)} \tag{3}$$

and

$$G(t') = \int_0^{t'} g(t)\,dt = \frac{F(t_0 + t') - F(t_0)}{1 - F(t_0)} \tag{4}$$

which is called the t' probability condition.

Based on the dispersion function F(t) and the extrapolation time t_0, we obtained G(t') function, the probability of the time occurence. As we have shown, $x_i = \log t_i$, F(t) function follows the logarithmic normal distribution, so $F(t) = \Phi(x) = \Phi(\log t)$.

We have used the χ^2 - test (Hsu Chung Tsi *et al.*, 1974). We hope to obtaine a normal distribution N(m,σ) for the above series. Find the mean value of x, and standard deviations:

$$\sigma^2 = \frac{\Sigma(x - \bar{x})^2}{n - 1} \tag{5}$$

so, we can use now $m = \bar{x}$. A number n - of x_i divided on k - groups. The probability of each group will be p_i, where (i=1,2,..k) and theoretical density for

TABLE 1. Earthquake in Albania in the period 1901-1988 ($M \geq 5.9$, $I_0=8$); Zone 1.

Nr.	YEAR	H	M	S	φ	λ	M	t_i	$x_i = \log t_i$
1.	01.06.1905	04	42	15.0	42.00	19.50	6.6	-	-
2.	04.08.1905	05	09	0.0	41.70	19.00	5.9	2.03	0.31
3.	18.02.1911	21	35	15.0	40.85	20.75	6.7	66.50	1.82
4.	0.02.1912	08	03	46.0	40.90	20.60	6.0	11.50	1.06
5.	22.12.1919	23	40	57.0	40.10	20.70	6.1	94.60	1.97
6.	26.11.1920	08	51	00.0	40.30	19.95	6.4	11.10	1.04
7.	18.12.1920	02	01	20.0	41.10	20.10	5.6	0.73	-0.13
8.	30.03.1921	15	05	30.0	41.70	20.50	5.8	3.40	0.53
9.	07.12.1922	16	22	10.0	41.80	20.60	6.1	20.20	1.30
10.	17.12.1926	11	39	55.0	41.30	19.50	6.2	48.30	1.68
11.	21.11.1930	02	00	27.0	40.20	19.60	6.2	47.10	1.67
12.	28.01.1931	05	55	15.0	40.60	20.70	5.9	02.20	0.34
13.	04.02.1934	09	35	30.0	41.25	19.60	5.6	36.20	1.56
14.	31.03.1935	03	21	31.0	41.25	20.25	5.7	13.80	1.14
15.	23.08.1940	00	39	56.0	40.60	19.60	5.5	64.70	1.82
16.	27.08.1942	06	14	11.0	41.65	20.45	6.0	23.90	1.38
17.	01.09.1959	11	37	40.0	40.85	19.80	6.2	203.6	2.31
18.	26.05.1960	05	10	11.0	40.60	20.70	6.4	08.80	0.94
19.	18.03.1962	15	30	32.0	40.70	19.60	6.0	21.60	1.33
20.	09.02.1967	18	08	18.0	39.95	20.35	5.6	58.70	1.76
21.	30.11.1967	07	23	50.0	41.35	20.40	6.6	09.70	0.98
22.	03.11.1968	04	49	34.0	42.00	19.30	5.5	11.90	1.07
23.	03.04.1969	22	12	22.0	40.50	19.95	5.6	05.00	0.70
24.	19.08.1970	02	01	53.0	41.15	19.80	5.5	16.50	1.22
25.	15.04.1979	06	19	40.1	41.94	19.08	7.2	100.9	2.00

TABLE 2. Earthquake series for zone 2, ($M \geq 5.9$, $I_0=8$)

Nr.	YEAR	H	M	S	φ	λ	M	t_i	$x_i = \log t_i$
1.	26.11.1920	08	51	00.0	40.30	19.95	6.4	-	-
2.	17.12.1926	11	39	55.0	41.30	19.50	6.2	72.70	1.86
3.	21.11.1930	02	00	27.0	40.20	19.60	6.2	47.10	1.67
4.	04.02.1934	09	35	30.0	41.25	19.60	5.6	38.20	1.58
5.	23.08.1940	00	39	56.0	40.60	19.60	5.5	78.60	1.89
6.	09.02.1967	18	08	18.0	39.95	20.35	5.6	317.6	2.50
7.	03.11.1968	04	49	34.0	42.00	19.30	5.5	20.80	1.31
8.	03.04.1969	22	12	22.0	40.50	19.95	5.6	05.00	0.70
9.	19.08.1970	02	01	53.0	41.15	19.80	5.5	16.50	1.22
10.	15.04.1979	06	19	40.1	41.94	19.08	7.2	100.9	2.00

TABLE 3. Earthquake series for zone 3, ($M \geq 5.9$, $I_0=8$)

Nr.	YEAR	H	M	S	φ	λ	M	t_i	$x_i = \log t_i$
1.	18.02.1911	21	35	12.0	40.90	20.75	6.7	-	-
2.	03.02.1912	08	03	46.0	40.90	20.60	6.0	11.50	1.06
3.	22 12 1919	23	40	57.0	40.10	20.70	6.1	94.60	1.97
4.	30.03.1921	15	05	30.0	41.70	20.50	5.8	15.20	1.18
5.	07.12.1922	16	22	10.0	41.80	20.60	6.1	20.20	1.30
6.	28.01.1931	05	55	15.0	40.60	20.70	5.9	97.70	1.99
7.	27.08.1942	06	14	11.0	41.65	20.45	6.0	139.0	2.14
8.	26.05.1960	05	10	11.0	40.60	20.70	6.4	213.0	2.33

TABLE 4. Earthquake series for zone 4, ($M \geq 5.9$, $I_0=8$)

Nr.	YEAR	H	M	S	φ	λ	M	t_i	$x_i = \log t_i$
1.	18.12.1920	02	01	20.0	41.10	20.10	5.6	-	-
2.	31.03.1935	03	21	31.0	41.25	20.25	5.7	171.4	2.23
3.	01.09.1959	11	37	40.0	40.85	19.80	6.4	293.0	2.46
4.	18.03.1962	15	30	32.0	40.70	19.60	6.0	29.6	1.47
5.	26.07.1963	04	17	12.0	42.00	21.40	6.1	16.2	1.21
6.	30.11.1967	07	23	50.0	41.35	20.40	6.6	52.1	1.71

each group is:
$m_i = np_i$, $\Sigma m_i = n \Sigma p_i = n$ and experimental density (observed) n_i, $\Sigma n_i = n$.

It is proved that quantity

$$\chi^2 = \Sigma_{i=1}^{k} \frac{(n_i - m_i)^2}{m_i} \tag{6}$$

with f = k-3 degree of freedom (Hsu Chung Tsi *et al.*, 1974) where: $\Sigma m_i = \Sigma n_i$, $m = \bar{x}$ follows approximately the χ^2 distribution with f = k-3 degree of freedom, so we have used 4 groups of m_i (Peçi, 1989)

To make use of the Gauss distribution, the theoretical probability for each grup is chose to be 0.25. From the tables, the interval for each series will be:

$$(-\infty, \bar{x} - 0.675\,\sigma); \quad (\bar{x} - 0.675\,\sigma, 0)\ ;$$

$$(0, \bar{x} + 0.675\,\sigma); \quad (\bar{x} + 0.675\,\sigma, +\infty)\ .$$

For the confidence level of 95%, the χ^2 value is 3.841, which will be found in most textbooks on statistics. Our values, in all cases that we have obtained, are less than 3.841 so, the statistical test shows that the difference between n_i and m_i is significant.

3. The Data Analyses

Based on our studies on the activity of some fault zones which are well know in our country (Aliaj Sh., 1983; Sulstarova E., 1986) and the main seismogenic zone and seismogenic lines, we have divided our territory in some zones as above, to see the probability of the reoccurrence interval for an earthquake with magnitude $M \geq 5.9$ ($I_0=8$, MSK-64) for the period 1901-1988.

t_1 serie belongs to events which have occured in the whole our country for the period 1901-1988, $M \geq 5.9$ ($I_0=8$, MSK-64), see Tab.1.

t_2 serie belongs to events which have occured in the Adriatic - Jonic seismogenic zone for the period 1901-1988, ($M \geq 5.9$). Tab.2.

t_3 serie belongs to events which have occurred in the Korca - Ohrid - Peshkopia seismogenic zone for the period 1901-1988, ($M \geq 5.9$). Tab.3.

t_4 serie belongs to events which have occured in the Lushnja - Elbasan - Diber, the main seismogenic line for period 1901-1988, ($M \geq 5.9$). Tab.4.

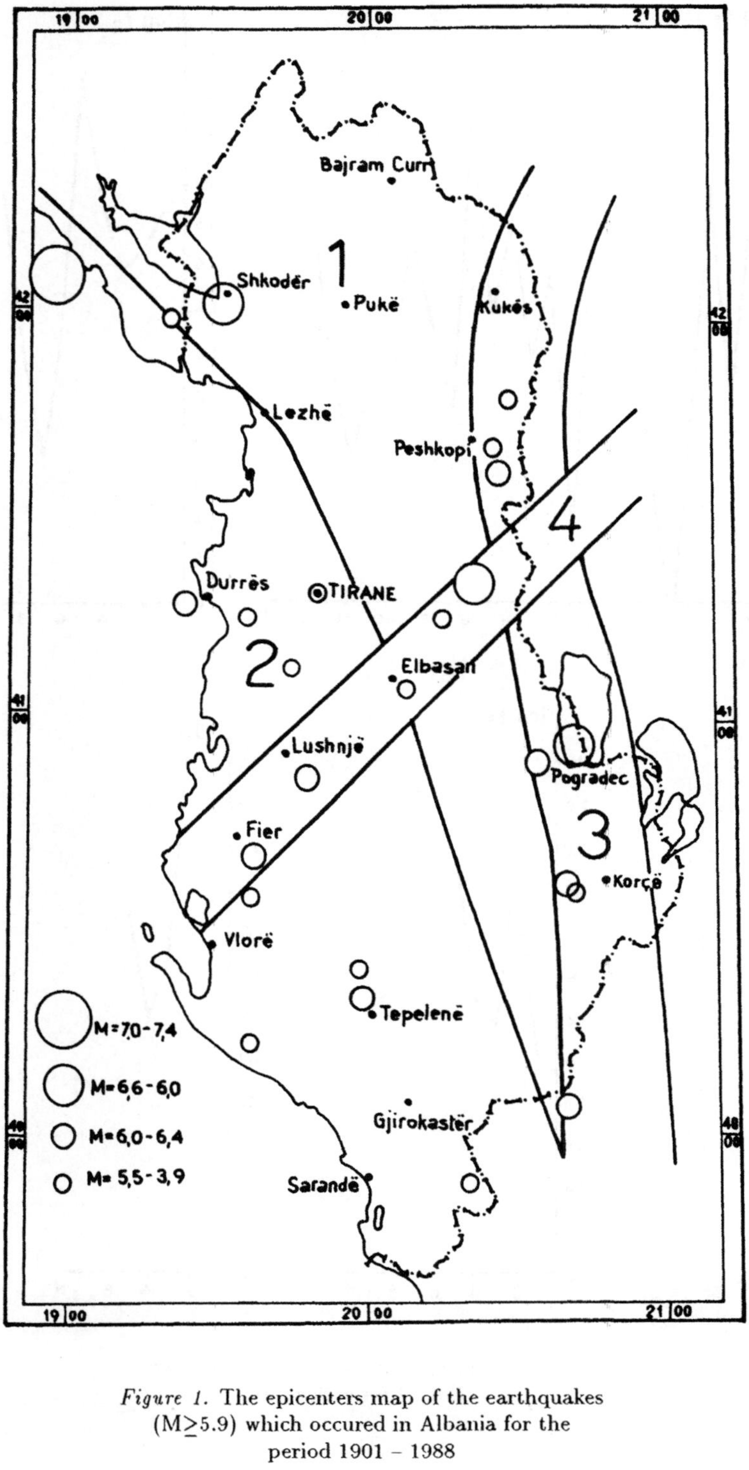

Figure 1. The epicenters map of the earthquakes ($M \geq 5.9$) which occured in Albania for the period 1901 – 1988

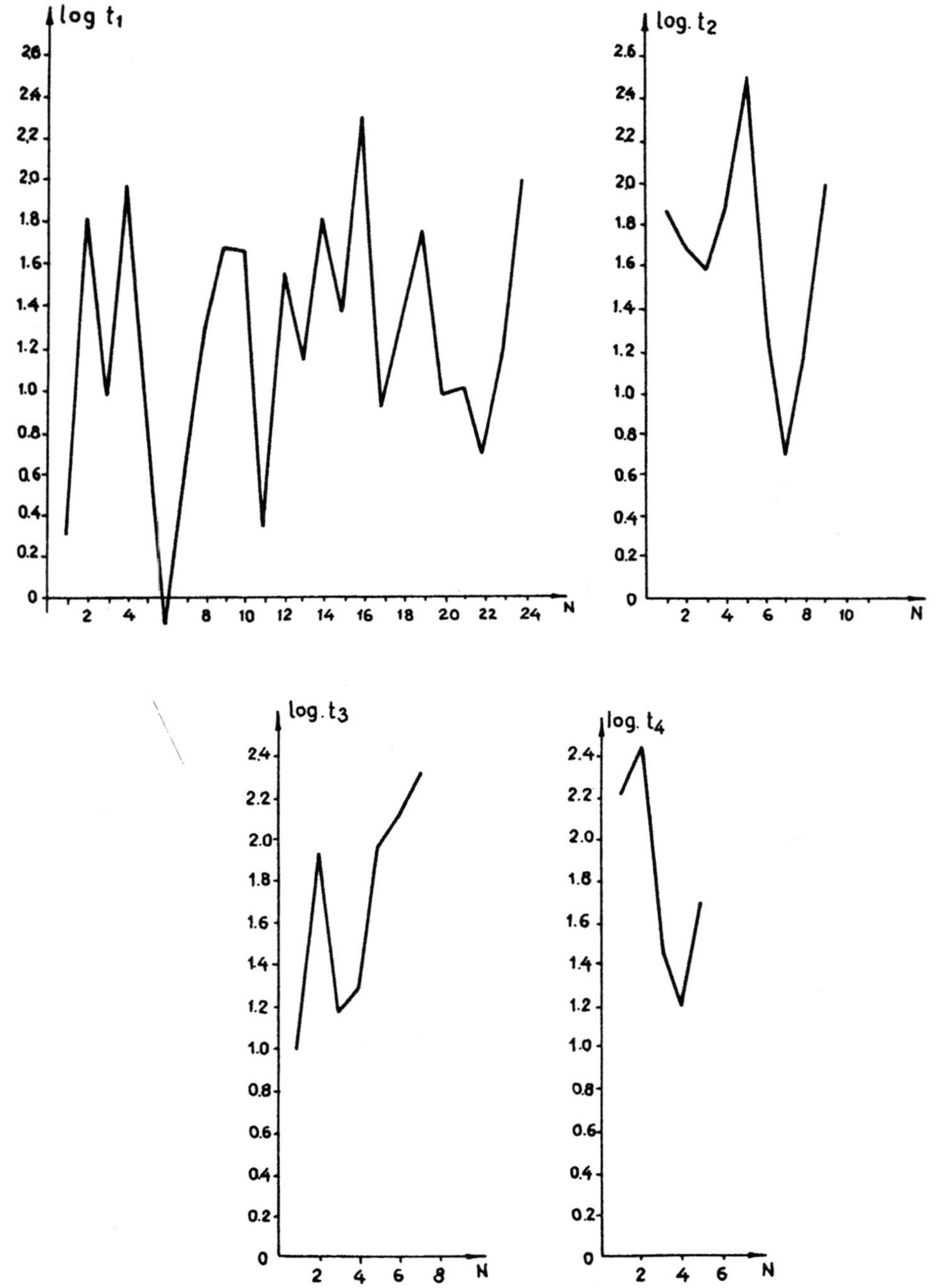

Figure 2. The graphic variation of time series ($\log t_i$) for each zones.

To analyse data we have made a program in BASIC for the microcomputer. All the results for each zone are showed in Tab.5.

Based on our historical and instrumental data for earthquakes $M \geq 5.9$, for period 1901-1988 in our territory, we have a different seismicity activity for the different zones and lines.

So the statistical estimation for the confidence level of 95%, for all this data shows that:

TABLE 5. Probability of the time of earthquake ($M \geq 5.9$, $I_0 = 8$ MSK-64) occurrence in some area of Albania

Series	$\bar{x}$	σ	Interval	χ^2	Reoccurrence time(on month)
t_1	1.24	0.61	0.83÷1.65	1.67	1.6 till 197.3 or 16.7 years
t_2	1.63	0.52	1.28÷1.99	0.33	4.2 till 455.3 or 37.9 years
t_3	1.71	0.52	1.36÷2.05	2.71	4.4 till 599.1 or 49.9 years
t_4	1.82	0.53	1.46÷2.17	2.2	4.0 till 1097.5 or 91.4 years

a - For zone 1, for all our teritory the expected time intervals that the next earthquake with magnitude $M \geq 5.9$ might occur is 2 till 197 months or approximately 0 till 17 years.

b - For zone 2, which is the Adriatic - Jonic seismogenic zone, the expected time intervals that the next earthquake $M \geq 5.9$ might occur is 4 till 455 months or 0 till 38 years.

c - For zone 3, which is the Korca-Ohrid-Peshkopia, the seismogenic zone, the expected time intervals that the next earthquake $M \geq 5.9$ might occur is 4 till 599 months or 0 till 50 years. d - For zone 4, which is Lushnja-Elbasan-Diber the main seismogenic line, the expected time intervals that the next earthquake $M \geq 5.9$ might occur is 4 till 1098 months or 0 till 91 years.

4. References

Aliaj Sh., (1983) Tiparet themelore sizmotektonike te Shqiperise, *Termeti i 15 Prillit 1979*, Shtepia botuse "8 Nentori", Tirane.

Hsu Chung Tsi *et al.*, (1974) Probability of the time of earthquakes occurrence, (1). *Acta Geophysica Sinica*, Vol.17, No.1,8.

Papazachos E.E. *et al.*, (1987) Probabilities of occurrence of large earthquakes in the Aegean and sorrounding area during the period 1986-2006. *Pageoph*, Vol. 125, No.4.

Peçi, V.(1989) *Prirja afatmesme e aktivitetit sizmik ne disa zona te vendit tone*, Disertacion i paraqitur per kerkimin e grades shkencore "Kandidat i Shkencave", Tirane.

Sulstarova E. and Kociaj S., (1975) *Katalogu i termeteve te Shqiperise.* Botim i Qendres Sizmologjike, Tirane.

Sulstarova E., (1986) *Mekanizmi i vatrave te termeteve dhe fusha e sforcimeve tektonike te sotme ne Shqiperi*, Disertacion per kerkimin e grades shkencore "Doktor i Shkencave", Tirane.

STATISTICAL ESTIMATION OF THE MAXIMUM MAGNITUDE AND ITS UNCERTAINTY FROM A CATALOGUE INCLUDING MAGNITUDE ERRORS.

B. OSHER
Joint Institute of Physics of the Earth,
B.Gruzinskaya 10, 123810, Moscow, Russia.

Abstract

Statistical methods of the maximum magnitude (M_{max}) estimation are based usually on the assumption that the earthquake magnitudes are known exactly. Actually there are errors in earthquake magnitudes of modern and historic events. Taking into account the standard deviations of errors in earthquake magnitudes results in modified distribution of observed magnitudes. The new magnitude distribution slightly differs from the Gutenberg-Richter's one for the large magnitudes and can explain non linear character of observed magnitude-frequency curves.

On a basis of the new distribution the formulas to obtain M_{max} confidence limits for large samples are derived. Numerical method for calculating exact M_{max} confidence limits for arbitrary sample size is also proposed.

Maximum likelihood estimates of M_{max} based on the new distribution are compared with the common estimates of maximum magnitude equal to the maximum of observed values. On the example of artificially generated catalogues the behaviour of the estimates for the different sample sizes and different levels of the magnitude errors is analysed. It is shown that the uncertainty in the M_{max} is usually much higher then the errors in initial magnitudes in catalogue.

As an example the M_{max} and its uncertainty was estimated for Caucasus region. We show that b value estimated with and without considering magnitude uncertainty are almost the same.

1. Introduction

To estimate the seismic hazard, a statistical model of earthquakes occurrence has to be selected. We restrict our analysis by the commonly assumed Gutenberg-Richter's magnitude-frequency relationship:

$$\lg(N) = a - bm, \tag{1}$$

V. Schenk (ed.), Earthquake Hazard and Risk, 25–37.

where N is the number of earthquakes in interval m $\pm$ Δm, a and b are parameters. Usually, this relationship is supposed to be valid up to maximum magnitude (M_{max}). In practice, the earthquake magnitudes are known from some threshold magnitude ($\underline{M}$). It is important that the value of $\underline{M}$ characterizes the completeness of our data and the equation (1) is still valid below $\underline{M}$ even for microfractures in rocks (see Scolz (1968)). Many methods of the estimation of parameter b in (1) were introduced. (See Bender (1983) for the comparison of different methods). The maximum likelihood estimates of M_{max} and its confidence limits was studied by Pisarenko (1992). All the estimates mentioned above are based on the assumption that the earthquake magnitudes are known exactly or grouped magnitude data are available (see Weichert (1980)).

The magnitudes of earthquakes are always defined with errors in real catalogues. Even instrumental magnitudes calculated by averaging are known with some uncertainty. Also there are round off errors. The magnitudes of the historic events are calculated from intensity data by inexact correlation formulas. Due to the lack of the completeness and reliability of the historic data, the errors in magnitude estimates of such earthquakes can be as much as one unit of magnitude and even more. However, we have to include all events in our consideration because, usually, the instrumental observation period is not long enough for the reliable maximum magnitude estimation and often the strongest events are the historic ones.

The model with magnitude errors was introduced in (Tinti and Mulargia, 1985). In the beginning of our considerations we start from this model. Then we get more simple and easy to use magnitude distribution. Unlike the Tinti and Mulargia model, our new distribution differs slightly from the Gutenberg-Richter's one only for the large magnitudes.

In the second part of the paper, we build the maximum likelihood estimates of parameters of the derived distribution taking into account the unequal observation periods for different magnitudes. Then we analyse how the variance and confidence limits of M_{max} estimators are depend on the sample size and the value of magnitude uncertainty. At the end, we estimate maximum magnitude and its uncertainty for the catalogue of Caucasus.

2. Modified Gutenberg-Richter's distribution of the observed earthquakes magnitudes

2.1. DERIVATION OF THE OBSERVED MAGNITUDES DISTRIBUTION

Our derivation of the new distribution is based on four assumptions:
a) the earthquake magnitudes obey Gutenberg-Richter's law;
b) the observed magnitudes are measured with error;
c) the magnitudes of successive events are independent;
d) the magnitudes are completely recorded from some threshold magnitude.

In the next three chapters we formulate these assumptions in mathematic form. As obvious from the above list, the Poisson or any other model of earthquake occurrence was not postulated. Such postulates are often used for M_{max} estimation (e.g. Kijko and Selevoll 1989,1992) but they are not necessary.

2.1.1. Gutenberg-Richter's model

To rewrite equation (1) in statistic form we denote by ξ the random value of magnitude. Then the probability density function for ξ is following:

$$f_\xi(m) = c \cdot \exp(-\beta m) \qquad M_1 \le m \le M_{max} \tag{2}$$
$$f_\xi(m) = 0 \qquad \text{otherwise}$$

where

$$c = \beta / \{\exp(-\beta M_1) - \exp(-\beta M_{max})\}, \tag{3}$$
$$\beta = b/\lg(e),$$

M_1 - denotes magnitude from witch (2) is valid.

(We introduce new notation M_1 to stress the new meaning of this parameter. For practical application of the model (2) without magnitude errors we can select M_1 equal to threshold magnitude $\underline{M}$).

2.1.2. Taking into account the magnitude uncertainty

Next step is to take into account the values of the magnitude errors. As in (Tinti and Mulargia, 1985) we assume that:

$$\langle\text{observed magnitude } \eta\rangle = \langle\text{true magnitude } \xi\rangle + \langle\text{random error } \varepsilon\rangle, \tag{4}$$

where random error ε is assumed to be normally distributed with zero mean and standard deviation equal to σ. This assumption is the simplest but it is possible to use any other magnitude error model. As follows from (4) the probability density function (p.d.f.) for the observed magnitudes $f_\eta(m)$ is equal to convolution of the function $f_\xi(m)$ defined by (2) with p.d.f. of normal distribution. The simple calculations result in the following expression:

$$f_\eta(m) = c_\eta \cdot g(m), \tag{5}$$

where

$c_\eta = c \cdot \exp(\sigma^2\beta^2/2)$, c is defined by (3),

$g(m) = \{\Phi[(M_{max}-m)/\sigma+\sigma\beta] - \Phi[(M_1-m)/\sigma+\sigma\beta]\} \cdot \exp(-\beta m)$,

$\Phi(x)$ is error function (see Abramowitz and Stegun, 1970) defined by:

$$\Phi(x) = 1/\sqrt{2\pi} \cdot \int_{-\infty}^{x} \exp(-t^2/2)\,dt \tag{6}$$

The p.d.f. $f_\eta(m)$ is not equal to zero for any value of magnitude m. In contrary to the initial model (2) this distribution has neither left no right cut-off magnitude. The p.d.f. defined by (5) is equivalent to the function introduced in by Tinti and Mulargia when it is considered that contrary to the usual practice they normalise their p.d.f to total number of earthquakes.

2.1.3. Taking into account the restricted interval of observations

Usually we know that the earthquake magnitudes are completely recorded from some threshold magnitude $\underline{M}$. To apply the distribution (5) to our observations, we use it only for $m \geq \underline{M}$. So we get the following p.d.f. $f_\eta(m)$:

$$f_\eta(m) = c_g \cdot g(m) \quad m \geq \underline{M} \tag{7}$$
$$f_\eta(m) = 0 \quad \text{otherwise,}$$

where g(m) is defined in (5) and norm factor c_g is given by

$$c_g = 1 / \int_{\underline{M}}^{\infty} g(m)dm \tag{8}$$

In the Gutenberg-Richter's model (2) the parameter M_1 can be treated as threshold magnitude $\underline{M}$. However in the model (7) the parameter M_1 has other meaning. This parameter characterizes the lowest magnitude for which the initial exponential law (2) is valid. But the exponential law (2) is valid even for microearthquakes. So, let $M_1 \ll \underline{M}$ in (7). As can be easily proved (see Appendix), the second term of function g(m) is much less then the first term for $m > \underline{M}$ and can be neglected. Thus we get new distribution of observed magnitudes $p_\eta(m)$:

$$p_\eta(m) = c_p \cdot \Phi[(M_{max}-m)/\sigma+\sigma\beta] \cdot \exp(-\beta m) \quad m \geq \underline{M} \tag{9}$$
$$p_\eta(m) = 0 \quad \text{otherwise,}$$

where norm factor c_p is given by formula:

$$c_p = \beta \,/\, \{\Phi[\frac{(M_{max}-\underline{M})}{\sigma} + \sigma\beta] \cdot \exp(-\beta\underline{M}) - \Phi[\frac{(M_{max}-\underline{M})}{\sigma}] \cdot \exp(-\beta M_{max} - \frac{\sigma^2\beta^2}{2})\}$$

Our new probability density function defined by (9) depends upon four parameters ($\underline{M}$, M_{max}, σ, β). Two of them: magnitude threshold $\underline{M}$ and the standard deviation of magnitude error σ are estimated for every earthquake, two other parameters: slope β and maximum magnitude M_{max} have to be estimated.

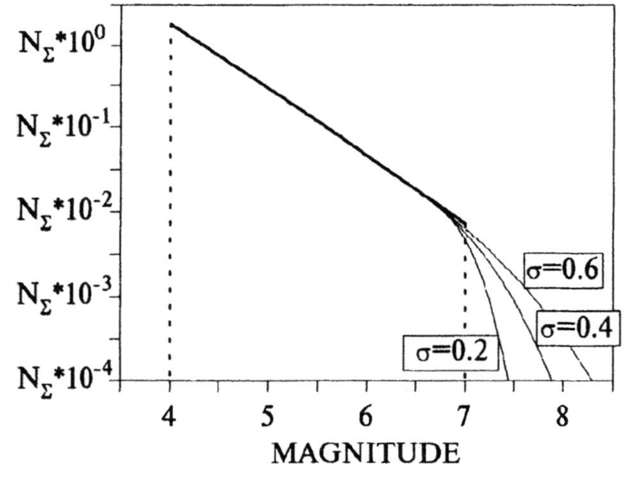

Figure 1. Three examples of magnitude-frequency curves defined by equation (9) for three values of magnitude errors: σ= 0.2, 0.4, 0.6 are shown by thin lines. The initial magnitude-frequency curve (2) is shown by solid line. ($\underline{M}$=4.0, M_{max}=7.0, β=0.8)

To see how the p.d.f (9) looks like we present it in graphic form. To draw the magnitude distribution function here and below we multiply it by the total number of earthquakes N_Σ and use logarithmic scale for ordinate axis. Such graphs are common for frequency- magnitude relations.

The figure 1 illustrates the typical example of distribution (9) for three values of standard deviation of magnitude errors. The initial distribution (2) is also shown on the same figure. Both distributions coincide up to approximately M_{max}-0.5. The greater is the value of the standard deviation σ the greater is probability to observe magnitudes more then M_{max}. The absence of the right cut-off magnitude for distribution (9) results from the selected normal noise model.

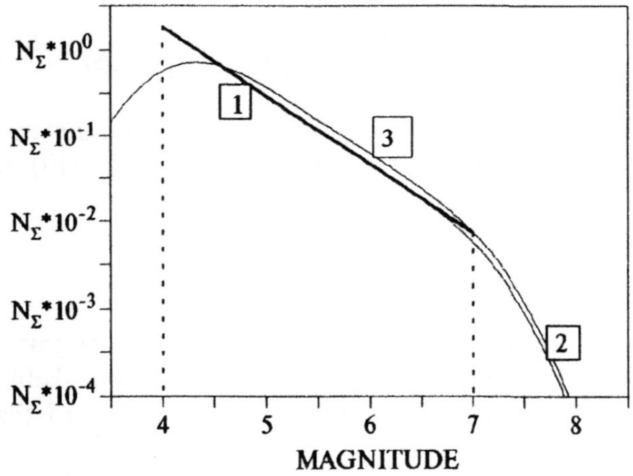

Figure 2. Three types of magnitude-frequency curves. The first curve is Gutenberg-Richter's defined by (2), the second is defined by equation (9) and the third is distribution of (Tinti and Mulargia, 1985) - equation (5). ($\underline{M}$=4.0, M_{max}=7.0, β=0.8, σ= 0.4)

2.2. COMPARISON OF DIFFERENT MAGNITUDE DISTRIBUTIONS

To compare the different magnitude distributions we show three of them in the figure 2. It is clear from the figure that the first and the second distributions differ slightly from each other only for the large magnitudes. The third distribution is shifted considerably from the first and the second ones. The shift is originated from the absence of left cut-off magnitude in the third model.

3. Estimation of the distribution parameters

Let our observed magnitudes $m_1,m_2...m_n$ are measured with errors $\sigma_1,\sigma_2...\sigma_n$ and completely recorded over magnitude $\underline{M}$. (We left the case when only group data available beyond the scope of this paper.) Regarding the magnitudes of successive events as independent the logarithm likelihood function for the model (9) is given by:

$$L(M_{max},\beta) = \sum_{i=1}^{n} \ln(p_\eta(m_i,\underline{M},M_{max},\sigma_i,\beta)) \tag{10}$$

Then we have to take into account the unequal observation periods for different magnitude thresholds. Let magnitude records are complete for $m \geq \underline{M}_1$ in the time interval from t_1 to T_1, magnitudes $m \geq \underline{M}_2$ are completely recorded in the time interval $[t_2,T_2]$ and so on. Selecting mutually disjoint time intervals for every group of completeness we get the unique time interval $[t_i,T_i]$ and the corresponding threshold magnitude $\underline{M}_i$ for every observed magnitude m_i. Thus we get the following logarithm likelihood function:

$$L(M_{max},\beta) = \sum_{i=1}^{n} \ln(p_\eta(m_i,\underline{M}_i,M_{max},\sigma_i,\beta)) \tag{10}$$

The maximum likelihood estimates of unknown parameters β and M_{max} are obtained from the maximizing of likelihood function (10) or (11). So

$$\{\boldsymbol{\beta_L}, \mathbf{M^L_{max}}\} = \arg\max\,(L(M_{max}, \beta)), \tag{12}$$

where

$\boldsymbol{\beta_L}$ is estimate of β, $\mathbf{M^L_{max}}$ is estimate of M_{max}.

Here and below we denote by bold letters the estimates of parameters.

For the Gutenberg-Richter's model (2) the maximum likelihood estimate of M_{max} is equal to the maximum observed magnitude and the estimate of the slope β is given by

$$\boldsymbol{\beta_{GR}} = \arg\max\left(\sum_{i=1}^{n} \ln(f_\xi(m_i,\underline{M}_i,M_{max},\beta)\right) \tag{13}$$

The numerical examples presented below show that usually the estimates $\boldsymbol{\beta_L}$ and $\boldsymbol{\beta_{GR}}$ are close to each other. This can be expected because two distributions differ only for the large magnitudes where only a few of the strongest events are present. So, common methods for estimation of slope β (see Bender (1983)) can be successfully used even if the earthquake magnitudes are recorded with errors.

4. Confidence limits and variance of the maximum magnitude estimators

4.1. METHODS OF ESTIMATION

It is possible to build confidence limits for both unknown parameters of our distribution (9). In the present paper we focus our attention on the study of uncertainties in maximum magnitude estimates and suppose that slope β can be estimated exactly. This simplification can result in the decrease of the maximum magnitude uncertainty. In real situation the value of the slope β is controlled by the large number of small earthquakes and so it can be estimated much more precisely then the maximum magnitude which is mainly controlled by the strongest events.

We use two approaches for estimation of the uncertainty associated to the maximum magnitude estimates. The first approach is based on the asymptotic properties of the maximum likelihood estimates $\mathbf{M}^{\mathbf{L}}_{\mathbf{max}}$. As shown in Cramer (1963), the estimator $\mathbf{M}^{\mathbf{L}}_{\mathbf{max}}$ follows the asymptotically normal distribution with mean equal to the true value M_{max} and variance defined by:

$$\mathrm{var}(\mathbf{M}^{\mathbf{L}}_{\mathbf{max}}) = 1/\mathrm{I}(\mathbf{M}^{\mathbf{L}}_{\mathbf{max}}), \tag{14}$$

$$\mathrm{I}(M_{max}) = \mathbf{E}\left[\partial L(M_{max},\beta) / \partial M_{max}\right] = \sum_i \mathrm{I}_i(M_{max}),$$

where

$$\mathrm{I}_i(M_{max}) = \mathbf{E}\left[\partial \ln(p_\eta(x_i, \underline{M}_i, M_{max}, \sigma_i, \beta)) / \partial M_{max}\right].$$

Simple calculations result in the following expressions:

$$\mathrm{I}_i(M_{max}) = \frac{c_p \exp(-\beta M_{max} - \sigma_i^2\beta^2)}{2\pi\sigma_i} \cdot K(\sigma_i\beta, u) - c_p^2 \cdot \Phi^2\left[\frac{(M_{max} - \underline{M}_i)}{\sigma_i}\right] \cdot \exp(-2\beta M_{max} - \sigma_i^2\beta^2)$$

$$u = (M_{max} - \underline{M}_i)/\sigma_i + \sigma_i^2\beta^2, \qquad K(\gamma, u) = \int_{-\infty}^{u} \frac{\exp(-x^2 + x\gamma)}{\Phi[x]}dx \tag{15}$$

This estimate of variation can be easily calculated but it is valid only for the large samples. We compare this estimate with the more precise one in the chapters 4.2 and 4.3.

The second approach is based on the Monte Carlo method. For a set of values M_{max} we generate N samples $\{m_1, m_2 \ldots m_n\}_j$ corresponding to the distribution (9). (The value of slope β is supposed to be the same for all values M_{max} and equal to its estimate from (12).) Then we calculate an estimate $\mathbf{M}_{\mathbf{max}}$ ($\mathbf{M}^{\mathbf{L}}_{\mathbf{max}}$ or $\mathbf{M}^{\mathbf{M}}_{\mathbf{max}}$) for every sample. Treating $\mathbf{M}_{\mathbf{max}}$ as random variable we can calculate the estimates of mean value $\overline{\mathbf{M}}_{\mathbf{max}}$ for every M_{max}. Setting up the confidence level p we can estimate the quantile $q_p(M_{max})$ defined by equation:

$$P\{\mathbf{M}_{\mathbf{max}} > q_p(M_{max})\} = p \tag{16}$$

The figure 3 illustrates the typical example of functions $q_p(M_{max})$ and $\overline{\mathbf{M}}_{\mathbf{max}}(M_{max})$. The value $\overline{\mathbf{M}}_{\mathbf{max}}$ is estimator of $E[\mathbf{M}_{\mathbf{max}}]$ (math expectation is calculated for fixed sample size n). If M_{max} is large, in almoust all random samples there are no observed magnitudes near or greater M_{max}, so as can be seen from the graph $M_{max} > \overline{\mathbf{M}}^{\mathbf{M}}_{\mathbf{max}}$ and $M_{max} > \overline{\mathbf{M}}^{\mathbf{L}}_{\mathbf{max}}$. If M_{max} is small, there is big probability that some observed magnitude is greater then M_{max} due to observation errors, so in this case $M_{max} < \overline{\mathbf{M}}^{\mathbf{M}}_{\mathbf{max}}$ and $M_{max} \approx \overline{\mathbf{M}}^{\mathbf{L}}_{\mathbf{max}}$. What is 'small' and what is 'large' also depends upon sample size n and slope β. Thus estimators $\mathbf{M}^{\mathbf{L}}_{\mathbf{max}}$ and especially $\mathbf{M}^{\mathbf{M}}_{\mathbf{max}}$ are biased and can be used with caution. However it is often possible to build confidence limits for both estimators.

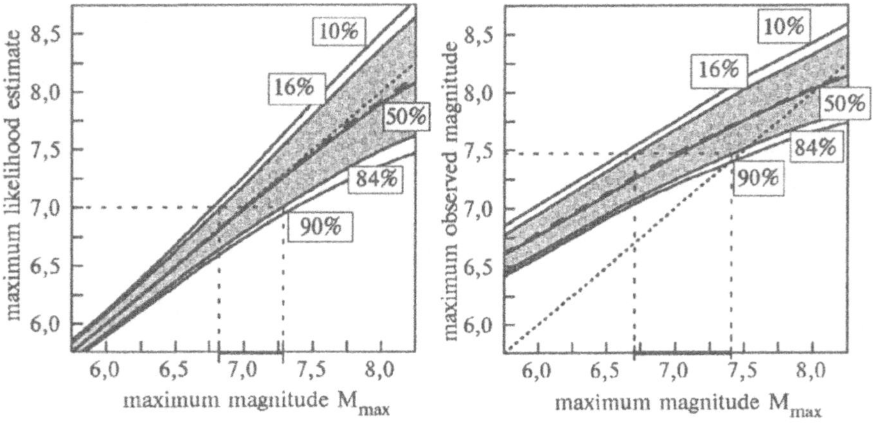

Figure 3. The examples of functions $q_p(M_{max})$ (solid lines) and $\overline{\mathbf{M}_{max}}$ (M_{max}) (dash lines) for two types of maximum magnitude estimates. (a) - maximum likelihood estimate $\mathbf{M}^{\mathbf{L}}_{\mathbf{max}}$; (b) - maximum observed magnitude $\mathbf{M}^{\mathbf{M}}_{\mathbf{max}}$. The values of the confidence levels p are shown on the graph. Bold horizontal segments show 68% confidence intervals for $M_{max}=7$. The 68% confidence area is shown by grey color. ($\underline{M}=4$, $\beta=0.8$, $\sigma= 0.4$, $n=500$)

If the function $q_p(M_{max})$ has inverse then the equation (16) can be rewritten as:

$$P\{M_{max} < q_p^{-1}(\mathbf{M_{max}})\} = p \tag{17}$$

The last equation shows that $q_p^{-1}(\mathbf{M_{max}})$ is equal to the upper confidence limit corresponding to probability p. Let p_1, p_2 are probabilities and $p_1 > p_2$. Then

$$P\{\, q_{p_1}^{-1}(\mathbf{M_{max}}) \le M_{max} < q_{p_2}^{-1}(\mathbf{M_{max}})\} = p_1 - p_2, \tag{18}$$

and $[q_{p_1}^{-1}(\mathbf{M_{max}}), q_{p_2}^{-1}(\mathbf{M_{max}})]$ is confidence interval corresponding to probability $r = p_1 - p_2$. The single difficulty in the practical application of this method is to build function $q(M_{max})$ with sufficient accuracy. According to Cramer (1963) the sampling quantile estimate is asymptotically normal distributed with mean equal to the true value of the quantile and standard deviation proportional to $1/\sqrt{N}$. So, the number of samples N has to be large enough to get the necessary accuracy. In the most cases considered below 2000 - 4000 samples were enough to get accurate estimates but in some cases more random samples were generated.

4.2. RANDOM EARTHQUAKE CATALOGUES

To study the influence of the magnitude uncertainty on the M_{max} estimates the series of random catalogues were generated. The example of confidence intervals for two types of maximum magnitude estimates are shown on the figure 3. It is clear from the graph that maximum observed estimate M^M_{max} results in much wider confidence intervals than maximum likelihood estimate M^L_{max}. It is also clear that the estimate M^M_{max} is considerably biased for M_{max} far from 8.

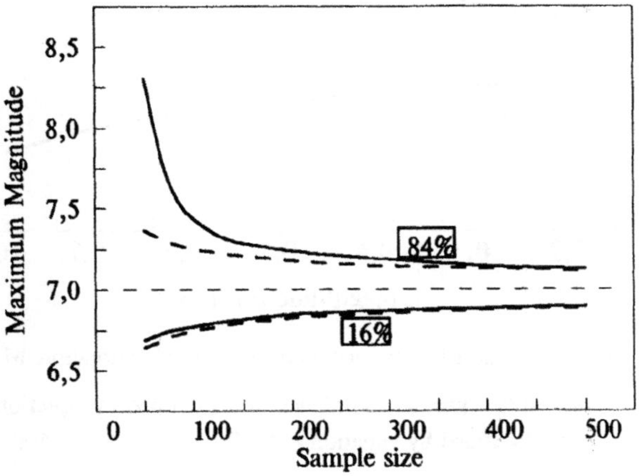

Figure 4. M_{max} confidence limits for the maximum likelihood estimation M^L_{max} as a function of sample size. Solid lines - precise confidence limits defined by equation (17). Dash lines - asymptotic confidence limits defined by equation (14). ($\underline{M}$=4, β=0.8, σ= 0.2, M_{max}=7)

There are two main factors affecting the M_{max} estimates uncertainty. The first factor is the size of sample. The second factor is the values of magnitude errors. To analyse this factors we evaluate the confidence intervals for a set of sample sizes n and a set of magnitude errors σ. The results are shown on the figures 4 and 5. As can be seen from the figure 4 the uncertainty in M_{max} increases dramatically as sample size decreases. In this example the 84% upper confident limit is 1.3 magnitude units higher then the maximum magnitude estimate when sample size is equal to 50. This can be explained by exponential character of initial magnitude-frequency law (2) as following. When the number of observations is small the probability to record magnitude near M_{max} is very low and so the uncertainty is very high. The asymptotic confidence limits defined by equation (14) tend to precise ones when the sample size increases.

As may be seen from the figure 5 uncertainty in M_{max} is much higher then the corresponding standard deviation of errors σ. (The 68% confidence area between upper and low limit in the figure corresponds to one standard deviation of normal distribution).

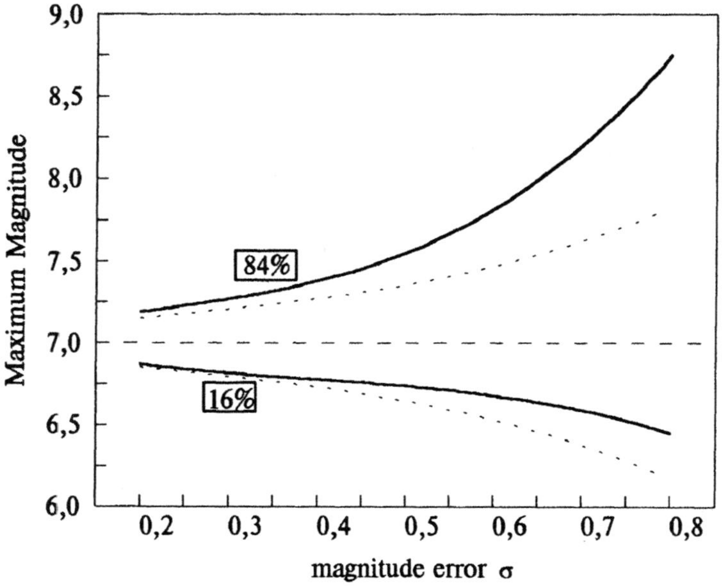

Figure 5. M_{max} confidence limits for the maximum likelihood estimation M^{L}_{max} as a function of magnitude error σ. Solid lines - precise confidence limits defined by equation (17). Dash lines - asymptotic confidence limits defined by equation (14). ($\underline{M}$=4, β=0.8, n= 500, M_{max}=7)

In real situation the two factors mentioned above acts together. The example of real catalogue is presented in the next chapter.

4.3. THE EARTHQUAKE CATALOGUE OF CAUCASUS

For illustration we estimated M_{max} for region including main part of Great and Lesser Caucasus ($\phi=40\text{-}44^0$, $\lambda=41\text{-}51^0$). We used data from two catalogues (New catalogue of strong earthquakes .. 1982 and General catalogue.. 1994). The parameters and results are presented in the table 1. The M_{max} was estimated for two sets of parameters (a) and (b). In the fist case (a) the strongest earthquake (t=1668, m=7.8, σ=0.5) and its aftershocks were excluded from estimation of M_{max}. In the second case (b) the strongest earthquake and its aftershocks were included in the estimation. As it can be seen from the table, the maximum likelihood estimate M^{L}_{max} is almost the same in both cases. The estimates β_L and β_{GR} are close to each other.

The figure 6 illustrates the confidence levels and observed magnitudes with error bars ±σ (in both cases the levels are practically the same). As can be seen from the figure the 68% confidence range for M_{max} is 6.79-7.42, the 90% range is 6.67-7.96. (The M_{max} standard deviation calculated by asymptotic formulas (14)-(15) is equal 0.27.)

TABLE 1. Two variants of b and M_{max} estimation. Parameters and results.

(a)

M_i	t_i	T_i	Total number
3.75	1962	1990	215
4.25	1950	1962	68
4.75	1900	1950	85
5.75	1800	1900	17
6.75	1700	1800	1

$M^{L}_{max} = 7.01$, $M^{M}_{max} = 6.9$,

$\beta_L = 0.72$, $\beta_{GR} = 0.71$

(b)

M_i	t_i	T_i	Total number
3.75	1962	1990	215
4.25	1950	1962	68
4.75	1900	1950	85
5.75	1800	1900	17
6.75	1650	1800	5

$M^{L}_{max} = 7.02$, $M^{M}_{max} = 7.8$,

$\beta_L = 0.72$, $\beta_{GR} = 0.71$

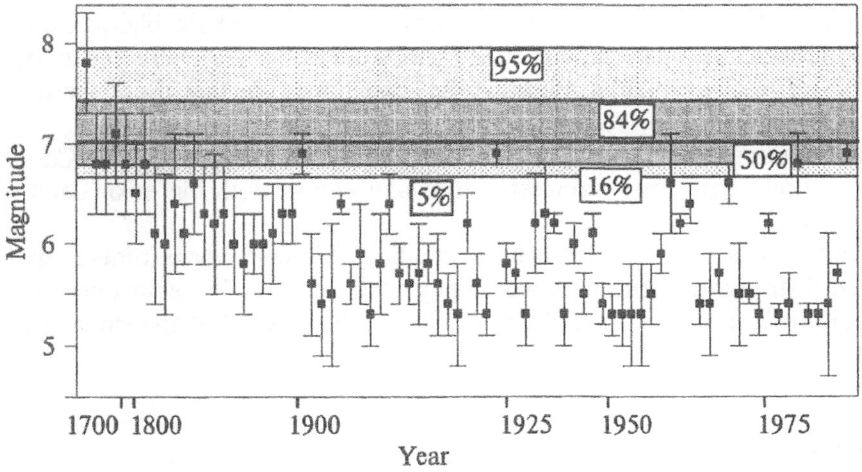

Figure 6. M_{max} confidence limits for the maximum likelihood estimation and initial magnitudes with error bars for region (ϕ=40-44^0, λ=41-51^0). Parameters of estimation are given in Table 1. Probability levels are shown on the graph. Only m>5 are shown. (We use non linear scale for horizontal axis to make the graph readable).

5. The maximum magnitude uncertainty and seismic risk estimation

In the examples considered above, the significant values of maximum magnitude uncertainty were found. The increase of maximum magnitude by 0.5 of unit and more can result in considerable increase in probability of strongest quakes.

To calculate seismic risk, M_{max} have to be estimated in model (2). Similarly to Pisarenko (1991) it is possible to use $q^{-1}_{50}(M_{max})$, $q^{-1}_{84}(M_{max})$ or any other confidence limit defined by (17) as the estimate of M_{max} in (2) to take into account the magnitude uncertainty. It is possible to calculate several versions of seismic risk estimates corresponding to the different confidence limits.

6. Conclusions

A new magnitude distribution taking into account the magnitude uncertainty was derived. This distribution explains non linear character of magnitude-frequency curves for the large magnitudes.

The connection between the magnitude uncertainty and the uncertainty in the estimates of maximum magnitude was studied. It was shown that the maximum magnitude can be estimated much less precisely then the magnitudes of observed earthquakes. This result can be explained by the limited number of available observations and exponential character of the magnitude-frequency law. As well known, the maximum likelihood estimator is asymptotically efficient. So, the uncertainties in M_{max} are result of the nature of the problem and can't be explained by the used method of estimation.

The example of maximum likelihood estimation of M_{max} for Caucasus region shows that even strongest earthquake known with big error can't alter the results significantly.

Our analysis was restricted by Gutenberg-Richter's magnitude-frequency low in distributed form (1) however any other earthquake recurrence models and other forms of the magnitude errors can be studied at the same manner as above.

7. References

Abramowitz, M. and Stegun I. A. (1964). *Handbook of mathematical functions,* 9th ed., Dover, New York.

Bender, B. (1983) Maximum likelihood estimation of b values for magnitude grouped data, *Bull. Seism. Soc. Am*, 73, pp. 831-851.

Cramer, H. (1963) *Mathematical methods of statistic*, Princeton University Press, Princeton, N.J.

General catalogue of strong earthquakes of North Eurasia, Page I, (1994) Editor Shebalin, N.V. IPE report, Moscow.

Gutenberg, B. and Richter, C. F. (1954). *Seismicity of the Earth*, 2nd ed., Princeton University Press, Princeton, New Jersey.

Kijko, A. and Sellevoll, M.A. (1989). Estimation of earthquake hazard parameters from incomplete data files. Part I. Utilization of extreme and complete catalogs with different threshold magnitudes., *Bull. Seism. Soc. Am*, 79, pp. 645-654.

Kijko, A. and Sellevoll, M.A. (1992). Estimation of earthquake hazard parameters from incomplete data files. Part II. Incorporation of magnitude heterogeneity., *Bull. Seism. Soc. Am*, 82, pp. 120-134.

New catalogue of strong earthquakes in the USSR from ancient times through 1977, (1982). World Data Center A for Solid Earth Geophysics. Report SE-31.

Pisarenko, V.F. (1992). Statistical estimation of maximum possible earthquakes. *Physics of The Solid Earth*, v.27, No. 9, pp.757-763.

Tinti, S. and Mulargia, F.(1985) Effects of magnitude uncertainties on estimating the parameters in the Gutenberg-Richter frequency-magnitude law, *Bull. Seism. Soc. Am*, 75, pp. 1681-1697.

Tinti, S., Rimondi, R. and Mulargia, F. (1986). On the frequency-apparent magnitude relations, *Ann. Geophys.*, 4, Ser. B, No. 4, pp. 465-472.

Scolz, C.H. (1968). The frequency-magnitude relation of microfracturing in rock and its relation to earthquakes, *Bull. Seism. Soc. Am*, 58, pp. 399-415.

Weichert, D.H. (1980). Estimation of earthquake recurrence parameters for unequal observation periods for different magnitudes. *Bull. Seism. Soc. Am*, 70, pp. 1337-1346.

Appendix

To estimate ratio $R=\Phi[(M_1-m)/\sigma+\sigma\beta]/\Phi[(M_{max}-m)/\sigma+\sigma\beta]$ for $m>\underline{M}$ we introduce the follows variables:

$$u = (M_{max}-m)/\sigma+\sigma\beta \quad \text{and} \quad \delta = (M_{max}-M_1)/\sigma.$$

Then $R = \Phi[u-\delta]/\Phi[u]$. From equation (6) we have:

$$\Phi(u) = 1/\sqrt{2\pi}\cdot\int_{-\infty}^{u}\exp(-t^2/2)dt = 1/\sqrt{2\pi}\cdot\int_{-\infty}^{u-\delta}\exp(-(t+\delta)^2/2)dt$$

Then for any $t<u-\delta$:

$$1/\sqrt{2\pi}\exp(-t^2/2) < 1/\sqrt{2\pi}\exp(\delta\cdot(u-\delta/2))\cdot\exp(-(t+\delta)^2/2)$$

Integrating the last equation from $-\infty$ to $u-\delta$ we have:

$$\Phi[u-\delta] < \exp(\delta\cdot(u-\delta/2))\cdot\Phi[u]$$

Then $R<\exp(\delta\cdot(u-\delta/2))$. Setting up realistic constrains of parameters:

$$\sigma\le 1,\ \beta\le 1,\ 1\le M_{max}-\underline{M}\le 5,\ \underline{M}-M_1>10,\ m>\underline{M}$$

we have:

$$\delta\ge 11,\ (u-\delta/2)\le(M_{max}-\underline{M})/(2\sigma)-(\underline{M}-M_1)/(2\sigma)+\sigma\beta<-1.5.$$

So, $R<\exp(-1.5\cdot 11)\approx 7\cdot 10^{-8}$ for any $m>\underline{M}$.

ESTIMATION OF UPPER BOUND MAGNITUDE IN EARTHQUAKE HAZARD ASSESSMENT IN SLOVENIA

J. K. LAPAJNE and B. ŠKET MOTNIKAR
Seismological Survey of Slovenia
Kersnikova 3, 61000 Ljubljana, Slovenia

Abstract

Estimates of upper bound magnitudes of seismic sources based on historical seismicity are commonly used as scaling parameters in areas that lack identified causative faults. The analysis of the earthquake catalogue of Slovenia and surrounding areas showed that for the region of Slovenia a doubly truncated exponential frequency-magnitude relationship may serve directly for this purpose, taking into account the following assumptions:

(1) The value of the upper bound magnitude, for which the corresponding least-squares estimator of the decay rate has a minimum standard error, is an optimal historical seismicity estimate of the upper bound magnitude.

(2) The difference between the estimated upper bound magnitude and the largest observed magnitude in an area with a well-defined earthquake catalogue is an appropriate increment (typically 0.1 to 0.2) for estimating upper bound magnitudes of seismic sources in the same area with the incremental technique.

There is an indication that such least-squares estimates are similar to or a bit greater than Kijko and Sellevoll's maximum likelihood estimates (1989). Both might be used with reasonable weights as alternatives for upper bound magnitudes in seismic hazard assessment in Slovenia.

Key words: *upper bound magnitude, least-squares estimate, maximum-likelihood estimate*

V. Schenk (ed.), Earthquake Hazard and Risk, 39–48.

1. Introduction

In earthquake hazard assessment, estimates of maximum expected magnitudes are based on geological and geophysical constraints and/or on earthquake catalogues. Estimates of maximum magnitudes based on geological and geophysical constraints are generally preferred (Doser and Smith, 1982; Krinitzsky and Slemmons, 1990; Salyards, 1991; Schwartz and Coppersmith, 1984; Wells and Coppersmith, 1994; Wesnousky, 1986; Youngs and Coppersmith, 1985). Nevertheless, estimates based on historical seismicity are the most commonly used scaling parameters in areas that lack identified causative faults (e.g., Krinitzsky and Slemmons, 1990).

A common way to use an earthquake catalogue is to estimate seismic parameters from frequency-magnitude relationship, e.g., a doubly truncated exponential (Youngs and Coppersmith, 1985):

$$N(m) = N(m_o)\,(1 - \exp(-\beta(m_u - m_o)))^{-1}\,(\exp(-\beta(m - m_o)) - \exp(-\beta(m_u - m_o))), \qquad (1)$$

where $N(m)$ is the cumulative number of earthquakes per year equal to or greater than magnitude m, $N(m_o)$ is the total number of earthquakes per year equal to or greater than the lower bound magnitude m_o, m_u is the upper bound magnitude, and β is the decay rate.

A statistically proper method of estimating β is the method of maximum likelihood (Aki, 1965; Page, 1968; Utsu, 1966; Weichert, 1980). The earthquake activity rate $N(m_o)$ can simply be the total number of events above m_o that equals the threshold of catalogue completeness (Weichert, 1980) or can be estimated (Kijko and Sellevoll 1989). Upper bound magnitude m_u is usually estimated from independent considerations. Using the assumption that the largest observed magnitude equals the largest expected magnitude in the time span of the catalogue, Kijko and Sellevoll (1989) found the maximum likelihood solution for m_u as well.

2. Least-Squares Estimate of Upper Bound Magnitude

The conventional method of least-squares is a reasonable approximation for well-defined earthquake data (Weichert, 1980). This is clearly demonstrated by **Fig. 1**, which gives maximum likelihood (ML) and least-squares (LS) doubly truncated exponential curves. The ML parameters and curve have been calculated according to the Weichert (1980) method, grouping events into magnitude classes of $m \pm 0.05$. (In the earthquake catalogue magnitudes are given to the first decimal point.)

In Fig. 1 the complete part of an aggregated catalogue consisting of the earthquake catalogue of Slovenia and the catalogues of bordering regions of the neighboring countries of Italy, Austria, Hungary and Croatia has been used as the earthquake data set (Živčić, 1992). The aggregate catalogue covers the period 567-1990 A.D. and an area of approximately 70,000 km^2. The magnitude m_{LH} defined by Karnik (1968) has been selected as a uniform measure of earthquake size (Živčić, 1992). The maximum historical event has been determined to be 6.5. The total number of events for m_{LH} equal to or greater than 3.5 after removing aftershocks is 842. The catalogue may be considered complete from the year 1870 for m_{LH} equal to or greater than 3.7. This value has been taken as the lowest applicable threshold of catalogue completeness. The total number of events with a magnitude of 3.7 and greater is 549, and the maximum observed magnitude in the period 1870-1990 is 6.1.

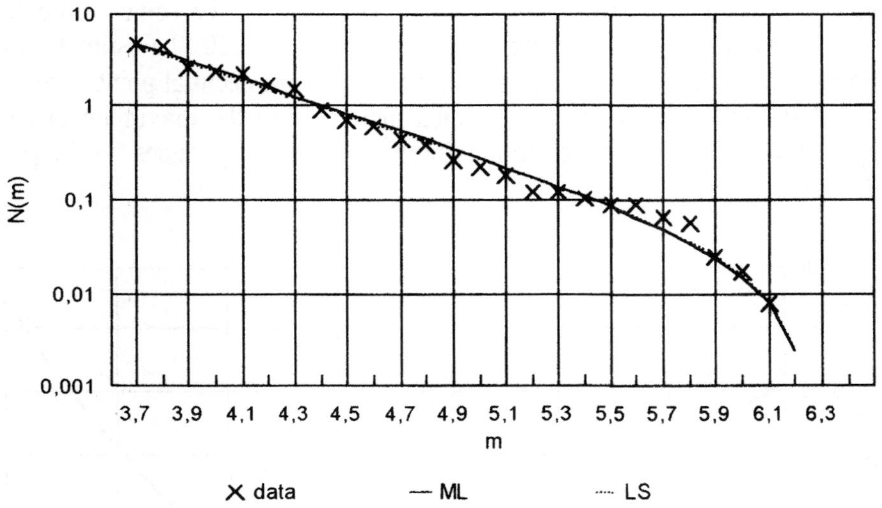

Data: Period: 1870-1990, No. of events = 549; m_o = 3.7;
ML: $N(m_o) = 4.54$, $\beta = 2.09 \pm 0.10$, $m_u = 6.25$;
LS: $N(m_o) = 4.20 \pm 0.06$, $\beta = 2.04 \pm 0.04$, $m_u = 6.25$.

Figure 1. Estimation from the complete part of the catalogue.

In the case of a well-defined data set for seismic sources or regions where seismicity is not clearly associated with tectonic structures (as in the case of the area under consideration), let the possible estimation of m_u rests on the following presumed criteria:

(1) The value of m_u for which the corresponding least-squares estimator of β has a minimum standard error is taken as the optimal historical seismicity estimate of m_u. (In this procedure, the estimators of the activity rate $N(m_o)$, the

decay rate β, and the upper bound magnitude m_u are mutually dependent. They are calculated by an iterative procedure.)

(2) The differences between estimated upper bound magnitudes and largest observed magnitudes for sources or regions with well-defined data sets are used in the incremental technique for estimating the m_u of seismic sources in the same area.

Fig. 2 shows the standard error of the least-squares estimator of β (SE(beta)) as a function of m_u for the earthquake data used in Fig. 1. The minimum standard error is at $m_u = 6.25$. Fig. 2 also shows that the curve of the squared standard error of the estimator of $\log N(m)$ (denoted as "mean square error") has a similar course and has its minimum at the same value for m_u. The value 6.25 has been used as the upper bound magnitude in the calculation of ML and LS curves in Fig. 1. (It should be pointed out that the ML curve is relatively insensitive to the value of m_u.) The difference between 6.25 and the maximum observed magnitude is 0.15. Similar calculations of m_u for some well-defined subsets of events have given values between 0.10 and 0.20. To estimate upper bound magnitudes for seismic sources in the area to one decimal point, a rounded value of 0.2 might be used as a regional increment to the maximum observed magnitudes. It should be added to maximum observed magnitudes in the period 567-1990.

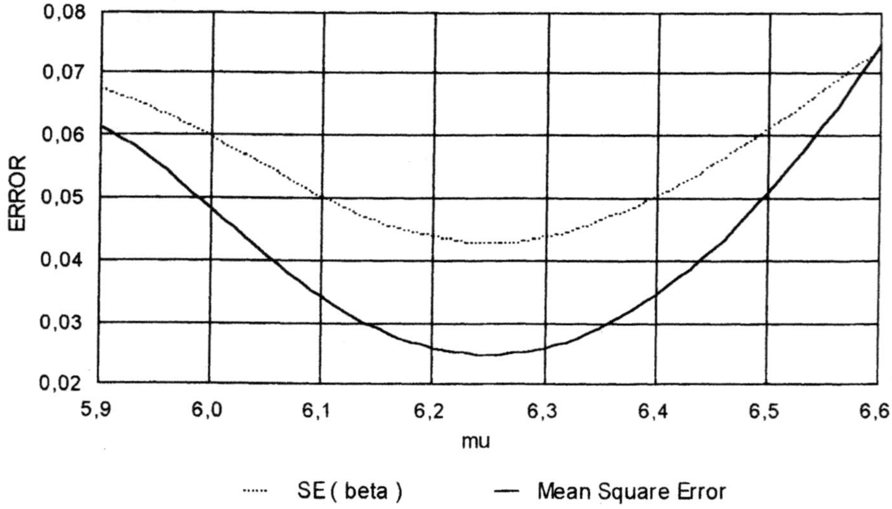

Figure 2. Least-squares errors.

For comparison, **Figs. 3** and **4** show ML and LS curves for poorly defined data sets of two selected seismic sources A1 and A2 from the bordering region of Slovenia and Croatia. The calculation procedure is the same as in Fig. 1. The differences are obvious. In the case of source A2, the optimal LS estimate of the upper bound magnitude is found to be less than the maximum observed value 5.7 (caused by the absence of magnitudes between 5.0 and 5.7). Therefore, 5.7 has been imposed as the m_u.

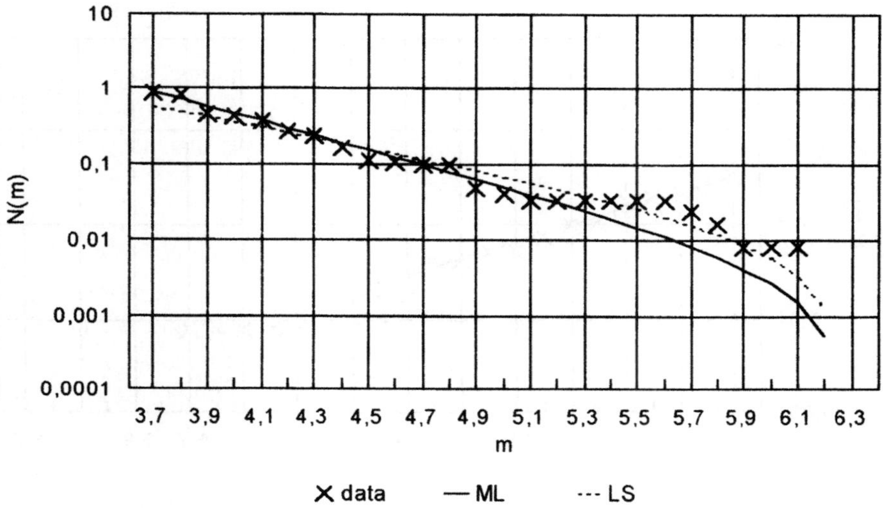

Data: Period: 1870-1990, No. of events = 110; $m_0 = 3.7$;
ML: $N(m_0) = 0.91$, $\beta = 2.19 \pm 0.22$, $m_u = 6.27$;
LS: $N(m_0) = 0.59 \pm 0.10$, $\beta = 1.56 \pm 0.07$, $m_u = 6.27$.

Figure 3. Estimation from the catalogue of seismic source A1.

In the above calculations, only the last part of the catalogue has been used. We have also taken into account older data using the Weichert method (1980). A summary of input data for the Weichert estimation procedure is given in **Table 1**. Events have been grouped into magnitude classes of $m \pm 0.25$.

TABLE 1. Iput data for the Weichert method.

m	No. of events	Period of completness
3.8	287	1870-1990
4.3	188	1870-1990
4.8	65	1800-1990
5.3	20	1750-1990
5.8	13	1750-1990
6.3	4	1300-1990

Fig. 5 shows ML curves for $m_0 = 3.8$ and two imposed upper bound magnitudes, 6.5 and 7.0. The estimate of β is practically the same as in previous calculations. (It should be pointed out that this needn't be the case.) For comparison, the LS curve is also given. The LS and ML curves and the estimates

of $N(m_o)$ and β are quite different, as the data in this case can not be regarded as well-defined. Nevertheless, the least-squares optimal estimate of m_u is larger by an increment of 0.2 than the maximum observed magnitude according to previous statement.

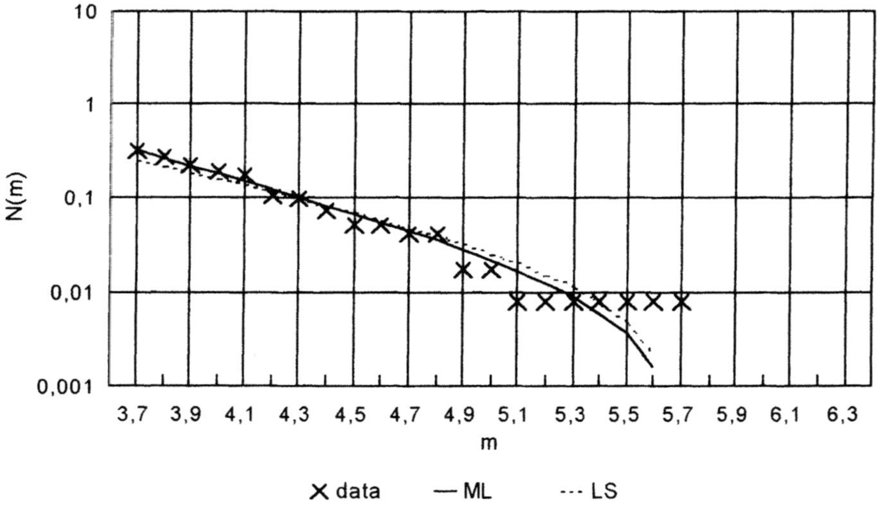

Data: Period: 1870-1990, No. of events = 39; $m_o = 3.7$;
ML: $N(m_o)$ = 0.32, $\beta = 1.86 \pm 0.36$, $m_u = 5.7$;
LS: $N(m_o)$ = 0.26 ± 0.10, $\beta = 1.49 \pm 0.08$, $m_u = 5.7$.

Figure 4. Estimation from the catalogue of seismic source A2.

Visual fitting, still used in some hazard studies, would lie somewhere between the ML and LS curves in Figs. 3, 4, and 5. Intuitively one would expect that the visual fitting should be closer to the least-squares fitting. It is known that Gutenberg and Richter (1944) invented a log-linear frequency-magnitude relationship in the least-squares sense.

3. Maximum Likelihood Estimate of Upper Bound Magnitude

For the maximum likelihood estimate of upper bound magnitude (as well as for estimates of $N(m_o)$ and β), the Kijko and Sellevoll method (1989) has been used. For this procedure, the catalogue has been divided into a complete part (1870-1990, as used in Fig. 1) and an extreme part (567-1869). A summary of input data is given in **Table 2**. The three largest events given in the extreme part occurred in 1348 (6.3), 1511 (6.5), and 1700 (6.3).

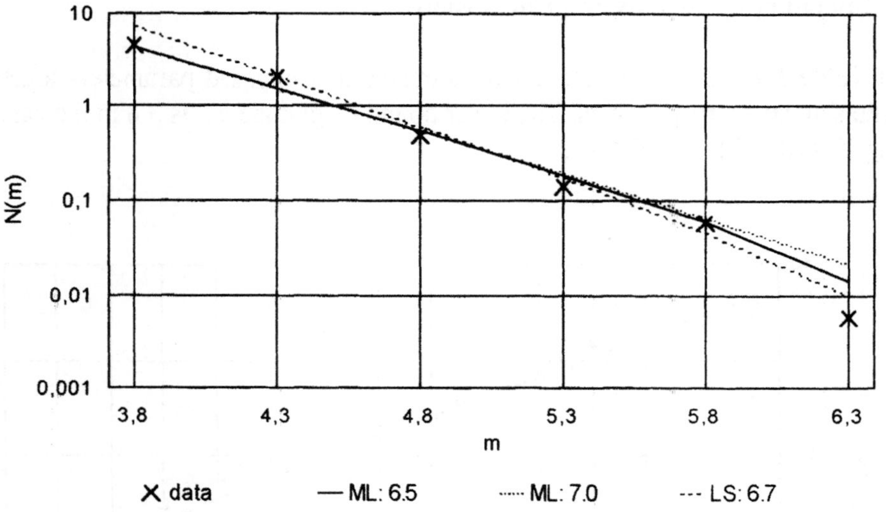

Data: Period (m_0, No. of events):
1870-1990 (3.8, 287), 1870-1990 (4.3, 188),
1800-1990 (4.8, 65), 1750-1990 (5.3, 20),
1750-1990 (5.8, 13), 1300-1990 (6.3, 4);
ML: $N(3.8) = 4.3$, $\beta = 2.05 \pm 0.07$, $m_u = 6.5, 7.0$;
LS: $N(3.8) = 7.10 \pm 1.20$, $\beta = 2.46 \pm 0.10$, $m_u = 6.7$.

Figure 5. Estimation from complete subcatalogue - Weichert method.

TABLE 2. Input data for the Kijko and Sellevoll method.

	Complete part	Extreme part
Period	1870-1990	567-1869
No. of events	549	3
Magnitudes	3.7-6.1	6.3, 6.5, 6.3
Average magnitude	4.12	

Fig. 6 shows ML curve and parameters calculated according to the Kijko and Sellevoll procedure. $N(m_o)$ is practically the same as in Fig. 1 (due to the influence of the complete part of the catalogue). The rounded value of m_u (6.6) is a bit smaller than the corresponding LS estimate (6.7). Only the estimate of β is rather different from previous ML estimates (due to the influence of the extreme part of the catalogue). Let us mention that Slejko and Kijko (1991) estimated the decay rate in the frequency-intensity relationship according to the Kijko and Sellevoll method for some zones of the same area. Converting their values for the decay rate β of the frequency-magnitude relationship, we obtained values ranging from 2.2 to 2.6, similar to our estimate.

4. Summary of Results and Conclusions

In **Table 3** a summary of the above estimated and imposed parameters together with standard errors is given. The lower bound magnitude m_o is 3.8 in the case of Fig. 5, and 3.7 in all other cases.

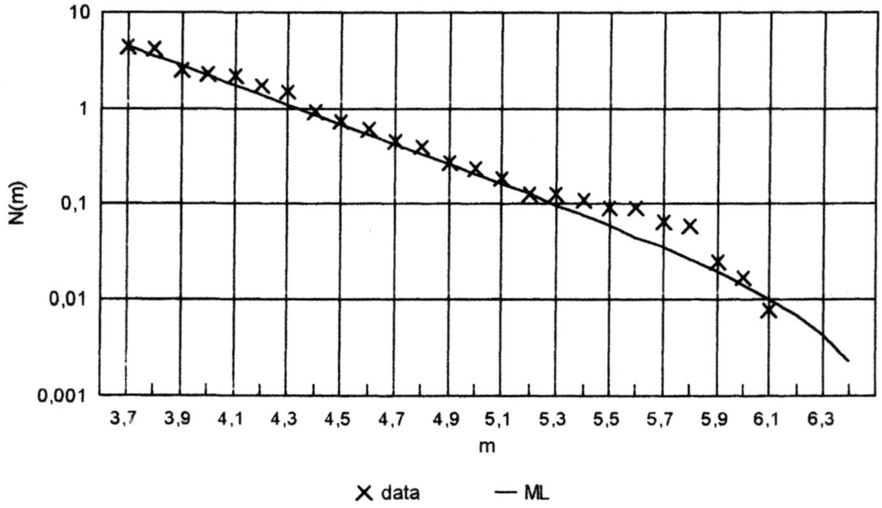

Data: Period (m_o, No. of events):
1870-1990 (3.7, 549), 567-1869 (6.0, 3);
ML: $N(3.7) = 4.52 \pm 0.19$, $\beta = 2.37 \pm 0.09$, $m_u = 6.55 \pm 0.03$.

Figure 6. Estimation from complete and extreme parts of the catalogue - Kijko and Sellevoll method.

TABLE 3. Summary of estimated and imposed parameters.

		$N(m_o)$	ß	m_u
Fig. 1	ML	4.54	2.09 ± 0.10	6.25
	LS	4.20 ± 0.06	2.04 ± 0.04	6.25
Fig. 3	ML	0.91	2.19 ± 0.22	6.27
	LS	0.59 ± 0.10	1.56 ± 0.07	6.27
Fig. 4	ML	0.32	1.86 ± 0.36	5.7
	LS	0.26 ± 0.10	1.49 ± 0.08	5.7
Fig. 5	ML	4.30	2.05 ± 0.07	6.5, 7.0
	LS	7.10 ± 1.20	2.46 ± 0.10	6.7
Fig. 6	ML	4.52 ± 0.19	2.37 ± 0.09	6.55 ± 0.03

Most of the standard errors are rather small. It should be stressed that uncertainties in the input parameters have not been taken into account in the above calculations.

Disregarding estimates from poorly defined data sets, representative estimates of β rounded to one decimal point would be 2.1 (from Figs. 1 and 5) and 2.4 (from Fig. 6). Both values could be used as regional estimates together with subjectively assigned weights.

Least-squares estimates of the upper bound magnitudes have been obtained 0.1 to 0.2 units larger than the observed maximum magnitudes. Thus, it would be appropriate to add an increment of 0.2 to the maximum observed magnitude in each defined seismic source in the area under consideration to obtain the corresponding upper bound magnitude. Such a least-squares estimate and Kijko and Sellevoll's maximum likelihood estimate of the upper bound magnitude might be used with reasonable weights as alternatives for the upper bound magnitude in seismic hazard assessment in Slovenia.

Incremental techniques usually involve increments of 0.5 to one magnitude unit (e.g., Basham et al, 1979). The basis for large increments is the assumption that a rare "maximum magnitude earthquake" has most probably not occurred during recorded history. Furthermore, large uncertainties in the evaluation of historic earthquakes justify large increments. Upper bound magnitudes which are assigned 0.5 to one unit larger than the largest historical events might be used as highest value alternatives in areas that lack identified causative faults.

References

Aki, K. (1965) Maximum likelihood estimate of b in the formula $\log N = a - bM$ and its confidence limits, *Bull. Earthq. Res. Inst. Tokyo Univ.* **43**, 237-239.

Basham, P. W., D. H. Weichert, and M. J. Berry (1979) Regional assessment of seismic risk in eastern Canada, *Bull. Seism. Soc. Am.* **69**, 1567-1602.

Doser, D. I. and R. B. Smith (1982) Seismic moment rates in the Utah region, *Bull. Seism. Soc. Am.* **34**, 525-551.

Gutenberg, B. and C. F. Richter (1944) Frequency of earthquakes in California, *Bull. Seism. Soc. Am.* **34**, 185-188.

Karnik, V. (1968) *Seismicity of the European Area, Part 1,* Academia, Czechoslovak Academy of Sciences, Praha.

Kijko, A. and M. A. Sellevoll (1989) Estimation of earthquake hazard parameters from incomplete data files. Part I. Utilization of extreme and complete catalogs with different threshold magnitudes, *Bull. Seism. Soc. Am.* **79**, 645-654.

Krinitzsky E. L. and D. B. Slemmons, Editors (1990) Neotectonics in earthquake evaluation, *Reviews in engineering geology* **VIII**, Geol. Soc. Am., Boulder.

Page, R. (1968) Aftershocks and microaftershocks of the great Alaska earthquake of 1964, *Bull. Seism. Soc. Am.* **58**, 1131-1168.

Salyards S. L. (1991) A preliminary assessment of the seismic hazard of the southern Rio Grande rift, New Mexico, *New Mexico Geological Society Guidebook*, 42nd Field Conference, Sierra Blanca, Sacramento, Capitan Ranges.

Schwartz, D. P. and K. J. Coppersmith (1984) Fault behavior and characteristic earthquakes: Examples from the Wasatch and San Andreas fault zones, *J. Geophys. Res.* **89**, 5681-5698.

Slejko, D. and A. Kijko (1991) Seismic hazard assessment for the main seismogenic zones in the Eastern Alps, *Tectonophysics* **191**, 165-183.

Utsu, T. (1966) A statistical significance test of the difference in b-value between two earthquake groups, *J. Phys. Earth* **14**, 37-40.

Weichert, D. H. (1980) Estimation of the earthquake recurrence parameters for unequal observation periods for different magnitudes, *Bull. Seism. Soc. Am.* **70**, 1337-1346.

Wells, D. L. and K. J. Coppersmith (1994) New Empirical Relationships among Magnitude, Rupture Length, Rupture Width, Rupture Area, and Surface Displacement, *Bull. Seism. Soc. Am.*, **84**, 974-1002.

Wesnousky, S. G. (1986) Earthquakes, quaternary faults, and seismic hazard in California, *J. Geophys. Res.* **91**, 12587-12631.

Youngs, R. R. and K. J. Coppersmith (1985) Implications of fault slip rates and earthquake recurrence models to probabilistic seismic hazard estimates, *Bull. Seism. Soc. Am.* **75**, 939-964.

Živčić, M. (1992) Earthquake catalogue for Probabilistic assessment for seismic hazard at Krško Nuclear Power Plant. Report of the Institute of Structural and Earthquake Engineering, Department of Civil Engineering, University of Ljubljana, 92 pages.

THE ANALYSIS OF THE SPATIAL-TEMPORAL STRUCTURE OF SEISMICITY IN THE BAIKAL RIFT ZONE

A.V. SOLONENKO, N.V. SOLONENKO, V.I. MELNIKOVA, E.A. SHTEIMAN
Institute of the Earth's Crust, Siberian Branch of Russian Academy of Sciences, Irkutsk, 664033, Russia

Abstract

To study the fundamental regularities of seismic process in the Baikal rift zone (BRZ) an attempt was made to investigate the spatial-temporal structure of seismicity of the region, using the method of fractal dimensions (Mandelbrot, 1982).

The analysis of selfsimilarity of the zone in the spatial distribution of earthquakes was made. "The completion of development" of earthquake epicentral fields of all parts of BRZ using the analysis of temporal variations of fractal dimensions in spatial distribution of events was investigated.

The resulting criteria for evaluation of "completion of development" of earthquake epicentral fields in BRZ may be used in estimations of average long-term parameters of seismic regime of the territory.

1. Introduction

The Baikal seismic zone is one of the most active continental rift zones on the Earth. About hundred thousand shocks of various intensity have been recorded here during the last 30 years.

The map of earthquake epicentres with $M \geq 2.2$ presented in Fig.1 gives an impression of high seismic potential of the zone. The data on earthquakes recorded by the regional seismic network of Pribaikalia show a complex mozaic structure of seismicity distribution of the Baikal rift zone (Misharina and Solonenko 1976; Misharina and Solonenko. 1990; Pshennikov and Fomina, 1964 and so on).

The key features of distribution of earthquake epicentres in this area are their concentration in more or less wide bands of predominantly north-eastern strike and a clear discontinuity of epicentral field (transverse to the strike of the zone), which is manifested in alternation of large regions of high and low density of epicentres.

The reliability of quantitative estimates of seismic hazard from seismological data depends heavily on the fact whether the currently available information on sesmicity distribution of the area under study is sufficient for the prognostic esti-

V. Schenk (ed.), Earthquake Hazard and Risk, 49–62.

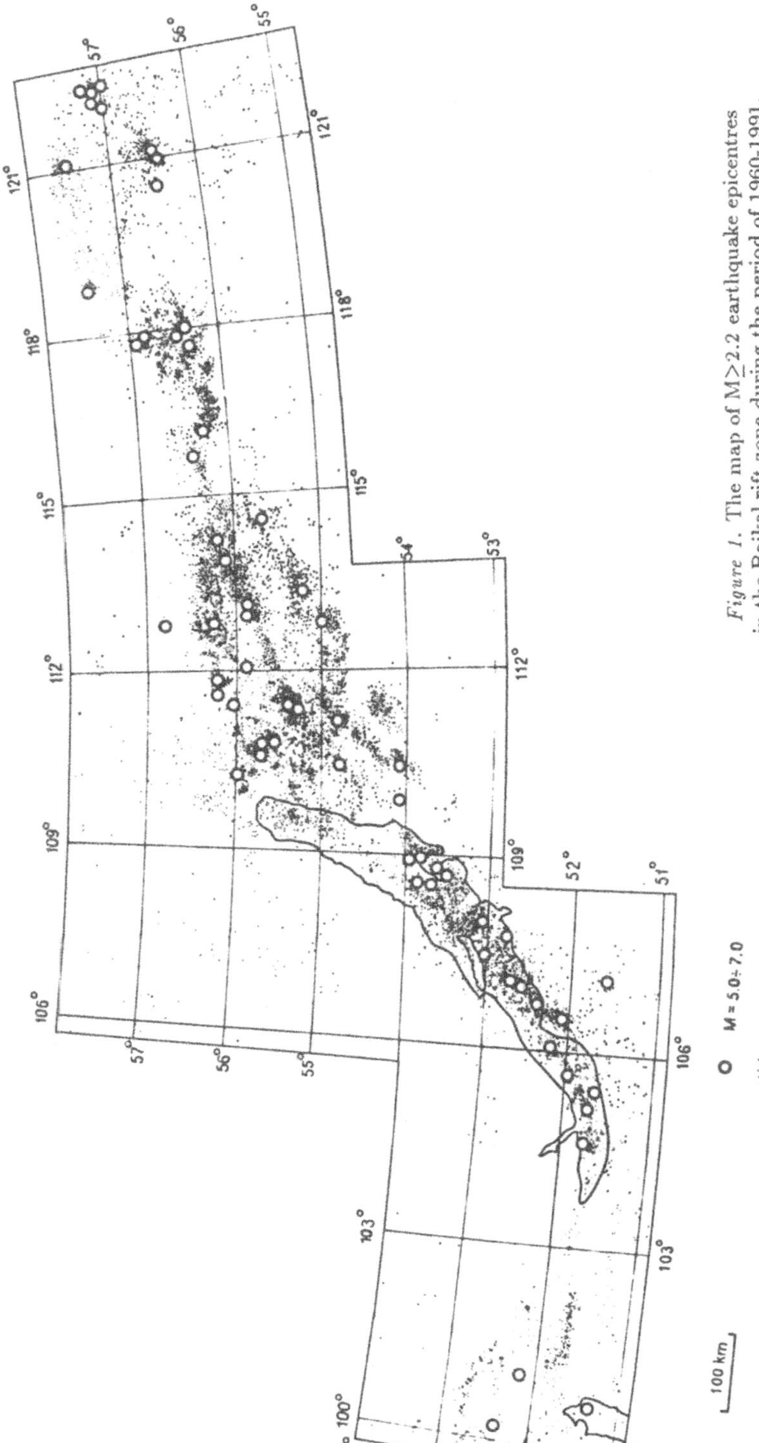

Figure 1. The map of $M \geq 2.2$ earthquake epicentres in the Baikal rift zone during the period of 1960-1991.

mates. And it is also necessary to be sure that evidences on earthquake epicentral field of the region, obtained commonly during a short period of instrumental observations, reflect its real structure adequately. The complete formation of this structure appears to need much more time.The criteria for evaluation of "completion of development" of earthquake epicentral field in Pribaikalia are presently unknown. Owing to lack of these criteria the calculations of average long-term parameters of the seismic regime of the area may involve significant errors, distorting the notion on its true seismic potential.

The peculiarities of the internal structure of the region and its stress state are reflected in earthquake epicentral field. Thus, investigation of regularities of formation of epicentral field is of great importance in the complex of study associated with the knowledge of regularities of development of seismic process.

Currently, the theory of multi fractals (Mandelbrot, 1982) has a wide application to study some peculiarities of seismic regime, manifested as spatial heterogeneity of seismic field. Investigation of temporal variations of spatial fractality of epicentral field seems to clarify the problem of evaluation of its "completion of development".

The epicentral field the development of which "has been completed" is such a field, in which new seismic events agree with its previously developed structure and do not change essentially its outline. If new earthquakes make significant changes in the previously developed structure of the epicentral field, it is believed that the development of such a field "has not been completed".

If we deal with the fractal structure of the epicentral field, the development of which "has been completed", the subsequent shocks will occur in most cases in portions where seismic events were observed before. Therefore the fractal dimension of epicentral field will not vary in time. The epicentral field the development of which "has not been completed" will have a dimension varying in time.

In practice the estimation of earthquake epicentral field involves the following operation (Golubeva *et al.*, 1987). The study area is divided into 4 equal parts (Ns=4). The average number of earthquakes in each area and the standard deviation from the average number of earthquakes is estimated. Then the number of areas (Nx), in which the number of earthquakes exceeds the standard deviation from the average number of earthquakes is calculated. Thus "the noise" background of epicentral field dependent, in particular, on errors in estimation of earthquake coordinates, is precluded. At the second stage the size of small areas decreases according to the certain law and the procedure of estimation is repeated. Division of the initial small area into more and more small-sized areas is performed until area sizes become commensurable with the errors in determination of earthquake coordinates. Dimension of epicentral field Ds is determined by angle of inclination of the plot lg Nx(lg L) where L= (1/Ns)1/2 is a minimum dimension of small area Ns (Nicolis and Prigozhin, 1990).

To estimate the epicentral field dimensions of various regions of the Baikal seismic zone we used the following change of the small area dimensions (Ns=4, 9, 16, 25, 36, 49, 64).

To analyse the epicentral fields, the development of which "has been completed", and their selfsimilarity in various parts of the Baikal seismic zone, 12 regions as large as 0.8*1.6 (from this point on the area dimensions are given in degrees), in longitude and latitude, respectively, were chosen (Fig.2). The symbols of the regions and their geographic coordinates are given in Table 1.

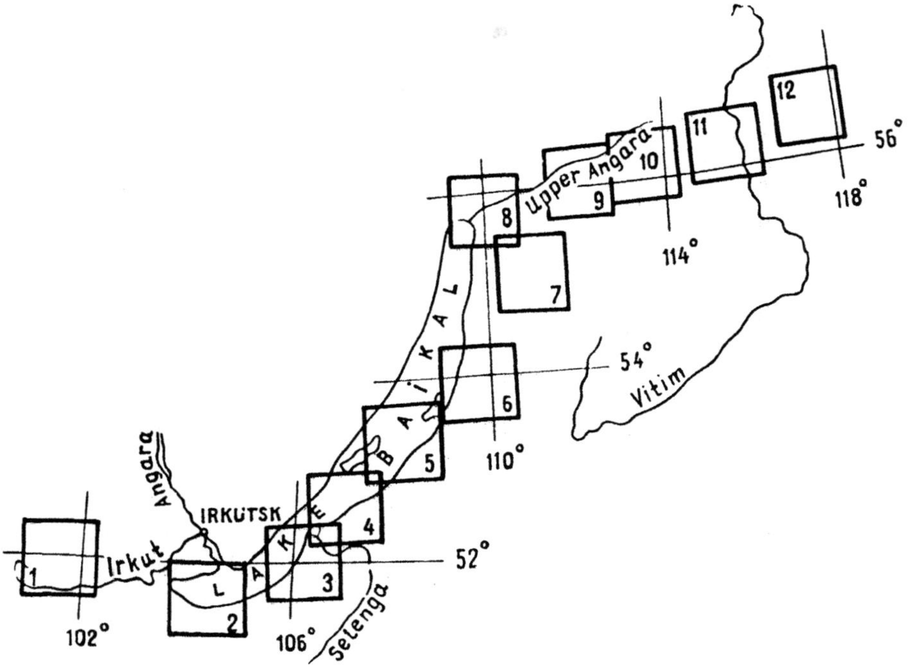

Figure 2. Scheme of location of the regions for which "the completion of development" and selfsimilarity of earthquake epicentral fields was analysed. The region boundaries and their names are given in Table 1.

In Table 1, N is the number of earthquakes with M≥2.2 recorded within the region during the period of 1962-1991. The last two columns (Ds and k) are discussed below. Earthquakes with M≥2.2, recorded practically over the whole territory under study, (Golenetsky, 1985) were analysed (besides instances specially indicated in the text).

TABLE 1. Regions of BRZ for which " the completion of development" and selfsimilarity of earthquake epicentral fields for the period 1962-1991 were analysed.

n	Name of the region	Location φ^o N	Location λ^o E	N	Ds	k
1	Tunka	51.5-52.3	100.8-102.4	518	1.63	7.3
2	Kultuk	51.2-52.0	103.8-105.4	531	1.89	10.2
3	South Baikalsk	51.6-52.4	105.4-107.0	805	1.66	10.3
4	Ust-Selenga	52.2-53.0	106.2-107.8	1237	1.75	8.1
5	Srednebaikalsk	52.9-53.7	107.4-109.0	1819	1.91	3.2
6	South Barguzin	53.5-54.3	109.0-110.6	624	1.58	5.9
7	North Barguzin	54.7-55.5	110.2-111.8	2211	1.68	3.0
8	North Baikalsk	55.4-56.2	109.3-110.9	780	1.80	6.6
9	Upper Angara	55.6-56.4	111.4-113.0	850	1.70	5.0
10	North Muya	55.7-56.5	112.9-114.5	2364	1.60	5.0
11	Muya	55.8-56.6	114.6-116.2	659	1.53	12.4
12	Udokan	56.1-56.9	116.6-118.2	1391	1.64	8.5

The results of estimates of dimensions of earthquake epicentral field are illustrated by the example of Srednebaikalsk (5) and North Muya (10) regions during the period of 1962-1991 (Fig.3).

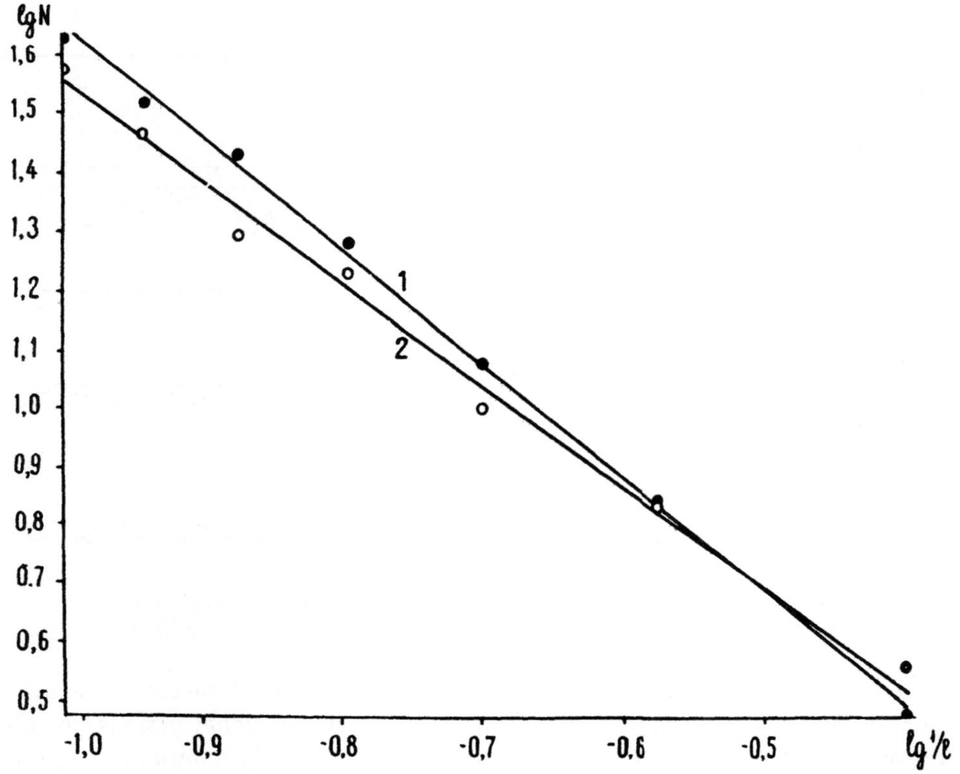

Figure 3. Determination of dimensions of earthquake epicentral fields of Srednebaikalsk (5) (the straight line 1) and North Muya (10) (the straight line 2) regions of BRZ.

As can be seen from the above data, earthquake epicentral field of the regions considered have fractal dimensions (Ds=1.91 and 1.60, respectively), i.e., represent a selfsimilar structure of the type of "Sierpinski carpet" (Feder, 1991) with a regular "hole" structure in a wide range of spatial hierarchy. As is seen from Fig.4, epicentral fields of the remaining 10 regions have also fractal dimension. Epicentral field dimensions are given in Table 1.

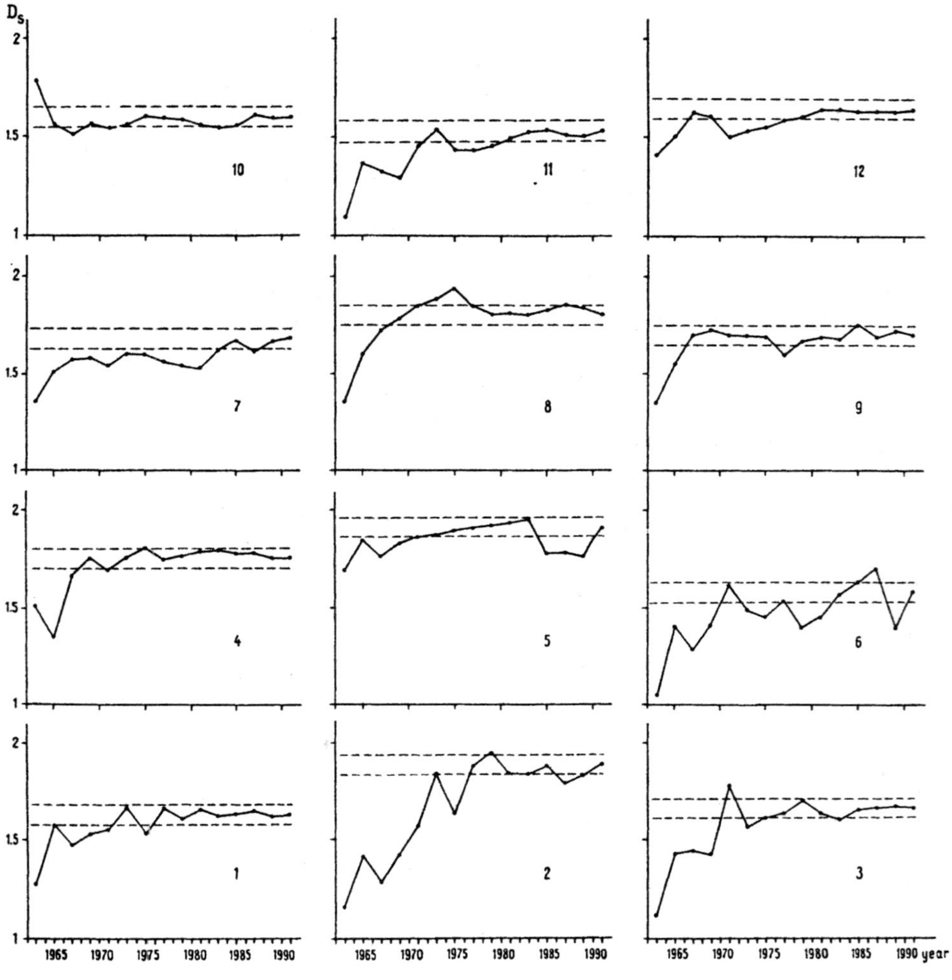

Figure 4. Temporal variations of dimensions of earthquake epicentral fields in various regions of BRZ. Plot numbers refer to region numbers in Table 1.

As can be seen from Fig.4, which was built by means of successive summation of earthquake epicentral fields from two-year observational periods, for the regions 1,3,4,8-12 the curves Ds(t) in the 1980's practically approached asymptotes. Therefore, there is reason to think that the main structure of earthquake epicentral fields of these regions has been essentially completed. In Kultuk (2), Srednebaikalsk (5), South Barguzin (6) and, to a lesser degree, North Barguzin

(7) regions the process of epicentral fields structure formation appears to be still in progress. In this connection, an extreme care (only seismological data should be used) is required to predict the development of seismic process in these regions, because it is not improbable that in future earthquakes will occur here and in those areas where they were not observed yet. It is reasonable to suppose that it is in these cases, when seismologists are forced to predict basing on insufficient data, "unexpected" large earthquakes may occur in areas of underestimated seismic risk. In particular, the largest earthquake for the past 30 years in the Baikal seismic zone (the Kyakhta earhquake of 13.05.1989, $M \geq 5.8$) occurred in the region of extremely low seismic activity for which shocks of such energetic level were not typical.

An essential feature of earthquake epicentral field, as indicated in some publications (Golubeva *et al.*, 1987; Sadovsky and Pisarenko, 1991), is its rather clear selfsimilarity. To analyse selfsimilarity in the structure of local seismicity of BRZ the procedure given in these publications was used. The catalog of earthquakes, recorded by the regional seismograph network of Pribaikalia during 1962-1991, was analysed. The catalog was compiled in the Institute of the Earth's Crust, Siberian Branch of the RAS under the direction of Dr. S.I.Golenetsky.

To analyse the selfsimilar structure of epicentral fields of various regions of BRZ, each of the previously distinguished regions (Fig.2, Table 1) was divided into trapezia of three hierarchical levels as large as: 1) 0.4 * 0.8, 4 trapezia; 2) 0.2 * 0.4, 16 trapezia; 3) 0.1 * 0.2, 64 trapezia; The minimum dimensions of the trapeziua were chosen taking into account average errors in determination of earthquake epicentral coordinates over the territory under study. Due to small accuracy in determination of focal depth only space distribution of earthquakes was investigated.

For the trapezium in each hierarchical level the relative index of seismicity in respect to the general seismicity of the trapezium of the previous elder level, which involves the initial trapezium, was estimated as in (Golubeva *et al.*, 1987). If the numbers of earthquakes in the trapezia of the first level of hierarchy are denoted by n(1), n(2), n(3), n(4), then the indexes of relative seismicity are estimated by the formulae (Golubeva et al.,1987): $p(1) = n(1)/N$, $p(2) = n(2)/N$, $p(3) = n(3)/N$, $p(4) = n(4)/N$, where $N = n(1) + n(2) + n(3) + n(4)$. The indexes p(1), p(2), p(3), p(4) are arranged in decreasing order and plotted. Similar calculations are performed at the next levels of hierarchy, at the second level the indexes being averaged in all four available trapezia and in the third one in 16 trapezia. In Fig.5 are shown plots for all regions discussed above. The plot numbers in Fig.5 correspond to the region numbers in Fig.2 and in Table 1.

Usually (Sadovsky and Pisarenko, 1991) aftershocks of large earthquakes in earthquake catalogs are not taken into account. To estimate the effect of aftershock sequences on the nature of spatial distribution of earthquakes at various hierarchical levels we took the Udokan region, which is characterized by numerous recurrent shocks (Solonenko and Solonenko, 1987). The first version of calculations was made for the whole 30 - year period of observation. In this case each earthquake with $M>2.2$, recorded in this period, was taken into account (Fig.6.1).

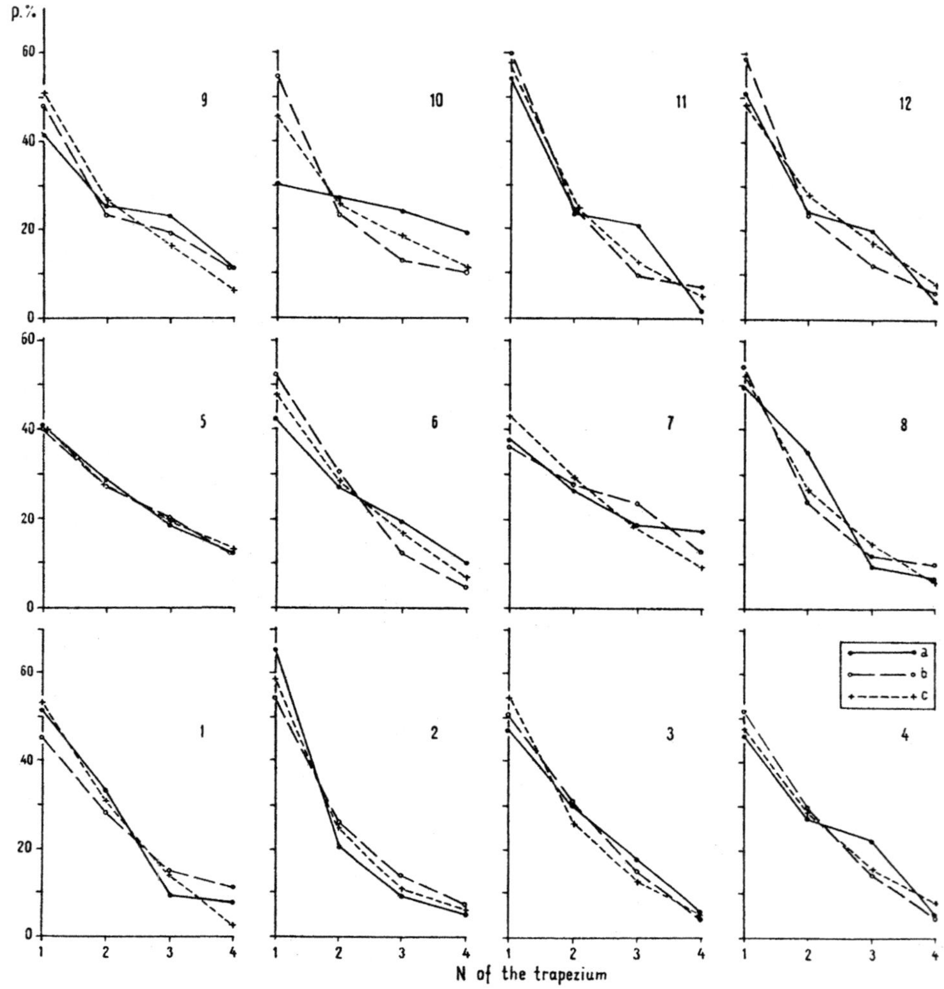

Figure 5. Proportions of seismicity for various levels of hierarchy.
Plot numbers refer to region numbers given in Table 1.
a – seismic percent in trapeziums of the 1st level;
b – averaging on trapeziums of the 2nd level;
c – averaging on trapeziums of the 3rd level.

The second version of calculations was made for the periods, in which the effect of aftershock sequences on the proportionality of spatial distribution of seismicity was minimum (Fig.6.2).

The period up to 1965, when remote aftershocks of the 1957 (M=7.6) Muya earthquake could be recorded, was eliminated. For the remaining three earthquakes (Kodar, 1970; Kalar, 1974; Udokan, 1981: each with M>5.4), which were accompanied by significant aftershock sequences, two-year time intervals after the main event of the sequence were eliminated. Practically the full coincidence of the

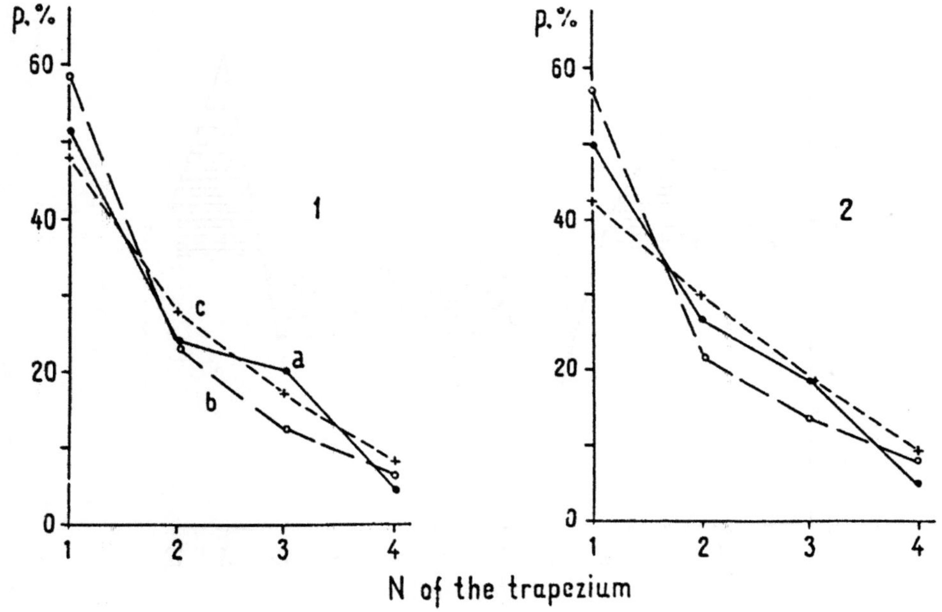

Figure 6. Proportions of seismicity of the Udokan region for various levels of hierarchy with not eliminated (1) and eliminated (2) aftershock sequences. Legend is identical with that of Figure 5.

curves in all levels of hierarchy (the correlation coefficient is 0.978 at the 1st level, 0.998 at the 2nd level and 0.990 at the third level) suggests that aftershock sequences do not essentially affect the proportionality of seismicity for various levels of hierarchy (at least in the Udokan region of BRZ).

Referring again to Fig.5, relatively stable level of proportionality in various hierarchical levels is kept in all regions (besides North Muya (10)). In the North Muya (10) region at the first level of hierarchy earthquake distribution on trapezia is significantly different from that at the 2nd and 3rd levels of hierarchy. This phenomenon is most likely related to the fact that two highly seismic structures (Upper Angara and Upper Muya basins), which behave independent at the first level of hierarchy, fall within the initial trapezium. Further division of the area shows that the parts of single seismic structures fall within smaller trapezia and selfsimilarity of seismicity field becomes more prominent.

To correlate the degree of contrast of seismicity field in various regions of BRZ the averaging of the data on proportionality of seismicity in three levels of hierarchy (for North Muya (10) in two low levels) was performed. Table 1 and Figure 7 present the corresponding data on the relation between k average (for three levels of hierarchy) percent content of seismicity in the most active trapezium and that in the less active one.

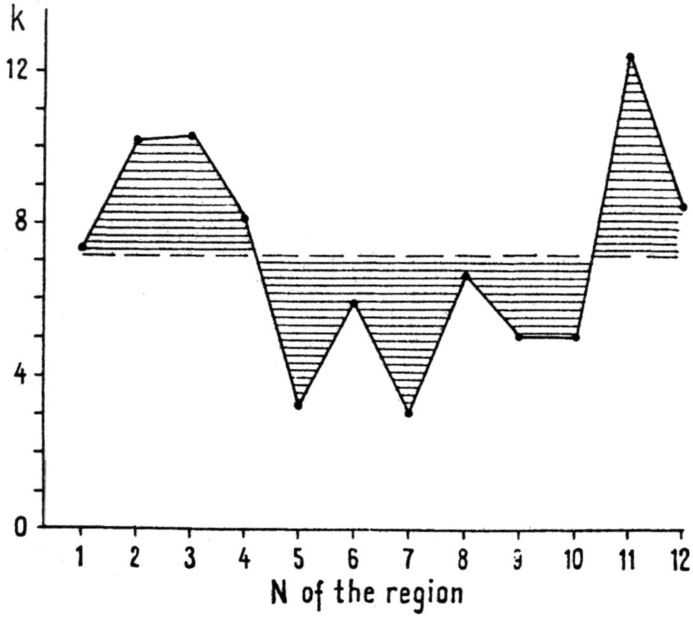

Figure 7. Values for k parameter for various regions of BRZ.

Average k meaning for all 12 regions is indicated by dashed line. One can recognize that k relation varies over a wide range: from 3.0 to 12.4. That is, at different levels of hierarchy the relation between seismicity of the most seismically active trapezium and that of less active averages between 3 and 12 and even more. The maximum k values are typical of the ends of seismic zone. In the central part of the rift the seismic field is less contrasting. Nevertheless, the level of seismic activity also varies here within small areas by a factor of 3-6.

It should be noted, that the stress character reflecting the internal structure of the region, is also heterogeneous. As is seen from Fig.8, normal faulting is predominant in the central part, the ends of the zone being characterized by the existence of strike-slip fault mechanism.

The analysis of selfsimilarity of seismicity field brings up the question: How much would the above conclusions on selfsimilarity of seismicity field be universal in an energetic sense. To settle this question, the data for four energetic levels were analysed, using examples of seismicity of North Muya (10) region, where all shocks with $M \geq 1.0$ are recorded. As shown in Fig.9, the curves for different magnitudes agree well, and in its turn indicate (as mentioned also in (Sadovsky and Pisarenko, 1991)) that magnitude-frequency relation is valid even on very small (about 10*10 km) areas, though the level of seismicity can grossly vary from area to area.

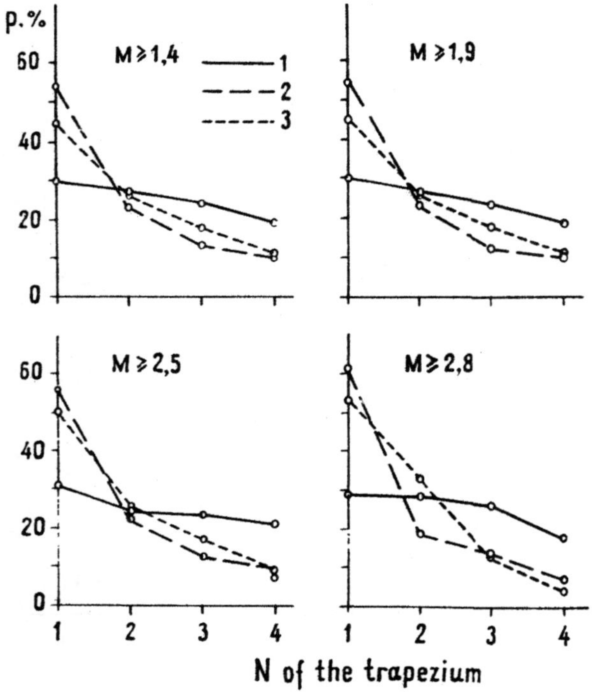

Figure 8. Earthquake focal mechanisms in the Baikal rift zone during the period of 1950 – 1990 (determinations are made on the Wolf grid in the projection of the lower hemisphere). Focal mechanism determinations of L.A.Misharina, N.V.Solonenko and V.I. Melnikova have been used in compiling the schematic map.
1 - Nodal planes
2 - P and T - stress axes
3 - Area of compressional waves
4 - Magnitude

Four versions of calculations of indexes of relative seismicity in time for North Muya region (10) were made (Fig.10) to follow their changeability. North Muya (10) region was chosen because a large earthquake swarm (more than 3500 shocks with $M \geq 0$), which was rather closely located in space (Kochetkov *et al.*, 1987) occurred here during the period of 1979-1983.

The local seismograph network in this region allows the estimation of epicentre coordinates with an error not exceeding 3 km in most cases. Occurrence of this swarm, as indicated above, slightly influences the proportionality of seismicity at high hierarchical level. At low levels even such a prominent event in seismicity of the region almost does not affect the contrast of its seismicity distribution, i.e. the spatial structure of the epicentral region of the swarm fits well into the general structure of selfsimilarity of seismicity field of the region.

The problem of stability in time of the relationship of seismic activity within individual trapezia at various levels of spatial hierarchy is of practical importance.

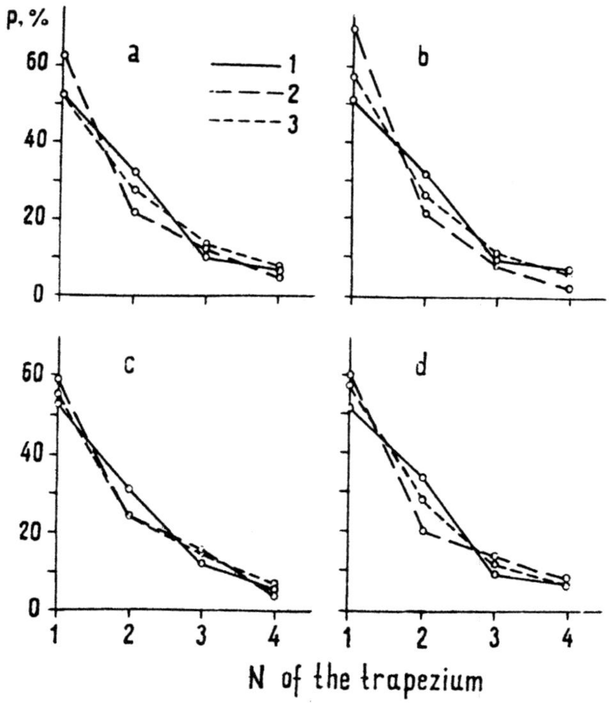

Figure 9. Proportions of seismicity for various levels of hierarchy (North Muya region (10), 1962-1991, earthquakes of various energetic ranges).

1 - seismic percent in trapeziums of the 1st level;

2 - averaging on trapeziums of the 2nd level;

3 - averaging on trapeziums of the 3rd level.

The results of estimation, made for North Muya region (10) in accordance with 10 year periods of observations (Fig.10), showed that at the first level of hierarchy the stability of ranging by seismic activity is kept at relatively low level and makes up 66.6 %. At low levels of spatial hierarchy, which are of most practical interest in estimation of local seismic activity, the stability of ranging is significantly higher and at the second level reaches 100 % for the most seismically active trapezia.

2. Summary

The results of the present studies are as follows: First, the structure of earthquake epicentral fields in some regions of BRZ (in accordance with the data of 30 - year period of regional seismological observations) continues to develop, changing its configuration from year to year.

Second, a clearly defined selfsimilarity of seismicity field at various spatial hierarchical levels in time and over a wide energetic range, which appears to be not disrupted by important aftershock sequences and earthquake swarms, is typical practically of all regioins of the Baikal rift.

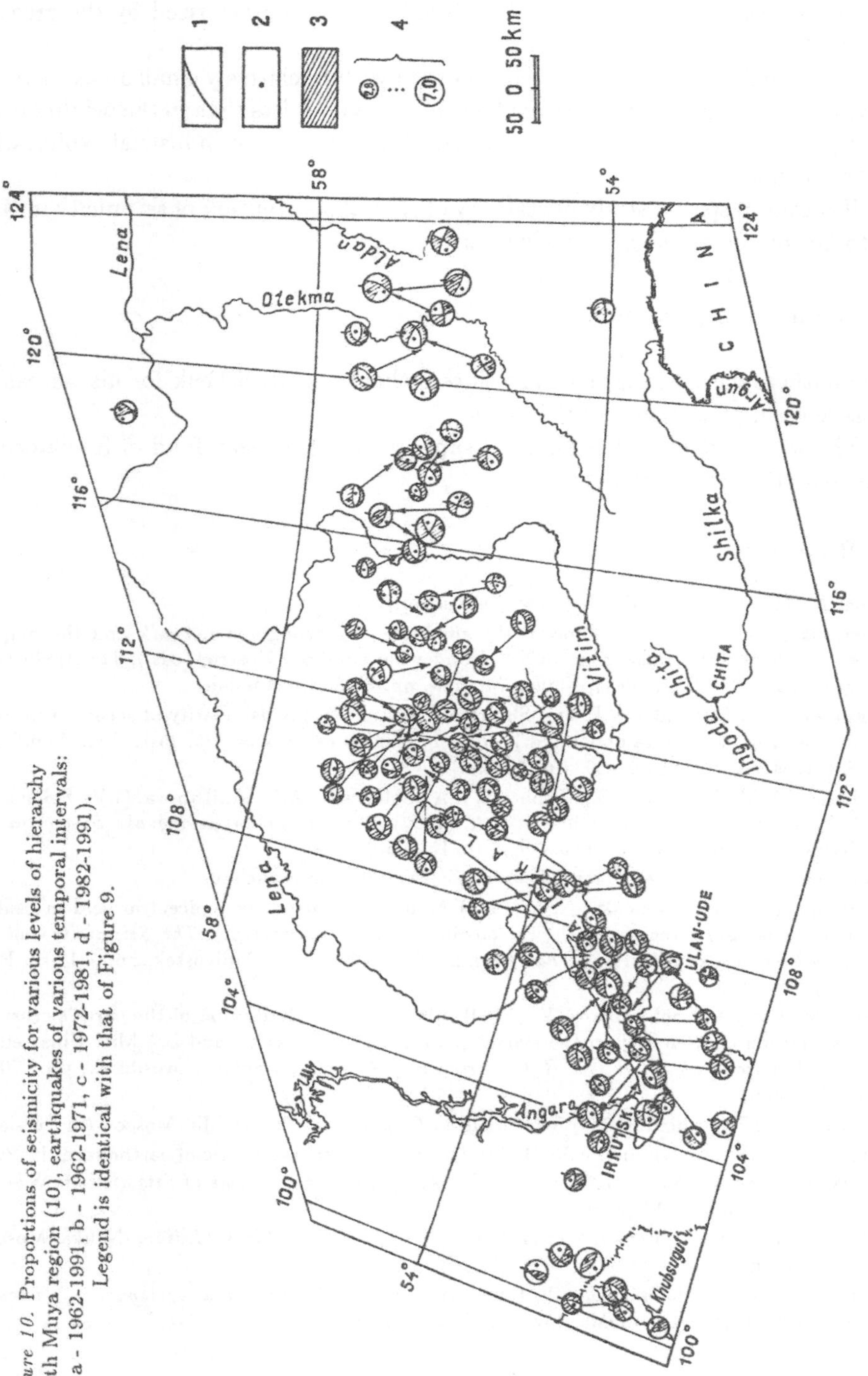

Figure 10. Proportions of seismicity for various levels of hierarchy (North Muya region (10), earthquakes of various temporal intervals: a - 1962-1991, b - 1962-1971, c - 1972-1981, d - 1982-1991). Legend is identical with that of Figure 9.

Third, the proportionality of seismicity at various levels of spatial hierarchy essentially varies over the zone, the rift sides being characterized by the greatest contrast.

And fourth, the study indicated that even within relatively small areas there are always portions in which the level of seismic activity is less than in the neighbouring parts by a factor of 3-12, that is of practical importance for industrial exploitation of the region.

However, a special study of stability of spatial distribution of seismicity in time is to be conducted in each specific case.

3. Acknowledgements

The authors wish to express their sincere thanks to Dr.Schenk for his unceasing assistance in preparation of the paper.

The work is done at the financial support of the Russian fund of fundamental research (Project 93-05-8143).

4. References

Feder, F. (1991) *Fractals*, Mir, Moscow, (in Russian).

Golenetsky, S.I. (1985) Estimation of the efficiency of seismograph network and the map of earthquakes in the Baikal zone, in N.V. Kondorskaya and B.V.Kostrov (eds.), *The Application of Computer in Seismology*, Nauka, Moscow, pp. 80-85 (in Russian).

Golubeva, T.V., Pisarenko, V.F. and Shnirman, M.G. (1987) Selfsimilarity of seismic processes in space, in A.M. Artamonov (ed.), *Recent Geophysical Researches*, Inst. Fiz. Zemli AN SSSR, Moscow, 1, pp. 61-69. (in Russian).

Kochetkov, V.M., Borovik, N.S., Misharina, L.A., Solonenko, A.V., Anikanova, G.V., Solonenko, N.V., Melnikova, V.I. and Gilyova, N.A. (1987) *The Angarakan Earthquake Swarm in the Baikal Rift Zone*, Nauka, Novosibirsk, (in Russian).

Mandelbrot, B. (1982) *The Fractal Geometry of Nature*, San Francisco.

Misharina, L.A. and Solonenko, N.V. (1976) Analysis of earthquake epicentral field in central and north-eastern regions of the Baikal rift, in S.L. Solovjev (ed.), *The Seismicity and the Deep Structure of Siberia and Far East*, DVNTS AN SSSR, Vladivostok, pp. 7-12 (in Russian).

Misharina, L.A., and Solonenko A.V. (1990) Influence of block division of the earth's crust on distribution of seismicity in the Baikal rift zone, in O.V.Pavlov and L.A.Misharina (eds.), *The Seismicity of the Baikal Rift. Forecasting Aspects.*, Nauka, Novosibirsk, pp. 70-78 (in Russian).

Nicolis, G. and Prigozhin I. (1990) *Knowledge on Complicated Things*, Mir, Moscow (in Russian).

Pshennikov, K.V. and Fomina, E.V. (1964) The energetic characteristic of earthquakes in Pribaikalia between 1952 and 1961, in A.A. Treskov (ed.), *The Problems of Seismicity in Siberia*, Novosibirsk, pp. 11-14 (in Russian).

Sadovsky, M.A. and Pisarenko, V.F. (1991) *Seismic Process in Block Medium*, Nauka, Moscow (in Russian).

Solonenko, N.V. and Solonenko, A.V. (1987) *Aftershock Sequences and Earthquake Swarms in the Baikal Rift Zone*, Nauka, Novosibirsk (in Russian).

CHANGES OF THE STATE OF STRESS AND DEFORMATION IN EPICENTRAL ZONES DURING THE AFTERSHOCK PERIOD

YU.P. SKOVORODKIN, T.V. GUSEVA, L.S. BEZUGLAJA
United Institute of Physics of the Earth RAS,
B. Gruzinskaja 10, Moscow, Russia, 123810

Abstract

Instrumental geophysical observations in epicentral zones of strong earthquakes during aftershocks are important for a study of seismotectonic processes and a prediction of strong aftershocks. The high precision geodetic, magnetometric and gravimetric repeated measurements after a main shock have been carried out in the epicentral zones of earthquakes with $M \geq 7$.

During a gradual descent of seismic activity, the trend and oscillatoring changes of geophysical fields, exceeding background values, occur. Before strong aftershocks ($K>12$), near the epicenters: the rate of deformation's change was (1-5) ppm/day, the amplitudes of local geomagnetic field variations were 7-17 nT and the gravity change was 50 mkGal.

The peculiarities of spatial-temporal change of fields reflect an unstability of deformation processes in zones of aftershocks concentration and show a character of stress redistribution at two stages of post-seismic activity: an intensive fissuring, leading to formation of blocks on a different scale (from metres to kilometres), and a consolidation of aftershock zone by smoothing of tectonic stress gradients at the block boundaries.

1. Introduction

Since the end of XIX century instrumental observation in seismically active regions and epicentral zones of strong earthquakes was carried out. Results of field measurements and modeling allowed to study the mechanics and dynamics of tectonic earthquakes, including determination of source mechanism, structure and medium properties, dynamics of stress field, creation of theory of source and preparation of earthquake. However, the theory of aftershock process does not exist on the same level.

Aftershocks are the final stage of the seismotectonic cycle. In view of these, instrumental observations during an aftershock activity can give new useful information about applicability and examination of some working hypothesis or theory.

V. Schenk (ed.), Earthquake Hazard and Risk, 63–76.

Our previous investigations in seismically active regions, epicenters of strong earthquakes, as well as field modeling of deformational processes by explosion and vibration showed that the change of stress-deformational structure of medium can be successfully studied by the optimal combination of geophysical methods, including seismological, geodetic, magnetic, gravity and other measurements.

This work presents results and qualitative analysis of observations during aftershocks of earthquakes with the magnitude M≥7 in Central Asia and Caucasus.

2. Tectonic situation in epicentral zones

The Gazli earthquake occurred on March 19, 1984, in the sand desert of Central Kizilcums (20 km from Gazli, Fig. 1). Instrumental focus parameters are: Φ=40.38°N, λ=63.36°E, H=15km, M_{LH}=7.2. The focus mechanism is strike-slip with the possible planes of disruption - striking submeridional and North-Eastern. The focus was situated in the zone of recent tectonic activisation of the Buchara-Guissar deep fault and coincided with the Gazli transverse uplift at the depths of 5 to 25 km. The stress field was characterized by near-horizontal compression and near-vertical tension axes. At the surface the rupture was not visible. The aftershock field was extended 60 km in E-W direction and 50 km in N-S direction. The depth of aftershock hypocenters was 10 - 15 km (Abdukadirov *et al.*, 1987).

The Rudbar earthquake of June 20, 1990, led to disastrous results in the Gilan province of North Iran. Instrumental focus parameters are: Φ=36.817 °N, λ=49.409°E, H=10km, M=7.3. The macroseismic epicenter was situated in the

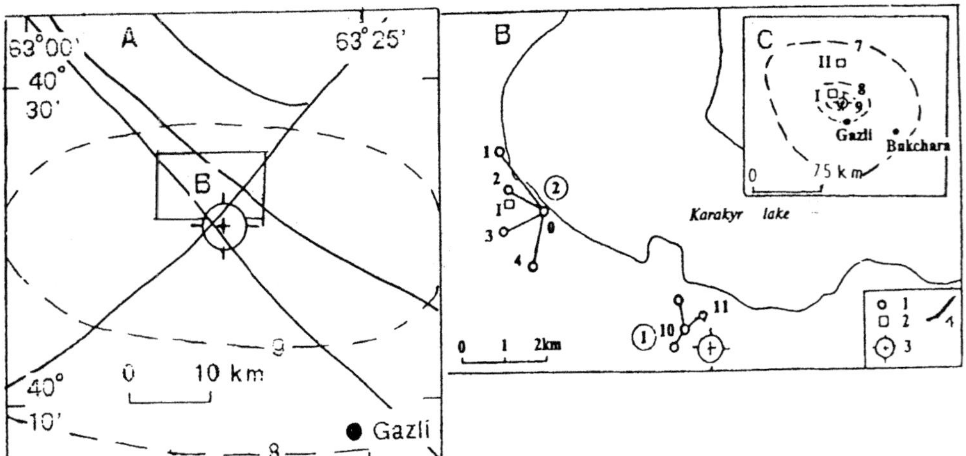

Figure 1. Isoseists of the 1984 Gazli earthquake (dotted lines on fig. A, C) and networks of observation (test areas 1 and 2 on fig. B): 1 - geodetic site, 2 - magnetic site, 3 - macroseismic epicenter, 4 - faults.

Elbrus range, near the Mangil city with MSK=10. The earthquake is likely connected with the Julfa fault, which crosses the region in SE-NW direction (Fig. 2). Epicenters of aftershocks (M>3) spread all over a territory of 90km in length along N60°W direction. The maximum aftershock activity was detected in the SE section of the epicentral zone (Tsukuda *et al.*, 1991)

The Racha (Racha-Djava) earthquake occurred on April 29, 1991, in the region of the South slope of the Great Caucasus. Instrumental focus parameters are: Φ=42.453°N, λ=43.673°E, H=17km, M=7.0. The mechanism of main shock is thrust, possible plane of disruption N264°E strike, declined to the North and inclined on 32° to the horizon, the vector of displacement formed an angle of 92° with the line of strike. The pleistoseiste zone was situated within several tectonic zones of N-W and E-W strike (Fig. 3). Here the large-scale deep Racha-Lechghum fault zone is marked. The aftershock zone aligned on 80 km and had a width of 10 km in the South-East and of about 20 km in the North-West. Hypocenters depth in average was 5 km. Peculiarities of the earthquake: absence of surface failure, anomalous high aftershock activity, relative quiescence before the strong Djava shock of June, 15, with MPLP=6.6 and seismic activisation after this shock (Arefyev *et al.*, 1993; Zakharova *et al.*, 1993).

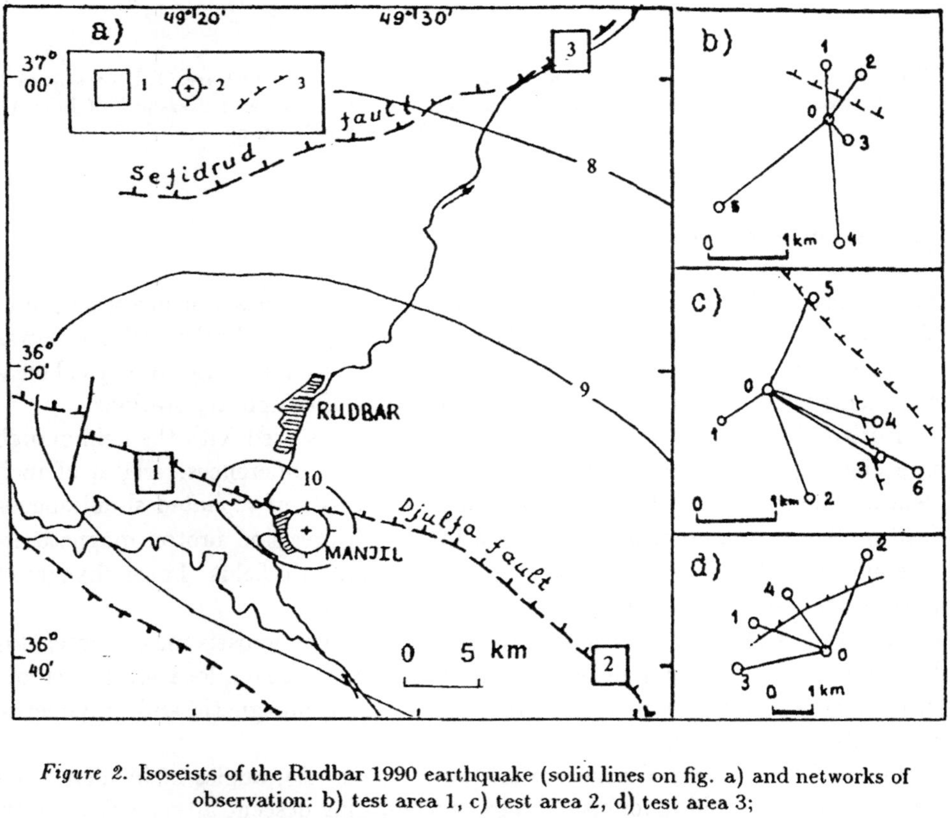

Figure 2. Isoseists of the Rudbar 1990 earthquake (solid lines on fig. a) and networks of observation: b) test area 1, c) test area 2, d) test area 3;
1 - test area, 2 - macroseismic epicenter, 3 - faults.

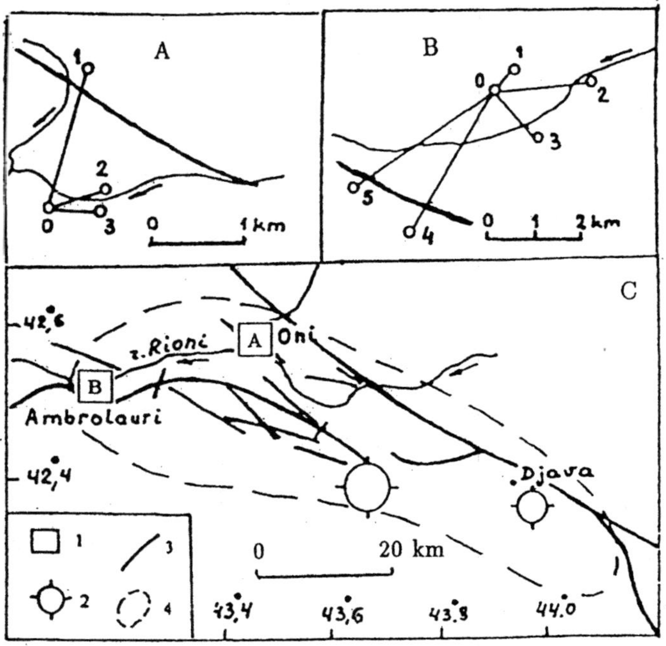

Figure 3. Epicentral zone of the Racha-Djava 1991 earthquake. A - test area 1, B - test area 2, C - pleistoseiste zone: 1 - test area, 2 - epicenters of the main shock and aftershock of June, 15, 3 - faults, 4 - boundaries of the pleistoseiste zone.

3. System of observations

We choose geodetic and magnetometric methods as the basis of investigation of the stress-deformational structure of medium during aftershocks. Where it was possible gravimetric and geoelectric measurements were made. Seismological measurements were carried out by means of a network of temporary stations.

Horizontal and vertical surface movements were observed with the help of high precision repeated measurements of distances, level differences, angles of incidence. The magnetometric method used here consists of repeated simultaneous field modulus measurements with the help of high precision proton magnetometers and permits to control changes of local geomagnetic fields. Local changes of gravity are observed with help of group of gravimeters.

As a rule, sites of observation are situated near the macroseismic epicenter, taking into account the spatial-temporal aftershock distribution, location of tectonic disturbances. As far as possible, sites of geodetic, geomagnetic and gravimetric observations are combined.

In the Gazli epicentral zone horizontal distances (D), angles of incidence (α) and geomagnetic field modulus (T) were made during descent aftershock activity.

The first network (1) was located in the macroseismic epicenter and another one (2) - in 4-5 km to the north-east (Fig. 1). On the network 1 measurements of three distances of 0.4-0.7 km were made in average every two hour with accuracy of 0.6 mm on the 11-14-th days after the shock. On the network 2 measurements of four distances of about 1.2 km (with accuracy of 4 mm) and vertical angles (with accuracy of 1") were made during one month (April, 13 - May, 13) twice a day (in the morning and in the evening). At the same period daily measurements of the geomagnetic field (T) were made in two magnetic sites (I, II) in the epicentral zone and also on the station "Garm", at 640 km from the epicenter.

From Aug. 4 to Sept. 26 geodetic and geomagnetic observations in the Rudbar epicentral zone were carried out. Near the macroseismic epicenter the geomagnetic station "Manjil" was situated, in the Djulfa fault zone the deformational networks "Gelmedari" (1) and "Sangrud" (2) were installed, in the Sefidrud fault zone the network "Imamzade-Hoshem" (3) was installed. Every network contains 5-7 sites in rocks, which allowed to record horizontal motions along and across of main fracture structure with the accuracy of 0.6 mm. Geomagnetic measurements made in the same sites, errors of determination of mean ΔT differences between the reference station and a site were lower than 2 nT on the networks 1, 3, and 3 nT - on the network 2 (Fig. 2).

In the Racha epicentral zone the combined network of observation was situated in zones of aftershock concentration: two test areas were made (Fig. 3). The magnetometric and gravimetric sites are placed near geodetic ones at distances of 10-30 m. Geomagnetic measurements were made in the reference station "Oni" (site 2 of the network 1) and in six sites of the network 2 every 1-3 day, for 2 minutes in each site. Errors of the ΔT mean were lower than 0.5 nT (Skovorodkin *et al.*, 1994).

The geodetic and magnetometric observations were performed also on the Goragorsk polygon, situated at the distance of 180 km from the epicenter. This test area has been installed near the city of Groznii in the epicentral zone of 3.08.89 earthquake (M=5.4). The measurement of distances and of the geomagnetic field modulus are constantly carried out since August 1989.

Repeated gravity measurements were carried out in the site GP "Oni" (near site 2 of the network 1) and in six sites of the network 2, one of which was the reference one - GP-0, by means of group of gravimeters GAG-3, GNU-KV and Scintrex. The square mean error of relative gravity (Δg) determination was less than 0.02 mGal for the individual measurement and 0.01 mGal - for the mean (m).

4. Results of observations

Temporal variations of distance D(t) in the Gazli earthquake epicenter are shown on Fig. 4a,b. It was found, that short-period sign-variable changes (variations) of D values took place on all azimuths with the period of 7 days and amplitude to 2.5 cm, i.e. linear deformation $\varepsilon=2\times10^{-5}$, relative to rectilinear trend. Changes of vertical angles ($\Delta\alpha$) and accordingly of level differences (Δh) also had the linear

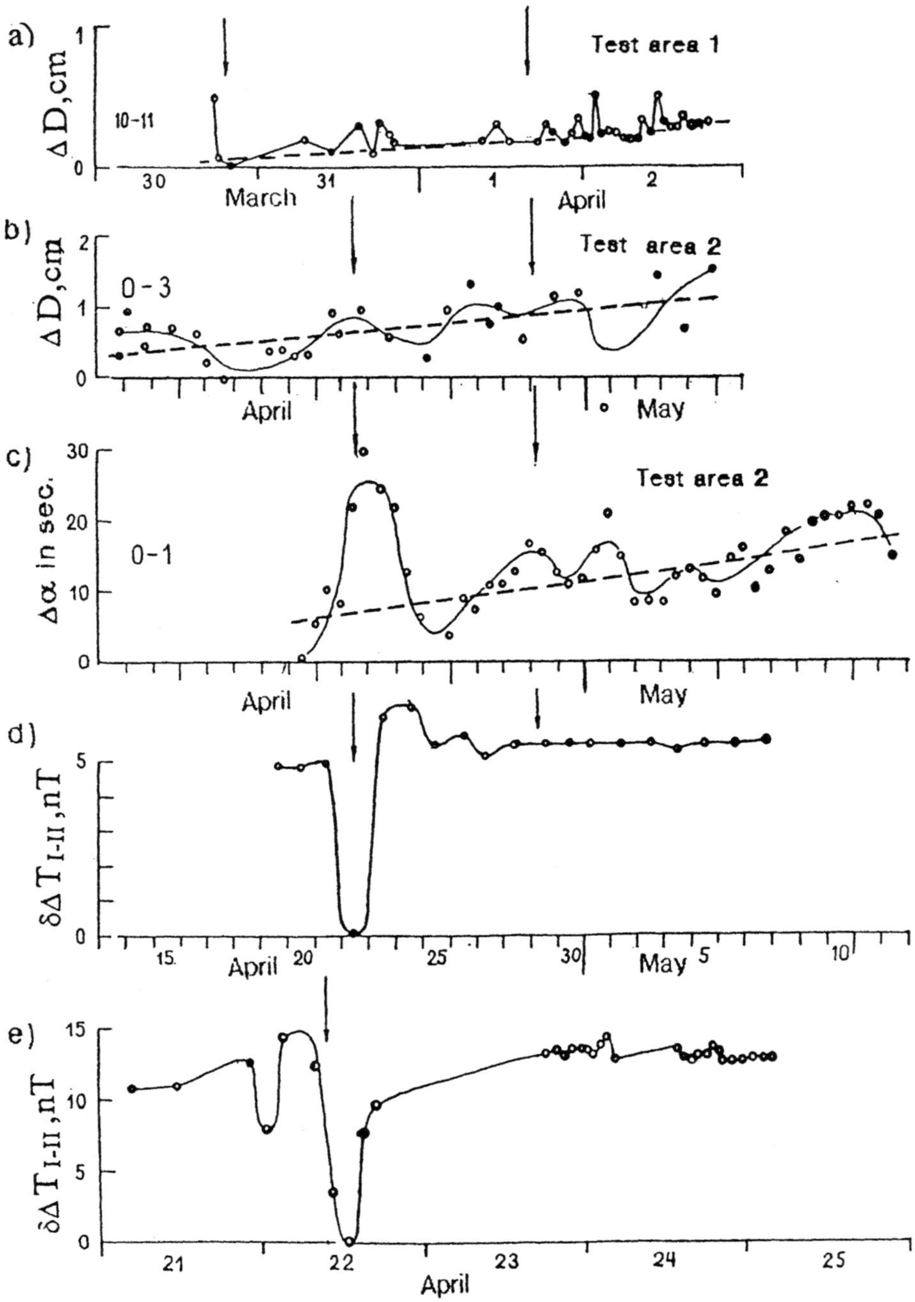

Figure 4. Surface deformations and geomagnetic variations of the Gazli 1984 aftershocks: a), b) - horizontal motions, c) - vertical motions, d), e) - changes (δ) of the local geomagnetic field difference (ΔT) between stations I, II. Moments of the aftershocks are showed by arrows.

trend (Fig. 4b). The maximum change value of Δh was 7 cm during 3 weeks. Variation of $\Delta\alpha$ with the maximum amplitude of 30" was observed along the 0-1 direction from April 20 to 24 and coincided with the aftershock of April, 22 (K=12.1), the source of which was situated on the test area. Temporal changes of $\Delta\alpha$ are correlated for different directions with linear correlation coefficients' r 0.6-0.76. In some cases, during measurements, the aftershocks with MSK=1.5-4 occurred. Before and after a shock sign-variable changes of D values with amplitudes 1-2 cm were recorded during several minutes. Vertical reversible motions of sites in the moment of a shock reached 3-8 cm.

As a result of geodetic measurements it was established that near the epicenter since April, 13 (25 days later than main shock) during one month the tension deformations were prevailing. Accumulated deformations had directions of main axes: tension 79° with a strain of $\varepsilon_1=5.2\times10^{-6}$, compression 169° with a strain of $\varepsilon_2=-1.8\times10^{-6}$.

Temporal changes of ΔT difference are shown on Fig. 4 (b, c). The maximum variation with the amplitude of 5 nT was observed for stations I and II during April, 21 to 24, what coincided with the aftershock (April, 22) with the epicenter close to the station I. Therefore, the decrease of the geomagnetic field T on the station I before the aftershock can be considered as a cause of the variation of stress. During another strong aftershock (K=12) with the epicenter on a distance 7 km from the station I, the ΔT level did not change. Temporal changes of the diurnal mean between stations I and "Garm" were of oscillatoring character. On the data of hourmean values between each pair of station I, II and "Garm" it was established that before the April, 22 earthquake on the station I a reversible decrease of T, and after it - a reversible increase of T on the station II occurred (Atrushkevitch *et al.*, 1986).

Comparison of geodetic and geomagnetic observations, resulting from the April, 22 shock, which occurred under conditions of near-horizontal sublatitude compression, show that vertical motions increased along 0-1, horizontal motions and magnetic moment decreased.

Results of geodetic observations of the Rudbar earthquake epicentral zone are shown on Fig. 5a. In comparison to the slow trend, the more short-period variations of D and ε values were observed. Maximum value of the trend constituent was recorded on the network 1 - $\varepsilon=4.3\times10^{-6}$ in 43 days. Maximum amplitude of ε variation was 22×10^{-6} on the network 2 at the middle of August, which coincided with the aftershocks of August, 12 and 13 (M=3.9). Variation amplitudes decreased with time.

In spite of general similarity, deformational processes had distinctions on different test areas. In approximation of quasi-uniform deformation of small test areas, directions of main axes were determined. Azimuths of the main lengthening axes were: on the test area 1 A=6-186° ($\varepsilon_1=6\times10^{-6}$, $\varepsilon_2=-1.6\times10^{-6}$), on the test area 2 A=160-340° ($\varepsilon_1=6\times10^{-6}$, $\varepsilon_2=0$), on the test area 3 A=58-238° ($\varepsilon_1=4.2\times10^{-6}$, $\varepsilon_2=-1.5\times10^{-6}$). On the test area 2 the direction of the fault motion was determined with the help of the tectonophysical method elaborated and applied by Gushtenko *et al.* (1993). This direction coincided with the main lengthening axis:

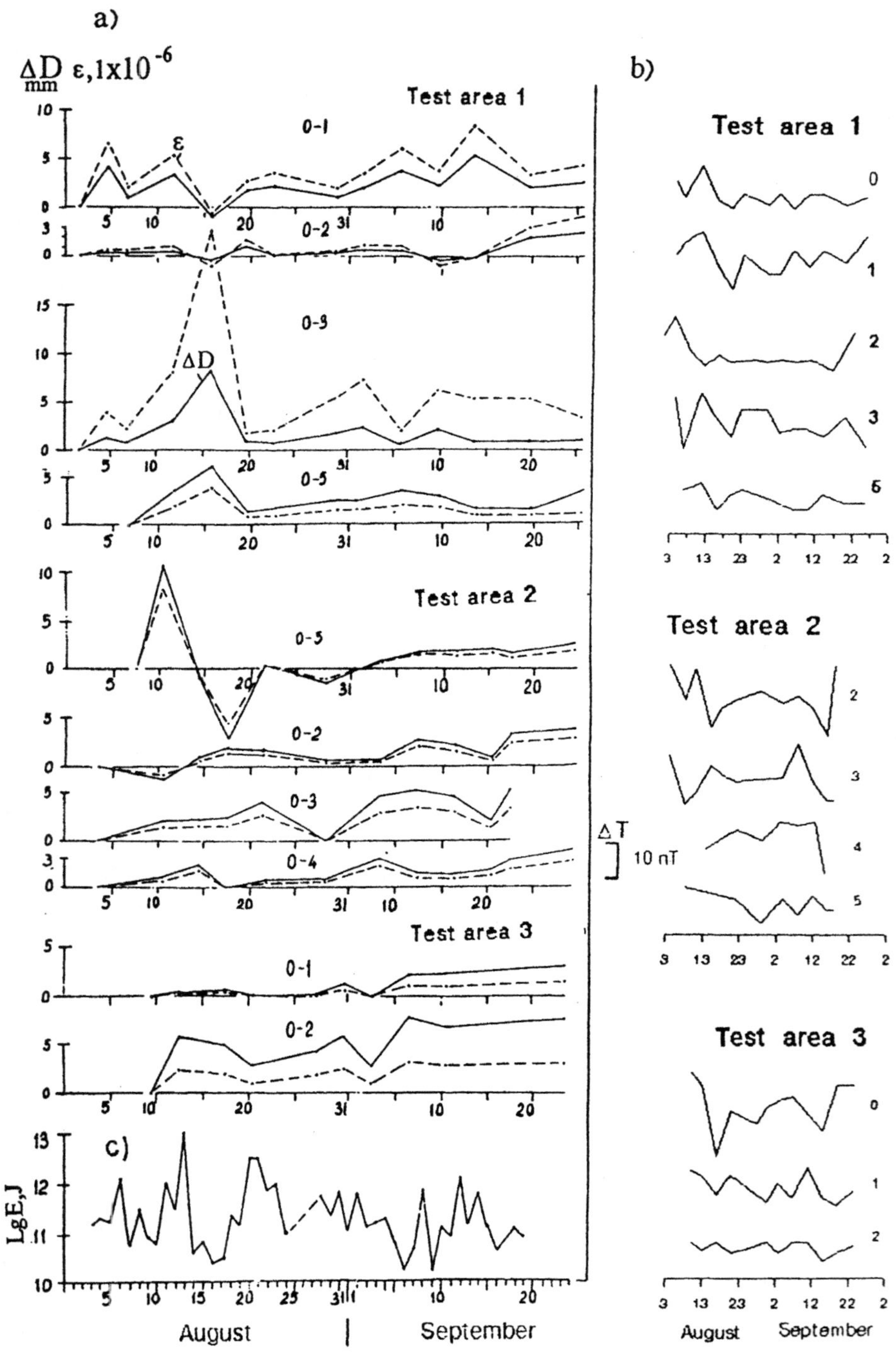

Figure 5. Surface deformations and geomagnetic variations of the Rudbar 1990 aftershocks: a) - horizontal motions on the test areas 1,2,3, b) - changes of the geomagnetic field differences on the test areas 1,2,3, c) - seismicity.

the difference between them did not exceed 6°.

Temporal changes of the ΔT difference are shown on Fig. 5b. On the test areas 1 and 2 two intervals can be distinguished by the character of ΔT changes. I) - in the August, 4 to 20, when relatively rapid variations with amplitudes of 5-7 nT were prevailing and not coinciding in a phase on different sites. II) - from the August, 20 to the September, 23, when changes not exceeded the maximum background. A linear trend was observed in the site 4 of the test area 1: ΔT decreased on 20 nT during all period of observations. On the test area 3, the most distant from the epicenter, ΔT changes in different sites did not coincide in a phase; maximum linear trend had 12 nT in the site 3.

The observed nonuniformity of ΔT changes within the limited test areas, situated on different distances and directions from the epicenter, as well as the distinction between main stress and strain axes, reflect non uniformity of the deformation of rocks volumes (blocks) at low depths.

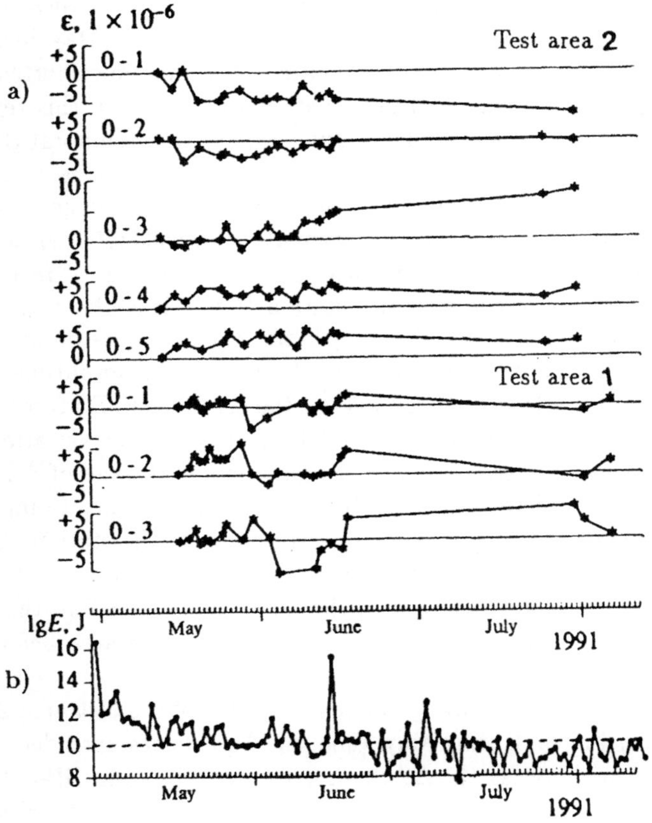

Figure 6. Surface deformations the Racha-Djava earthquake epicentral zone (fig. a) and seismicity (fig. b).

Results of the combined measurements in the Racha epicentral zone are shown on Fig. 6-8. According to the temporal changes of ε, in two zones of aftershock concentration the intensive horizontal motions occurred. On the test area 1 the distance changes were of oscillatoring character. The maximum amplitude value was 11.2×10^{-6} along the 0-3 direction; residual deformations during all observational period were not recorded; temporal changes of D values were interrelated for the lines 0-2 and 0-3, with the value of a linear correlation coefficient equal to 0.6. On the test area 2 oscillations of D values occurred also, the maximum amplitude value was 9.5×10^{-6} along the 0-3 direction. Moreover, uniform directional motions along the 0-1 and 0-3 lines can be approximated by linear relationships, with the rate of - 0.72×10^{-6}/day (shortening of the 0-1 line) and 0.11×10^{-6}/day (lengthening of the 0-3 line). During all periods of distance measurements a residual lengthening was observed on 0-3, 0-4, 0-5, and the distance 0-1 is characterized by residual shortening.

From May 12 to 25 the directional changes of distances were observed along each direction. On the test area 2 maximum value of correlation coefficient was 0.8 for the lines 0-4 and 0-5; temporal changes of ε along these distances had a mutual correlation with the value of coefficient 0.7. From May 25 on the test area 1 the sign of deformation changed: on each line a reversible shortening took place. It may result from a change of deformational processes in this region before the strongest aftershock, of M=6.3, on June 15. Spatial-temporal changes of accumulated deformations may reflect non uniformity of movements of the points on surface, especially a change of azimuths of compression-tension and a distinction of the deformation sign on the same azimuth (Atrushkevitch *et al.*, 1986).

Temporal changes of ΔT difference are shown on Fig. 7. In every site of the test area 2, relative to the reference station "Oni", irregular oscillations with the maximum swing from 3 to 7 nT and periods 3-5 days were observed during May-June. The reversible variation of "bay" form with the amplitude of 5-10 nT and duration about 10 days occurred also in the sites 0,1,3. Features of ΔT changes showed different characters of interrelation between the variations on different stages of aftershock activity: in July-August temporal series of ΔT revealed interrelation between different sites, contradicting to May-June. Comparison between the field changes in the epicentral zone and those in the station "Goragorsk", far from the epicenter, showed that in the aftershock zone from 4 to 9 June there was a decrease of field modulus T, with the amplitude of 4 nT in the station "Oni", that can be caused by seismotectonic processes in the area before the shock of June, 15.

Temporal changes of gravity difference in sites of the test area 2 and GP"Oni" are shown on Fig. 8. During May-June the Δg changes exceeded 3m values, and close correlation between the Δg changes in sites GP-4 and GP-5, GP-2 and GP-3 was found. At the same time , the long-period components of series Δg(t) differed in these sites. More important feature was an appearance of reversible variation with the amplitude of 0.05 mGal in sites GP"Oni" and GP-4, GP-5 of the test area 2 during the period from May 20 to 30, with the extremum on May 24-26. In July-August the Δg changes had less amplitudes. The maximum changes of Δg

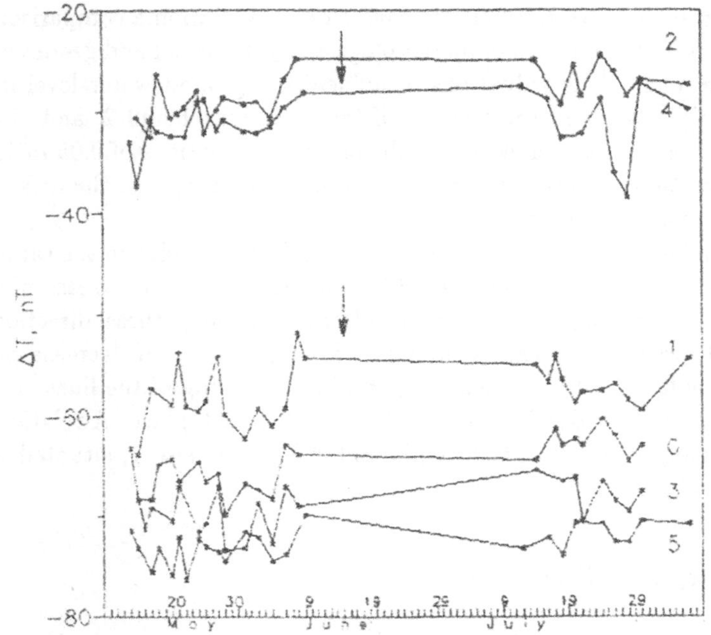

Figure 7. Magnetic field differences between the sites of the test area 2 and the reference station "Oni". The moment of the June, 15, aftershock is showed by arrows.

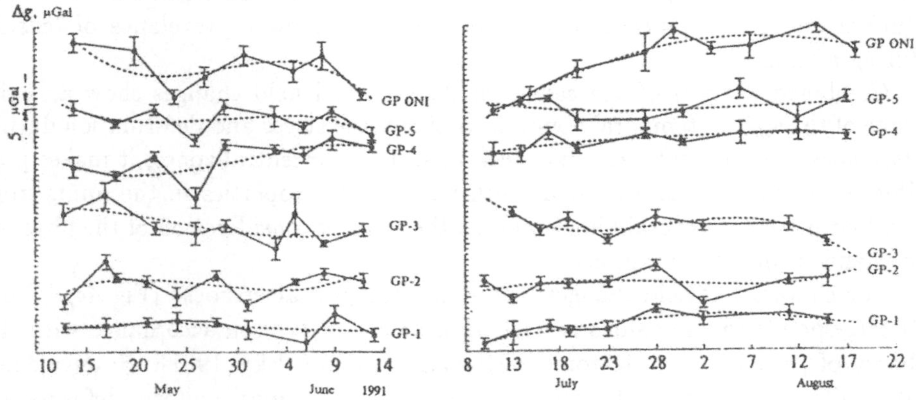

Figure 8. Gravity variations on the test area 2 and GP"Oni" (dotted lines - approximation by polynomial of degree 3).

values reached 0.05 mGal in GP"Oni" from the first half of July to the second half of August. To investigation the causes of Δg variation a comparison between gravimetric results and data of meteorological, hydrological and geodetic measurements were made. The valuations of influence of ground water level in the well "Oni", precipitation's in the vicinity of the test areas 1 and 2, and also vertical relative motions of sites showed that the observed variations of 0.05 mGal can not be caused by these. Thus, we conclude that density changes of the rocks cause the observed gravimetric effect.

A close relationship between the changes of different fields in the same observational sites was revealed. Thus, on 12-25 May, there was an increase of distances' 0-4, 0-5, corresponding to the strain deformation along these directions, and a decrease of gravity field values in sites GP-4, GP-5, i.e. a decrease of effective density. Simultaneously, at the same period a shortening of the lines' 0-2, 0-3 and increase of gravity field values in sites GP-2, GP-3 took place. A relation between temporal changes of Δg in GP "Oni" and ΔT in the site 1, situated besides of sites GP-0, also occurred.

5. Summary and discussion

The results of observation in epicentral zones show, that the significant changes of distances, gravity and magnetic fields occurs during aftershocks of earthquakes with $M \geq 7$. These changes are manifested by a type of short-period variations of geophysical parameters, often relative to the trend. The amplitudes of variations exceed the fluctuations of backgrounds observed in tectonically active regions during the seismic quite times and the periods of preparation of strong earthquakes. As a rule, the amplitudes of variations of the geophysical parameters decrease while the aftershock activity decreased. Results of geodetic measurements in the Gazli, Rudbar and Racha epicentral zones in the whole show a prevelance of tension during aftershocks.

Combined analysis of the causes of the observed field changes show an existence of the spatial-temporal nonuniformities of the stress and deformation during aftershocks. Taking into account network scale in epicentral zones, it makes possible to evaluate the size of areas with nonuniform properties in the limits from hundred meters to several kilometers, on the basis of consideration of the physical mechanisms of field variations.

The character of field changes before the strongest aftershocks (Fig. 4c, 4d, 6a, 7) corresponds to anomalous variations of precursor type in accordance with the theory of preparation of tectonic earthquake (Dobrovolsky, 1991). However, not all results can be interpretative in assumption of the quasi uniform deformation of solid medium. The local gravity changes with the amplitude of 50 mkGal in the Racha epicentral zone, at the end of May to the beginning of June (Fig. 8), could not occur because of changes of the rock density caused by compression with the value of ε equal to 1×10^{-5} (test area 1, Fig. 6a). The calculated values of the gravity changes were lower on 2 orders than those observed on the surface (Skovorodkin *et al.*, 1994). Also the difference of temporal changes interrelations of the parameters (fields) between different sites on the distances less than some kilometers, can not be explained.

The study of the spatial-temporal nonuniformities of variations made possible to suggest a new mechanism of the origin and development of non uniformities of the stress and deformation during aftershocks in superficial parts of the Earth crust. We assume that such non uniformities can appear as a result of block's formation, section boundaries of which are fissures (faults) of different rank. Origin of fissures and its parameters' changes, including opening and closing, is a direct consequence of processes at the moment of main shock and after it. It is known that the fissures' parameters are not uniform in a massif volume. Their number decreases along the direction from hypocenter to the earth surface. The fissurity depends also on lithological composition. However, at the moment there is no possibility to evaluate the distribution of the seismogene fissure system. Therefore, we are forced to do no more than following premises to evaluate the Δg changes, caused by the density changes in the fissured medium. 1) The change of the effective density ($\Delta\rho$) in a volume of rocks is a result of formation, opening and closing of fissures. 2) During the aftershock activity the dilatant fissures are formed and developed by means of fluctuations of external tectonic stress and residual stresses in blocks. 3) The fissures are distributed relatively uniformly and isotropy in focuses' volume ($\Delta\rho$=const). In accordance with the mean depth of the aftershock hypocenters, equal to 5 km, the calculations of the maximum values of anomalous variations were done. As a result, the observed change value Δg=-50 mkGal can appear at the following values of ratio $\Delta\rho/\rho$ (ρ=2.5 g/cm^{-3}): -0.05% - for the vertical ledge with the thickness of 1 km, -0.2% - for the horizontal circular cylinder with the radius of 1 km on the depth 5 km, -2% - for a sphere with the radius of 1 km on the depth 5 km. It is reasonable to suppose that the origin of fissurement in focal zone, which gives the degree of opening-closing of fissures, corresponding to such density changes. Evidently, it results from this model, that the amplitude of back-translation motions in focal zone is to exceed on 2 orders than those observed on surface. At this, a vertical distribution of tectonic stress can be unchanged or can be weakly changed. It is also noted that under our assumptions it is not important whether filling of fissures by water occurs.

6. Conclusion

Feature of the aftershock period, as a final stage of seismotectonic process, is relatively rapid change of the source structure in comparison with the period of earthquake preparation. Combined geophysical observations in the epicentral zones of the earthquakes with the M$\geq$7 made possible to distinguish at any rate two stages of the aftershock process and to reveal the peculiarities of the changes of the stress and deformation in geological medium.

We assume that the processes of the fissuring in the aftershock zone play a main part. On the first stage, at the time of high activity after the main shock, the blocks of different scale are formed in superficial parts of the Earth crust as a result of formation and development of fissures and its systems. Probably, the variations of the stress and deformation in these blocks are caused by interaction of changing external regional fields of tectonic stress with residual ones in forming blocks. Such interaction, resulting in stress redistribution in the medium, occurs due to the changing boundary conditions by the intensive formation of new fissures and closing-opening of preexisting ones. Loss of individual peculiarities of the stress and deformation in blocks and a decrease of variation amplitudes, is characteristic for the second stage, at the decrease of aftershock activity: the variations become correlated. In accordance with the temporal changes of the geophysical parameters, at that time the smoothing of the interblock stress gradients takes place under the influence of external regional tectonic forces, resulting from the consolidation of the aftershock zone. The fluctuation feature of the temporal stress changes remains in the aftershock zone.

Until now it is not possible for us to discuss the moment of the regular state's achievement because the present experimental evidence is not enough. It is noted that the seismotectonical process, in this case, corresponds to the intermediate type model of earthquake preparation and that the transition from the first stage to the second one is smooth, evolutionary (Dobrovolsky, 1991). It needs certainly a more detailed combined investigations of the source region (in a space and time) for a study of aftershock process.

7. References

Abdukadirov, A.A. *et al.* (1987) Gazli earthquake of March, 19, 1984, in *Zemletrjasenija v SSSR*, Nauka , Moscow, 67-85.

Arefyev, S.S. *et al.* (1993) The Racha earthquakes of 1991: results of field seismological observations, *J.Fizika Zemli RAN* **3**, 12-23.

Atrushkevitch, P.A. *et al.* (1986) Stress-strain medium state in the epicenter of Gazli earthquake of 1984, *J. Dokl. Acad. Nauk SSSR* **288**, N5, 1082-1085.

Dobrovolsky, I.P. (1991) *Theory of Preparation of Tectonic Earthquake*, IPE AN SSSR, Moscow.

Gushtenko, O.I. *et al.* (1993) The recent regional field of stresses and the mechanism of the lithosphere deformation of seismoactive East-Asia region: Abstract, in *7th Meeting of European Union of Geosciences*, Oxford, 259.

Skovorodkin, Yu.P. *et al.* (1994) Geophysical observation of aftershock of the April 29, 1991, Caucasus earthquake, *J.Fizika Zemli RAN* **1**, 59-69.

Tsukuda, T. *et al.* (1991) Aftershock distribution of the 1990 Rudbar, Northwest Iran, earthquake of M=7.3 and its tectonic implications, *Bull. Earthq. Res. Inst. Univ. of Tokyo* **66**, p.2, 351-382.

Zakharova, A.I. *et al.* (1993) Major source parameters of the Racha earthquake and its aftershocks, *J.Fizika Zemli RAN* **3**, 24-41

DISPERSION AND PROBABILITY DENSITY FUNCTION FOR FOCAL MECHANISM TENSORS

S.G. TYULENEVA, S.L. YUNGA
Joint Institute of Physics of the Earth,Russian Ac. Sci.,
B.Grusinskaya 10, Moscow 123810, Russia.

Abstract

Recently, large catalogues of earthquake focal mechanisms have become available. The question now arises of whether the statistical methods are applicable to this type of catalogues. In this paper we discuss the problem and introduce probability density function and dispersion for focal mechanisms set, where a focal mechanism tensor is treated as a random object. The procedure of numerical simulation of random focal mechanism tensors is proposed. We also describe the process of numerical simulation of a set of random focal mechanism tensors corresponding to the given probability density function.

1. Introduction

The earthquake focal mechanisms contain an important part of information necessary for the evaluation of earthquake hazard and risk. This is the information on development of tectonic process. The accumulation of focal mechanisms data provides us an opportunity to make the description of seismotectonic deformations with statistical methods. The application of statistical analysis methods to focal mechanisms sets gives the material for the tectonophysic interpretation. And thus we can make conclusions on the character of the deformation and stress state of seismically active regions.

Now the sufficient amount of data on the focal mechanisms of earthquakes with low and middle magnitude values is available and statistical methods may be applied.

The offset character in the seismic sources indicates on the variety of seismotectonic movements in even uniform geotectonic structures, whereas the statistic deals with homogeneous objects. Hence before proceeding any further the question of similarity in seismotectonic process should be resolved. The hypothesis about offset similarity in the sources of weak, medium and strong earthquakes was put forward and directly checked on experimentally determined focal mechanisms of earthquakes. The existence

V. Schenk (ed.), Earthquake Hazard and Risk, 77–84.

of a certain similarity was found out (Danilova et al., 1990, Lukk et al., 1989, Yunga, 1990).

2. Focal mechanism tensor

The classical theory of elasticity gives the description of an earthquake focal mechanism by a seismic moment tensor or a seismic moment density tensor (SMD tensor) for bounded size sources. Being interested only in the geometrical characteristics of earthquake source we use more simple object - so called earthquake focal mechanism tensor. It is a second rank tensor that differs from the SMD tensor by the zero determinant.

2.1 DEFINITION

Lets denote as **b** a unit length vector laying in a fault plane and indicating the direction of the offset and as **n** a unit length vector normal to the plane. Obviously **b** and **n** are orthogonal vectors. So the focal mechanism can be represented as tensor:

$$\mathbf{m} = \frac{1}{\sqrt{2}}(\mathbf{n}\otimes\mathbf{b} + \mathbf{b}\otimes\mathbf{n}) ,$$

where $\otimes$ is dyadic product of vectors.

This tensor is named a focal mechanism tensor. In a fixed coordinate system it is represented by the matrix:

$$\mathbf{m}_{ij} = \frac{1}{\sqrt{2}}(\mathbf{n}_i\mathbf{b}_j + \mathbf{b}_i\mathbf{n}_j),$$

where **i,j** = 1,2,3.

Lets introduce the scalar product of two matrices (i.e., tensors):

$$(\mathbf{m1}, \mathbf{m2}) = \sum_{i,j} \mathbf{m1}_{ij}\mathbf{m2}_{ij},$$

where **i,j** = 1,2,3. The norm of **m** is

$$\|\mathbf{m}\| = \sqrt{\sum_{i,j} \mathbf{m}_{ij}^2}$$

where **i,j** = 1,2,3.

The value of the scalar product can be considered as a distance (or angle cosine) between our matrixes.

2.2. PROPERTIES

From the definition one can easily deduce that focal mechanism tensor matrices have the following properties:

$$\text{tr}\,\mathbf{m} = \sum_{i=1}^{3} \mathbf{m}_{ii} = 0$$

$$\det \mathbf{m} = 0$$

int $\mathbf{m} = \sum_{i,j} \mathbf{m}_{ij}^2 = 1$, where i,j=1,2,3.

So eigenvalues is $\frac{1}{\sqrt{2}}, -\frac{1}{\sqrt{2}}, 0$, i.e. **m** is two-dipole matrix.

$\mathbf{m}_{ij} = \mathbf{m}_{ji}$
rank $\mathbf{m} = 2$

3. Average tensor of earthquake focal mechanisms set

One can judge about seismotectonic deformation directionality basing upon statistics of focal mechanisms. For example, an average tensor $\overline{\mathbf{m}}$ calculated for a seismogeneous volume of the earth crust can be used:

$$\overline{\mathbf{m}}_{ij} = \frac{1}{N}\sum_{k=1}^{N} \mathbf{m}_{ij}^{k}$$

where $(\mathbf{m}_{ij}^{k})$ is the matrix of k-th experimentally determined focal mechanism for the region under study and N is their number.

If we consider the focal mechanism tensor as a discrete random variable then an average tensor represents an estimate of its mathematical expectation.

4. Probability density function and dispersion.

For many geotectonic and geophysical problems it is not sufficient to have only an average tensor. So let us consider a question of a definition of probability density function (PDF) for tensors treated as random objects.

4.1. EXAMPLES OF PROBABILITY DENSITY FUNCTION.

Let **M** is fixed matrix, then the simplest probability density function is

$$y = C\ \mathbf{H}[(\mathbf{m},\mathbf{M})]\ , \tag{1}$$

where C is norming constant and **H** is Heaviside function:

$$\mathbf{H}(x) = \begin{cases} 1, x \geq 0 \\ 0, x < 0 \end{cases}$$

As matrices of our set have a unit intensity we can represent their as points on a unit hypershere. Then the formula (1) gives a uniform distribution of points only on the hemihypershere.

Another more complex example of PDF is

$$y = \mathbf{C}\ exp(-\mathbf{k}'(\mathbf{m}, \overline{\mathbf{m}}))\ , \tag{2}$$

where **C** is norming constant, **k'** is concentration parameter. It is easy to prove that this PDF can be represented as a product of two PDFs each of them is spherical PDF relatively specific matrix based on principal axes.

Both above mentioned PDF are depending only of a distance between fixed matrix and an arbitrary one.

4.2. DISPERSION

A dispersion for this case is introduced as follows:

$$D(m) = \left\langle \left\| m - \langle m \rangle \right\|^2 \right\rangle$$

where <**m**> is a mathematical expectation of a random variable **m**.

Let $\kappa = \left\| \langle \mathbf{m} \rangle \right\|^2$. Lets simplify the expression for **D(m)**:

$$\mathbf{D(m)} = \langle(\mathbf{m} - \langle\mathbf{m}\rangle, \mathbf{m} - \langle\mathbf{m}\rangle)\rangle = \langle (\mathbf{m},\mathbf{m}) - 2(\mathbf{m},\langle\mathbf{m}\rangle) + (\langle\mathbf{m}\rangle,\langle\mathbf{m}\rangle)\rangle =$$

$$= \left\langle \left\|\mathbf{m}\right\|^2 \right\rangle - \left\| \langle m \rangle \right\|^2 = 1 - \kappa$$

Having $\left\langle \left\|\mathbf{m}\right\|^2 \right\rangle = 1$ as only tensors with unit norm are considered.

4.3. EXAMPLE OF THE PDF CALCULATED FROM THE EXPERIMENTAL DATA

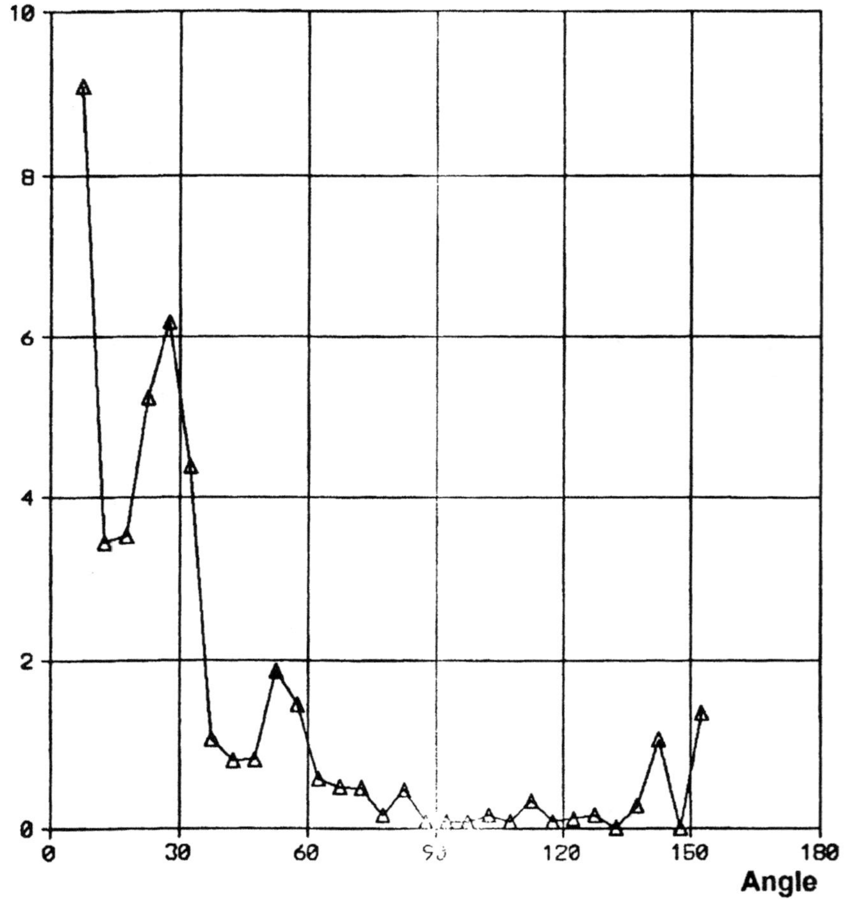

Fig.1 Probability density function for the experimental data set of focal mechanism tensor.

To illustrate the foregoing we selected an circular area in Central Tian-Shan with center in 43°N, 75°E and radius 2°. The catalogue of earthquake focal mechanisms in use is described in (Lukk, A..A. and Yunga, S.L., 1990). The sample consists of 108 events with magnitude M > 4 registrated from 1953 up to 1994. The calculated stress direction is close to meridianonal and tension one - to vertical, $\kappa = 0.4$.

The calculated probability density function is shown in the Figure 1.

5. Numerical simulation of random focal mechanism tensors

The focal mechanism tensor is completely defined by its principal axes. So the problem of random focal mechanisms simulation is equal to the one of simulation of three-dimensional space random rotations.

We use the representation of a three-dimensional rotation by four Euler parameters to construct a set of random uniformly distributed rotations of three-dimensional space (Moran, P.A.P., 1975, Kagan, Y.Y. et al., 1985 and Kendall, M.G. et al., 1963).

The most convenient representation of a rotation of space combined with an expansion or contraction is given in quaternion terms (Klein, 1932). In our case the absence of scale changes impose the additional restriction.The sum of four squared Euler parameters must be equal to 1, i.e. 4D point represented by Euler parameters lies on 4D unit hypersphere. So we have only 3 independent parameters.

Euler parameters for each rotation are obtained with a generator of random numbers as follows.

We generate random numbers on the interval (-1, 1) for each of four coordinate axes. These numbers are the coordinates of some point in 4D cube. We select from these points only those that fall into the solid sphere inscribed in the cube. Then each point is projected on the sphere along its radius. At last, the vectors, falling on the low hemisphere are transferred on the upper hemisphere.

The direction cosines of obtained vectors are accepted as Euler parameters for rotation matrix **A**. Matrix of tensor **m** in the new coordinate system is calculated by formula:

$$\mathbf{m}' = \mathbf{A}\mathbf{m}\mathbf{A}^*,$$

where $\mathbf{A}^*$ is transpose of **A**.

We generate a set of random matrices by random rotation of some initial matrix **m**.

To control the algorithm we estimate through Monte-Carlo simulation value of π by formula:

$$\pi = \sqrt{\frac{32\mathbf{p}}{\mathbf{q}}},$$

where **p** is the number of points within the unit hypersphere and **q** is the number of points within the unit cube in 4-dimensioned Eucledean space.

For series of 100 000 - 1000 000 random matrices we got two valid digits for π value.

6. Numerical simulation of a set of random focal mechanism tensors corresponding to the given PDF.

6.1. THE SIMILARITY COEFFICIENT AND DIRECTOR-MATRIX.

Let the resulting matrix $\overline{\mathbf{m}}$ is equal to the sum of mechanism matrices divided by their number. Then the similarity coefficient can be defined by formula:

$$\mathbf{K}=(\mathbf{m},\ \overline{\mathbf{m}})/\|\overline{\mathbf{m}}\|$$

The similarity coefficient **K** can be treated as the measure of likeness between some matrix m and the resulting matrix $\overline{\mathbf{m}}$.

Let $m' = \dfrac{\overline{m}}{\|\overline{m}\|}$, so $\|m'\| = 1$ and $\|m\| = 1$ by definition. Then

$$(m,m') < 1 \tag{3}$$

because m is two-dipole matrix and m' is three-dipole, i.e. m has two distinct from zero eigenvalues and m' - three.

The maximum of (m,m') for fixed m' is $\dfrac{m_1 - m_3}{\sqrt{2}}$, where m_1 is maximum and m_3 is minimum eigenvalue of m'.

To obtain equality in (3) we must introduce the director-matrix m* as follows:

$$m^* = \sqrt{2}\,\frac{m'}{m_1 - m_3}.$$

6.2. THE DIFFERENTIAL VOLUME ELEMENT DEFINITION.

To define the probability density function for random mechanism matrices we need to build volume elements in matrix space. This can be done easily when the differential volume elements correspond to increment of angle between some matrix and selected director-matrix. Let the total volume of matrix space is equal to one unit. We subdivide this volume into sufficiently big number of small elements by dividing the whole range of angles into small nonintersecting intervals. Then the volume of every small element can be estimated by Monte Carlo method as the ratio of a number of random matrices falling in this volume to the total number of generated random matrices. We realised this method by dividing the whole range of angles into 100 intervals.

6.3. CONSTRUCTION OF A SET OF RANDOM FOCAL MECHANISM TENSORS CORRESPONDING TO THE GIVEN PDF.

When the volume element in matrix space is built it is possible to generate a set of random matrix corresponding to the PDF (2). This was done in the following way. First of all we selected a director-matrix and the random matrix sequence was generated using the method described above. Then for every random matrix falling in a differential volume element the weight factor proportional to the volume of the element and proportional to PDF (2) was calculated. Next step was summation of the matrices with the corresponding factors and the calculation of the director-matrix.

It is not obvious that the calculated director-matrix is the same as initial director-matrix. Only the coincidence of the principl axis can be derived from the symmetry.

Our calculations proved that for all tested values of the concentration parameter **k'** in (2) the calculated director-matrix coincide with initial director-matrix.

Also we can assume that it is possible to estimate the concentration parameter **k'** on the base of experimental data.

7. Conclusion.

We introduced the probability density function for arbitrary focal mechanisms set where director-matrix is parameter of distribution. The procedure of numerical simulation of a set of random focal mechanism tensors corresponding to the given probability density function were examined in details.

It was shown that the estimate of director-matrix calculated from a generated set of mechanisms practically coincides with initial director-matrix. This result shows that our definition of the probability density function is correct. So it is possible to estimate parameters of the distribution on the base of experimental data.

8. Acknowledgements.

The work reported here was supported in part by a grant from International Science Foundation (ME8000).

9. References

Danilova, M.A. and Yunga S.L. (1990) Statistical properties of matrixes in problems of interpretation of earthquakes focal mechanisms, *Izv. Acad. Sci. USSR, Phys. Solid Earth* **2**, 42-48.

Moran P.A.P. (1975) Quaternions, Haar Measure and the Estimation of a Palaeomagnetic Rotation, *Perspectives in probability and statistics*, Academic press inc., London.

Kagan, Y.Y.and Knopoff, L. (1985) The first order statistical moment of the seismic moment tensor, *Geophys. J.R.astron. Soc.* **81**, 429-444.

Kendall, M.G. and Moran P.A.P. (1963) *Geometrical probabilities*, Hafner, New-York.

Klein, F. (1932) *Elementary Mathematics from an Advanced Standpoint. Vol. I. Arithmetic, Algebra, Analysis.* Macmillan, London.

Lukk, A.A. and Yunga, S.L. (1989) *Geodymnamic and stress strained state of lithosphere of Middle Asia*, Donish, Dushanbe.

Yunga, S.L. (1990) *Methods and results of study of seismotectonic deformations*, Nauka, Moscow.

Part 2:

EARTHQUAKE HAZARD AND SEISMIC RISK

PRELIMINARY SEISMIC HAZARD ASSESSMENT FOR THE ITALIAN SEISMIC CODE

D.SLEJKO
and GNDT WORKING GROUP "SEISMIC HAZARD"
Osservatorio Geofisico Sperimentale
P. O. Box 2011, 34016 Trieste, Italy

Abstract. The GNDT seismic hazard assessment project for a future seismic zonation of the Italian territory is described and the first preliminary results are presented. They have been obtained by applying two methodologies (the Cornell approach and the mixed method), and refer to two stages of elaboration: after having summarised the products performed during a previous national project, and after the first release of the products developed ad hoc for the present GNDT project. Maps of horizontal and vertical PGA referring to 100 and 500 year return periods and maps of exceedance probability for intensity VII MCS in 50 years are shown and commented.

1. The Italian Seismic Code

As Italy experienced repeatedly destructive earthquakes, the different governments which dominated parts of the territory faced the problem to preserve people and things from the earthquakes (Figure 1). The first legislative measures were taken by the Bourbon government in Calabria

V. Schenk (ed.), Earthquake Hazard and Risk, 87–124.

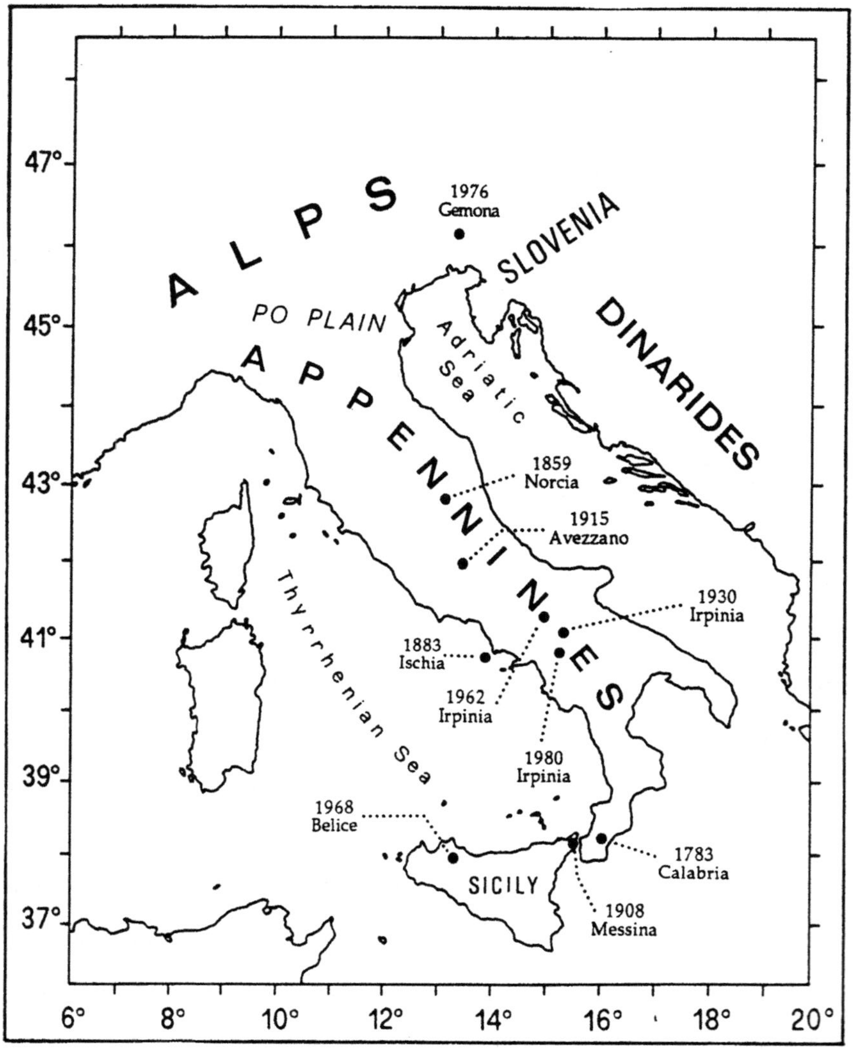

Figure 1. Index map of the study region and location of the earthquakes important for the evolution of the Italian seismic code.

after the 1783 earthquakes which caused more than 30,000 deaths. In the following years, the choice of the sites for the rebuilding, as well as the construction standards, were considered by the law and by the regulations issued from the Papal states after the 1859 earthquake in Norcia. After the union of Italy, all previous regulations lapsed and the Italian state was unprepared to manage the situation after the 1883 earthquake which ruined all villages of Ischia island. The quake which destroyed Reggio Calabria and Messina on December 28, 1908, causing 80,000 estimated deaths, was probably the strongest event in the Italian peninsula during the last ten centuries. Soon thereafter, the national seismic code was promulgated: it consisted of the list of the municipalities of Sicily and Calabria where technical rules for building, defined by Royal Decree, were applied. The seismic zonation was updated after every destructive earthquake simply by adding municipalities to the official list: it was therefore based only on the fact that a municipality had experienced damage because of earthquakes after 1908, without any scientific consideration on the Italian seismicity and had principally the purpose of public aiding (for more details see Petrini, 1991). In 1974, a new seismic code was promulgated (law 64/1974), which, because it only states the criteria for constructing in a set of given seismic zone categories, possesses an inherently dynamic character. The geographic description of the zones, intended as list of regulated municipalities, is established by decree, and, therefore, it can be easily updated, after every damaging earthquake, following the old philosophy or, better, when the increased knowledge of the Italian seismicity provides a revision of the zonation. Although the classified municipalities were inserted into two seismic categories according to the suffered damage, this distinction was rather fictitious and it is not yet solved.

After the 1976 Gemona earthquake, the regional Public Administration asked two public institutions (Osservatorio Geofisico Sperimentale of Trieste and Politecnico of Milano) to provide the scientific support for planning the reconstruction of the destroyed villages. This can be considered the first urban intervention based on seismic hazard studies (Faccioli, 1979; Giorgetti et al., 1980): on the basis of probabilistic studies performed for Friuli the criteria for distinguishing the two seismic categories were given.

Many different studies devoted to the knowledge of Italian seismicity started after the 1976 Gemona earthquake; those studies were developed by cooperation among geologists, geophysicists, and engineers in the framework of the "Progetto Finalizzato Geodinamica" (PFG) of the "Consiglio Nazionale delle Ricerche" (CNR). One of the products was the maps of shakeability of Italy (Gruppo di Lavoro Scuotibilità, 1979; Petrini et al., 1981) on the basis of which the CNR's proposal of seismic zonation was based (see Petrini, 1980; Servizio Sismico del Consiglio Superiore dei Lavori Pubblici, 1986; Petrini et al., 1987). That proposal was accepted completely by the Italian government, particularly following the 1980 Irpinia earthquake which caused 3,000 deaths, and translated into a series of decrees by the Ministry of Public Works between 1980 and 1984. On the basis of probabilistic studies the municipalities indicated as hazardous were inserted into the second category, leaving the already classified municipalities in their old position. In addition, a third category was defined for some municipalities of southern Italy, where even minor shaking could produce severe damage. A limitation of this seismic code consists in its applying mainly to new buildings, without giving reinforcement provisions for existing buildings.

Generally speaking, it can be said that the present Italian seismic code, comprising both zonation and technical rules fixed by laws, takes into account the seismic risk in its three ingredients (hazard, vulnerability, value/exposure). Already mentioned is the third zonation category for very vulnerable buildings. In addition, the technical rules evaluate vulnerability by defining the response spectrum for new buildings (the amplitude of the spectrum dependent only on the seismic zone category), and evaluate value/exposure by introducing a coefficient of seismic protection according to the importance of the building. In future, the seismic code should be the result of a national global risk analysis, where seismic hazard, vulnerability and value/exposure are assessed adequately, and a rigorous strategy of seismic risk mitigation is defined.

2. Main Features of the GNDT Project for the Revision of the Present Zonation

More than ten years have passed since the CNR's proposal for a seismic zonation, and the research activity of the CNR's "Gruppo Nazionale per la Difesa dai Terremoti" (GNDT) in the last years has been defined in order to reach two main goals: 1) a proposal for a new seismic zonation for the national territory, and 2) the definition of the methodologies to be used for seismic risk estimation and the testing of strategies for its reduction. The increase in seismotectonic knowledge (see e. g. Albini and Barbano, 1991 and GNDT, 1991 for the state-of-the-art) has supported the updating of the zonation. A project for the seismic hazard assessment in Italian territory has been being conducted since 1989 by GNDT, and a detailed analysis of the computational approaches as well as of input data set was performed before assessing the hazard. A schematic representation of the GNDT project (see the description in Slejko, 1992) is shown in Figure 2; the five main topics, whose tasks have almost been concluded, and among which there is a continuous transfer of information, are: 1) collection and revision of the seismological data (earthquake catalogue and intensity maps); 2) analysis of the seismotectonic information and preparation of a seismotectonic / seismogenic zonation of the whole of Italy; 3) definition of adequate attenuation relations (for macroseismic intensity and horizontal peak ground acceleration, PGA) for the Italian earthquakes; 4) definition and application of the most suitable probabilistic approaches for Italian conditions; and 5) modelling of the seismic hazard by deterministic approaches (see Costa et al., 1992) for checking the results obtained by probabilistic approaches.

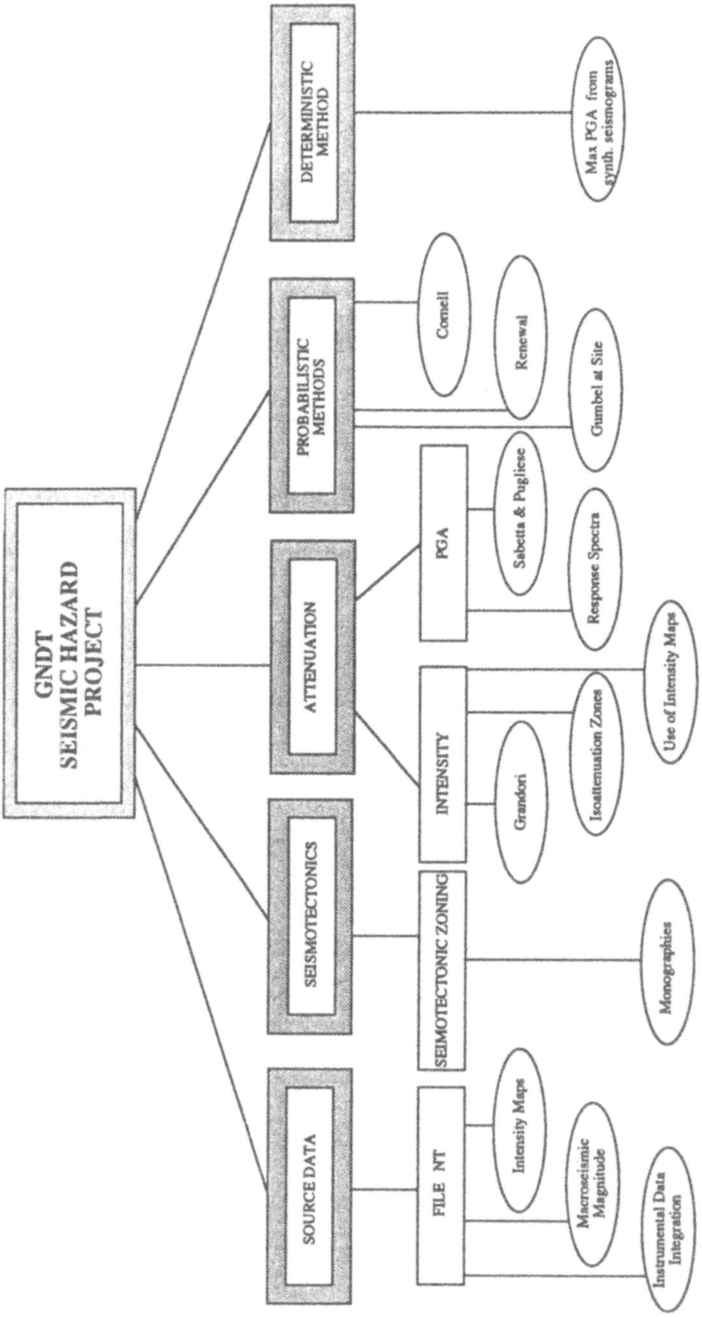

Figure 2. Schematic representation of the GNDT seismic hazard project.

3. The Probabilistic Approaches in the GNDT Project

Three probabilistic approaches should have been applied: 1) a simple statistical analysis of the actual shakings at the site; 2) a memoryless technique for seismicity recurrences, given homogeneous seismogenic zones (SZ's) following the method developed by Cornell (1968); 3) a memory-dependent mixed procedure based on the renewal process (Grandori et al., 1991). At present, only the Cornell (1968) and the mixed methods have been applied, the latter with the memoryless option for a direct comparison of the results obtained with the two methods. The first approach will be applied since a long time documentation of felt shakings is available for most Italian cities (see an example in Monachesi et al., 1994), it has not been applied yet on the whole Italian territory because the input information needed has to be still improved in quantity and homogenised in formats.

The Cornell (1968) approach is based on two work hypotheses: the earthquake recurrence times follow the Poisson distribution (the arrivals are, then, independent and the process is stationary in time) and the magnitude is exponentially distributed (the Gutenberg - Richter relation holds for magnitude and number of events in each magnitude class). In addition, the seismicity is considered uniformly distributed over the SZ. This approach was firstly translated into computer code by Algermissen et al. (1976) and McGuire (1976); the code SEISRISK III (Bender and Perkins, 1987) represents the most recent public domain product and has been used for the present hazard elaboration: its main advantage with respect to the previous codes (e. g.: RISK4a, Algermissen et al., 1976) is that it can account for a probabilistic location of the SZ boundaries and, therefore, also a sort of epicentral accuracy can be taken into account in the processing.

The mixed method (Grandori et al., 1984, 1991) is a variance of the renewal approach and it combines the hypothesis of homogeneous and uniform seismicity over the SZ's, as in the Cornell (1968) approach, with the use of the epicentral locations as reported in the catalogue. This is possible because the distribution of the seismological parameter (macroseismic intensity) and that of the recurrence time are computed separately. More precisely, the distribution of the site intensities is computed, as in the Cornell (1968) approach, by attenuating the different seismicity contributions that characterise each SZ; the recurrence times are, on the contrary, computed from the catalogue epicentral parameters considering all the shakings regardless of their intensity at the site but exceeding a threshold value. It is possible, in such a way, to model non - Poisson processes, by introducing the information about the last event through adequate recurrence time distribution functions. In the GNDT project, the mixed approach has been applied in its simplified form, that is under the Poisson condition for the recurrence times. This is motivated by the fact that, when applied in a wide context, site occurrency behaviour is Poissonian even if source is not. The SZ characteristic seismicity is described by the epicentral intensity distribution function and by the rate, normalised in time and space, of earthquakes exceeding a threshold value (VI Mercalli - Cancani - Sieberg, MCS).

4. Input Data for the GNDT Project

The input data needed in seismic hazard assessment change according to the approach considered. When epicentral parameters are used, as in the present application, an earthquake catalogue, one or more attenuation relations, and, sometimes, a seismogenic zonation are requested. In the present case, the Cornell (1968) approach has been applied considering PGA as seismological parameter and, consequently, spatial definition of the SZ's, parameters of their characteristic seismicity, and PGA attenuation relation are needed. The mixed approach has been applied considering intensity as seismological parameter and, consequently, catalogue, SZ's, intensity isoattenuation zonation, and intensity attenuation relations are needed.

Two earthquake catalogues have been used: the one prepared in the frame of the PFG activities which lists 37211 earthquakes in the Italian territory during the period 1000 - 1980 (Postpischl, 1985b), and the earthquake file prepared by the GNDT's Working Group "Macroseismology" which lists 3555 main events in Italy and bordering regions during the same time period (see Stucchi et al., 1993 for its description). This earthquake file is based on a complete revision of data on the major earthquakes (epicentral intensity Io larger than VII MCS) affecting Italy since 1000 by retrieval and analysis of all the information already used and cited in the earthquake catalogues (see GNDT Macroseismic WG, 1992). The macroseismic parameters have been derived from intensity point maps compiled since 1985 by several investigators or agencies, including GNDT. In the mean time, an assessment of instrumental magnitude for 20th century earthquakes was made (Margottini et al., 1993), together with a preliminary determination of magnitude/intensity relation for Italian earthquakes (Rebez, 1993) developed directly on the data of the GNDT earthquake file. This relation assigns larger magnitude values to historical events than that by Karnik (1969) especially for medium - low quakes: this tendency agrees with the re - evaluation of the magnitude of historical Italian earthquakes through seismic moment computation (Westaway, 1992). The magnitude assessment is crucial in hazard estimates as the majority of the catalogue contents come from historical seismicity.

A seismotectonic model for Italy has been proposed (Scandone et al., 1991) on the basis of a complete analysis of geological and geophysical data, which led to the definition of a kinematic evolution model. It consists of 58 areas (Figure 3a) of different seismotectonic behaviour, which have been used in the hazard analysis as SZ's, and for which the characteristics of the seismicity (frequency versus magnitude and intensity relations) have been computed. A revision of the seismotectonic / seismogenic zonation (Figure 3b) has been recently done (Scandone, 1993, personal communication): it consists in slight changes of the SZ geometry (see the Alpine sector) and in the definition of some SZ's external to the Italian territory which can contribute to the hazard in Italy. In fact, for long-period ground motion parameters, say 1 Hz, distant large magnitude sites may be important out to several hundred kilometres. In Figure 4 the major seismicity (Io ≥ VI MCS) from the PFG catalogue (Figure 4a) and from the GNDT earthquake file (Figure 4b) is reported: a good agreement between seismicity and zonation can be seen.

For PGA the attenuation relation established by Sabetta and Pugliese (1987) was chosen: it is based on the data of 17 earthquakes of magnitude 4.6 to 6.8 recorded in Italy in the period 1976 - 1984. The relation refers to rocky sites and epicentral distances are considered. It is a single isotropic relation which is considered valid for all the SZ's with the probable exception of the volcanic zones. As the relation refers to two kinds of magnitude according to the size of the earthquake, the magnitude available in the catalogue has been eventually converted to the proper one by the Margottini et al. (1992) formulas for Italian earthquakes. The Sabetta and Pugliese (1987) relation has been extrapolated to values lower than 4.6 (lower limit of the data set used for assessing the relation) when needed. For macroseismic intensity the Grandori et al. (1987) relation, which is based only upon the size of the equi-intensity areas and describes also eccentric and elliptical forms (Drei et al., 1988), was chosen and its parameters were calculated (Petrini, 1993, personal communication) separately for 42 isoattenuation zones (Figure 5) using the available

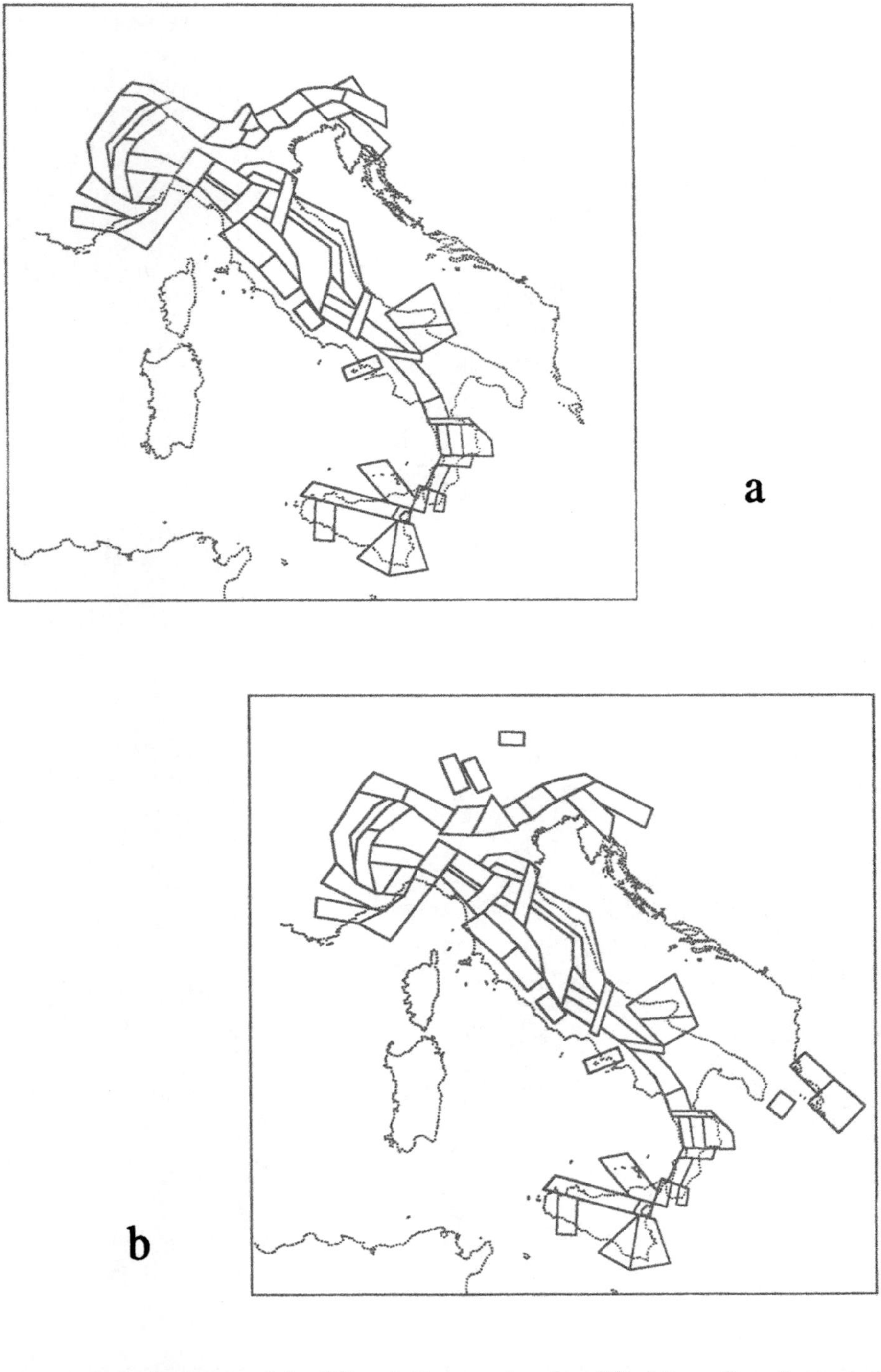

a

b

Figure 3. Location map of the SZ's: a) first version (modified from Scandone et al., 1991); b) revised version (Scandone, 1993, personal communication).

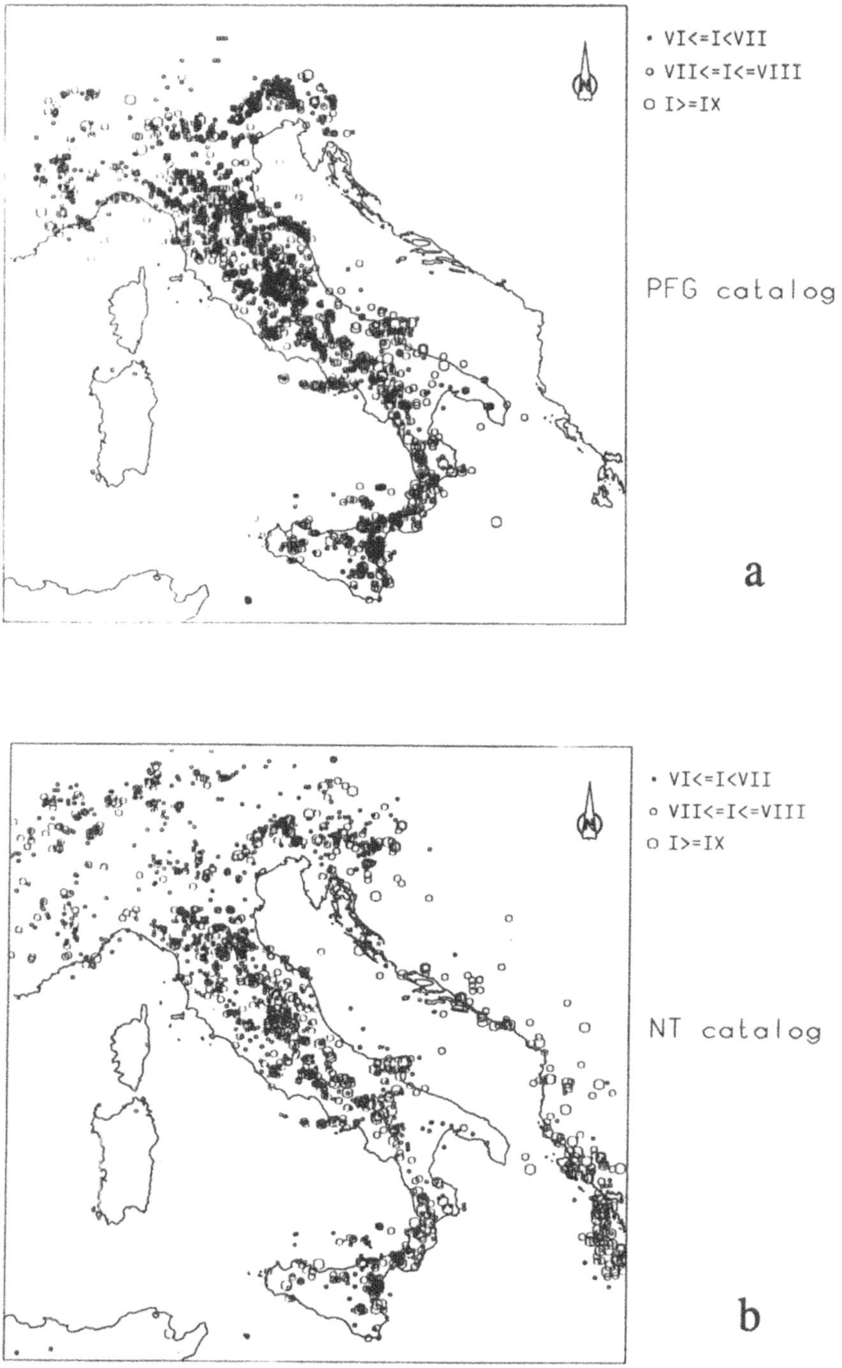

Figure 4. Epicentre map of the major earthquakes (Io ≥ VI MCS): a) taken from the PFG catalogue, b) from the GNDT earthquake file.

intensity maps (Postpischl, 1985a); the remaining background seismicity has been treated with average parameters.

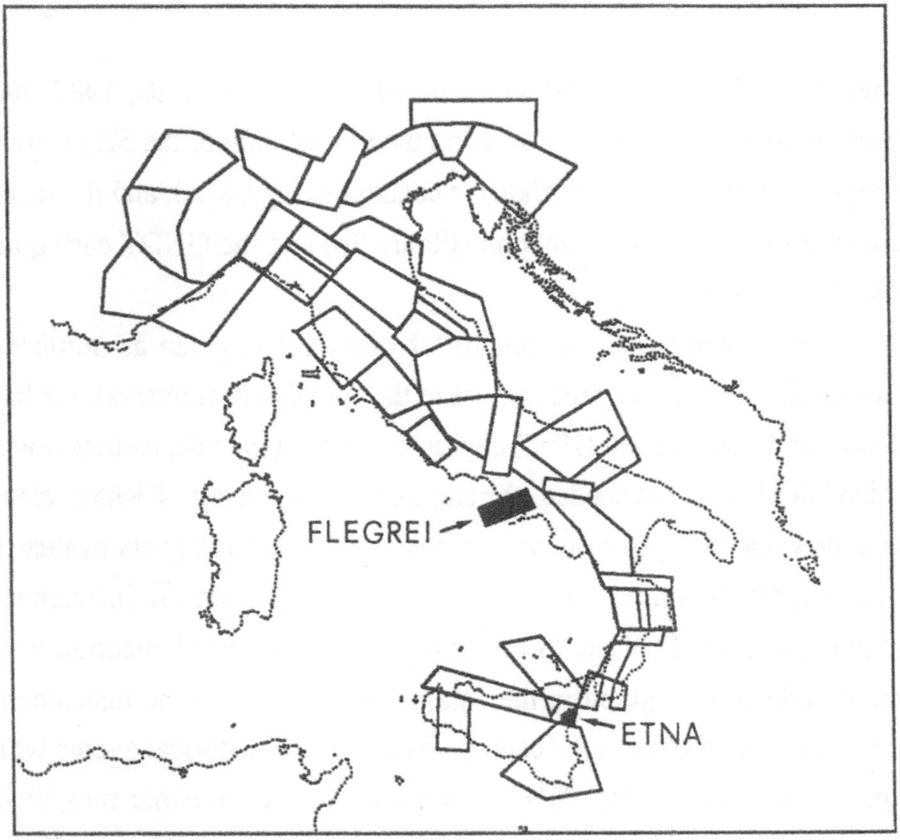

Figure 5. Location map of the intensity isoattenuation zones (Petrini, 1993, personal communication).

5. GNDT Probabilistic Hazard Assessment for the Seismic Code

Most of the elaborations expected by the GNDT project have been already done. Some basic results are described in the following sections: they have to be considered as preliminary as some comparisons are still missing.

5.1. THE CORNELL APPROACH

Two runs of the code SEISRISK III (Bender and Perkins, 1987) have been performed: the first considering the first version of the SZ geometry (Figure 3a) and the PFG earthquake catalogue (Figure 4a), and the second considering the revised zonation (Figure 3b) and the GNDT earthquake file (Figure 4b).

The characteristic seismicity of every SZ is given as number of earthquakes in each magnitude class of the completeness interval; the basic condition in the Cornell (1968) approach of an exponential distribution for seismicity is then not needed (the slope of the Gutenberg - Richter relation and the value of the maximum magnitude are consequently not evaluated). The magnitude values have been taken directly from the instrumental magnitude column of the PFG catalogue as no further indication of the magnitude type is given in the catalogue itself. When the instrumental magnitude is missing in the catalogue (i.e. for the historical events which are the majority in the PFG catalogue) the macroseismic magnitude, computed from the epicentral intensity by the Karnik (1969) formula, has been used. The Karnik (1969) formula does not surely hold for the volcanic areas (Campi Flegrei in the southern Apennines and Mount Etna in eastern Sicily, see Figure 5) where, having no specific relation available, a correction to the Karnik (1969) macroseismic magnitude has been applied simply subtracting 0.5. As only independent events have to be considered, aftershocks have been removed by a space-time filtering algorithm defined on the basis of the Gardner and Knopoff (1974) table. The removal of foreshocks and secondary events in swarm sequences has been manually checked for the major events (i.e. those with Io larger than

VIII MCS). The years when the seismological data collection improved have been pointed out on historical basis (Stucchi, 1993, personal communication), they are: 1000, 1224, 1458, 1691, 1782, 1899, 1915, 1936, and 1971. For each of these intervals the number of earthquakes normalised to 100 years for every magnitude class has been computed (see methodology in Peruzza and Slejko, 1993) considering the magnitude step 0.2. The choice of the preferable rate has not been always easy because of some relevant oscillations in the graph: a subjective opinion has been, therefore, necessary. The chosen values are marked with large squares in Figure 6: the choice of the large magnitude event rate is crucial, while that of medium - small events does not influence strongly the results. It implies that the magnitude estimate of strong historical quakes is very important. Considering the long time span of the earthquake catalogue used, the maximum observed magnitude has been considered representative of the maximum expected magnitude. Deep earthquakes ($h > 50$ km) have not been considered in the hazard assessment because they give low contribution to the estimates. A fixed uncertainty of 15 km has been accounted for the boundaries of the SZ's and the few earthquakes external to the identified SZ's have been referred to three large background SZ's. According to the specifics of the seismic Eurocode in preparation at present, the produced hazard maps refer to 100 and 500 year return periods. The first map (Figure 7) shows the mean value of the expected PGA with 100 year return period. The most hazardous regions are the eastern Alps (absolute maximum of the map 0.12 g), whole Apennines with the exception of limited areas in the southern sector, and part of the northeastern coast of Sicily. The map referred to 500 year return period (Figure 8) shows that the maximum PGA (which reaches 0.20 g) is expected to occur in the southern Apennines and Calabrian Arc. Slightly lower values are reached in central Apennines and the easternmost sector of the Alps displays values as high as 0.18 g. The hazard increases of

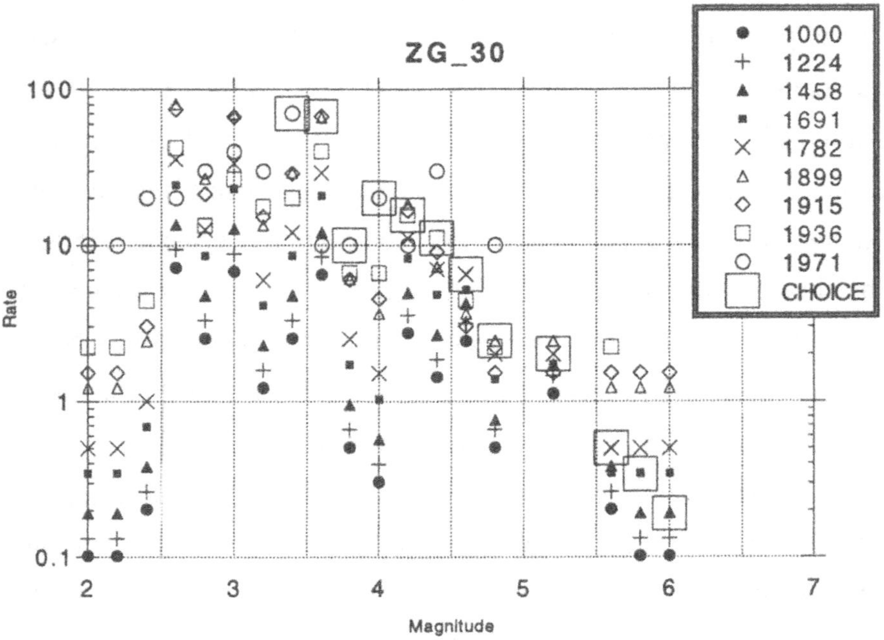

a

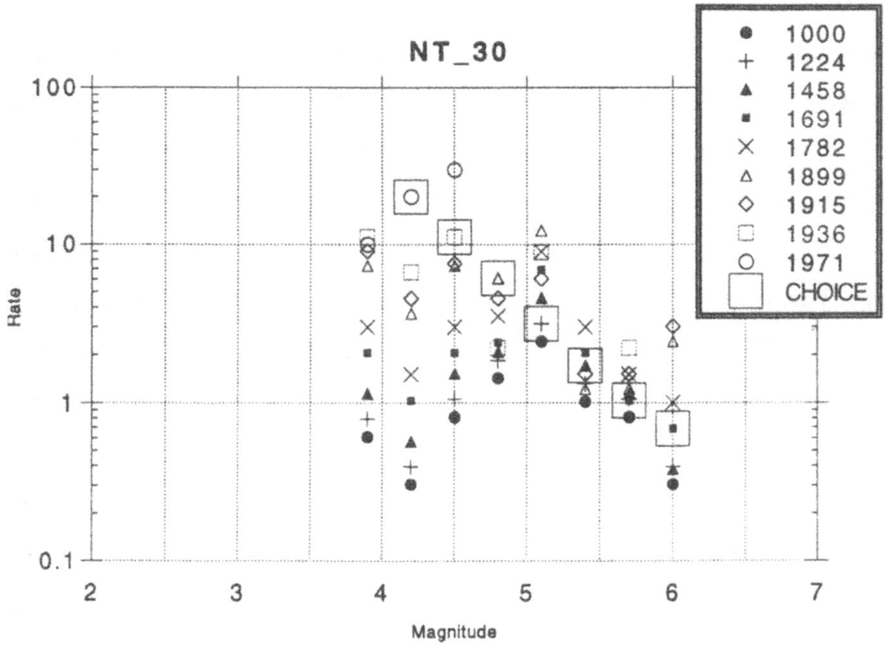

b

Figure 6. Seismicity rate (number of events in 100 years) for a SZ in central Apennines: a) from the PFG catalogue; b) from the GNDT earthquake file.

0.02 g in average when the standard deviation of the PGA attenuation relation (Sabetta and Pugliese, 1987) is taken into account (Figure 9) and values larger than 0.22 g are encountered in the southern Apennines and in a limited area of the eastern Alps. Almost the same features of the previous map are shown by the map obtained without weighting the SZ boundary

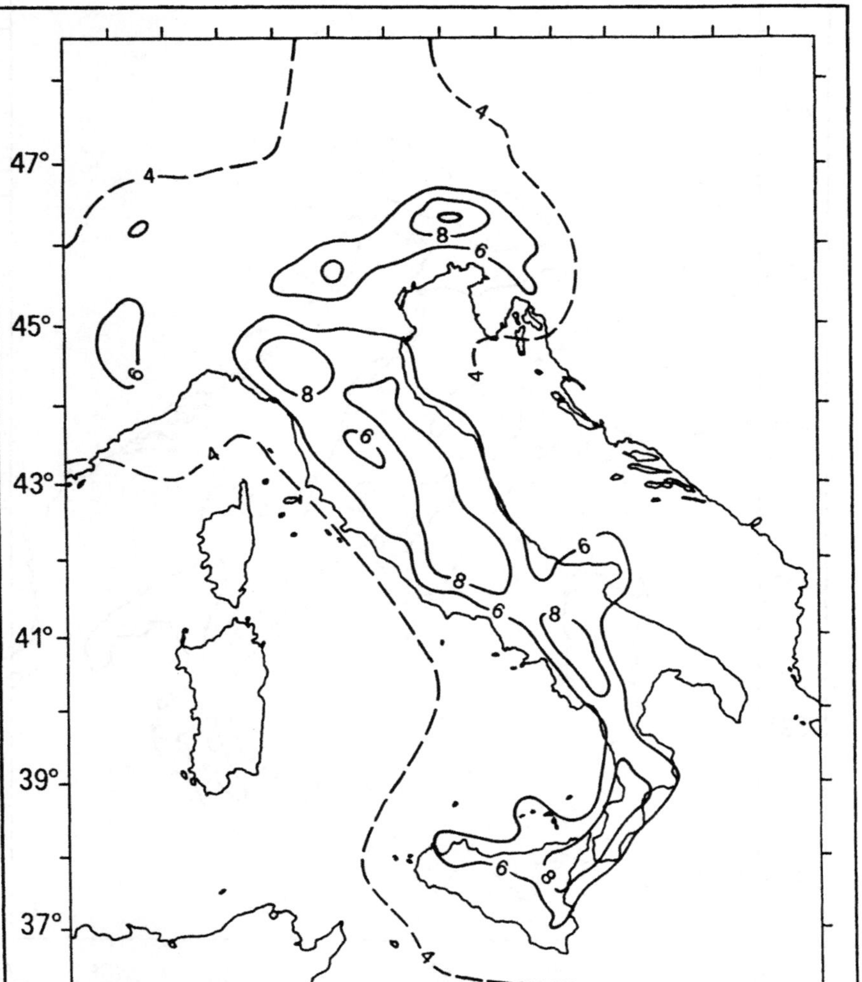

Figure 7. Cornell approach: horizontal PGA (in hundreds of g) with 100 year return period, soft boundaries for the SZ, background seismicity added, PFG catalogue used.

locations (Figure 10): the SZ geometry is here more evident and values slightly higher are reached. Specific analysis of the influence of the background SZ's in the results has revealed that larger values are reached only in aseismic areas, while no difference can be seen in areas influenced

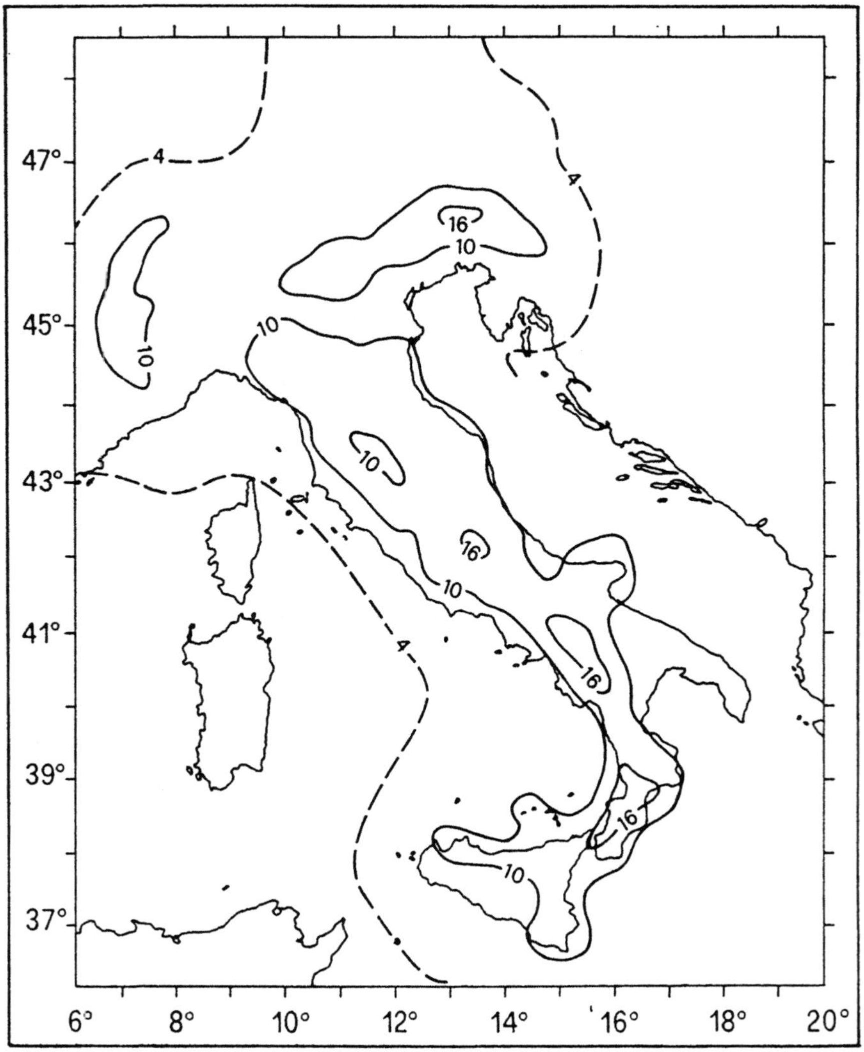

Figure 8. Cornell approach: horizontal PGA (in hundreds of g) with 500 year return period, soft boundaries for the SZ, background seismicity added, PFG catalogue used.

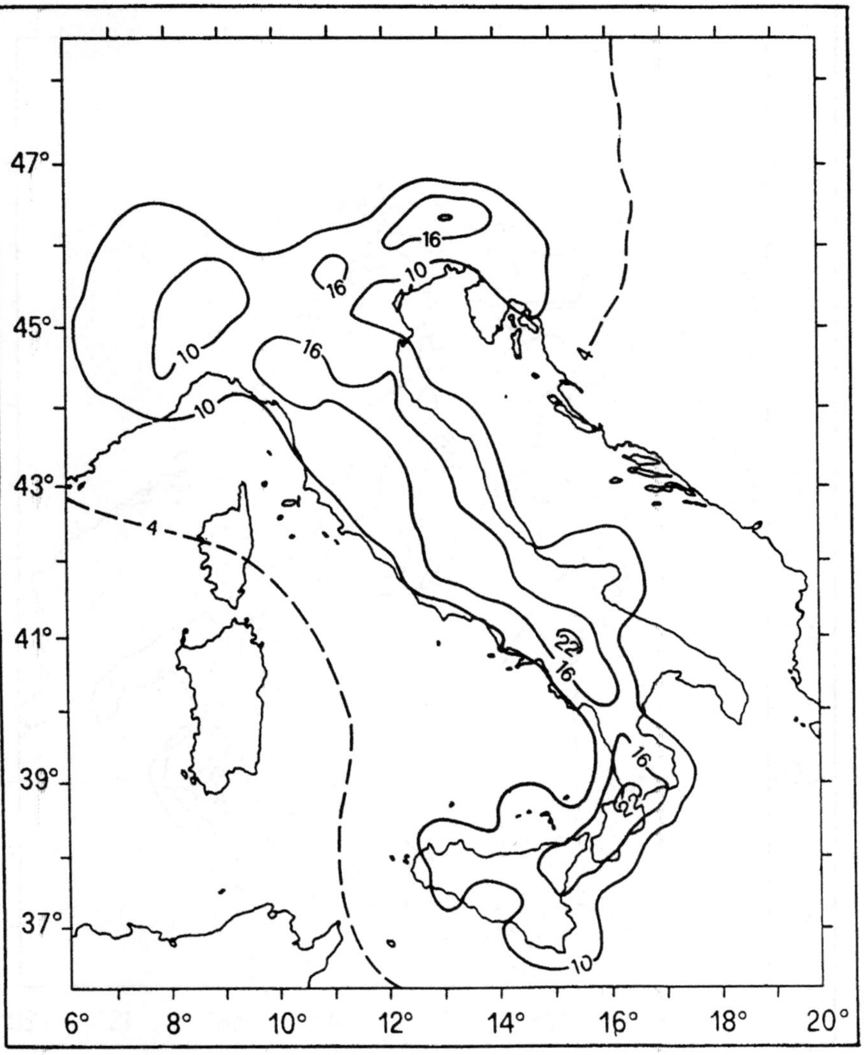

Figure 9. Cornell approach: horizontal PGA (in hundreds of g) with 500 year return period, soft boundaries for the SZ, background seismicity added, PFG catalogue used, standard deviation of the PGA attenuation relation considered.

by SZ seismicity. Another quantity recently considered for engineering purposes is the vertical PGA. The values expected with 100 year return period are shown in Figure 11; in this case the attenuation relation proposed by Pugliese and Sabetta (1993) for vertical PGA has been

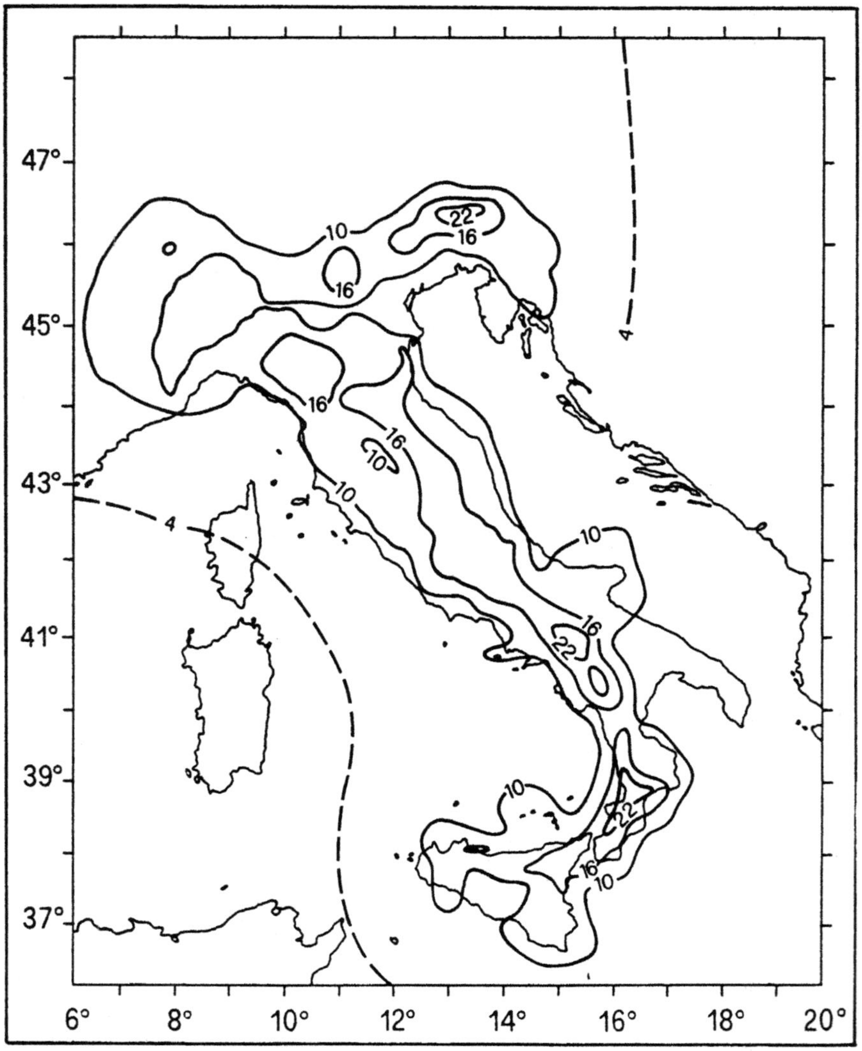

Figure 10. Cornell approach: horizontal PGA (in hundreds of g) with 500 year return period, hard boundaries for the SZ, background seismicity added, PFG catalogue used, standard deviation of the PGA attenuation relation considered.

considered and, again, the background seismicity and 15 km soft boundaries have been taken into account. Figure 11 shows that the maximum values, larger than 0.04 g, are encountered in the eastern Alps and in the southern Apennines and along the Calabrian Arc: they are about

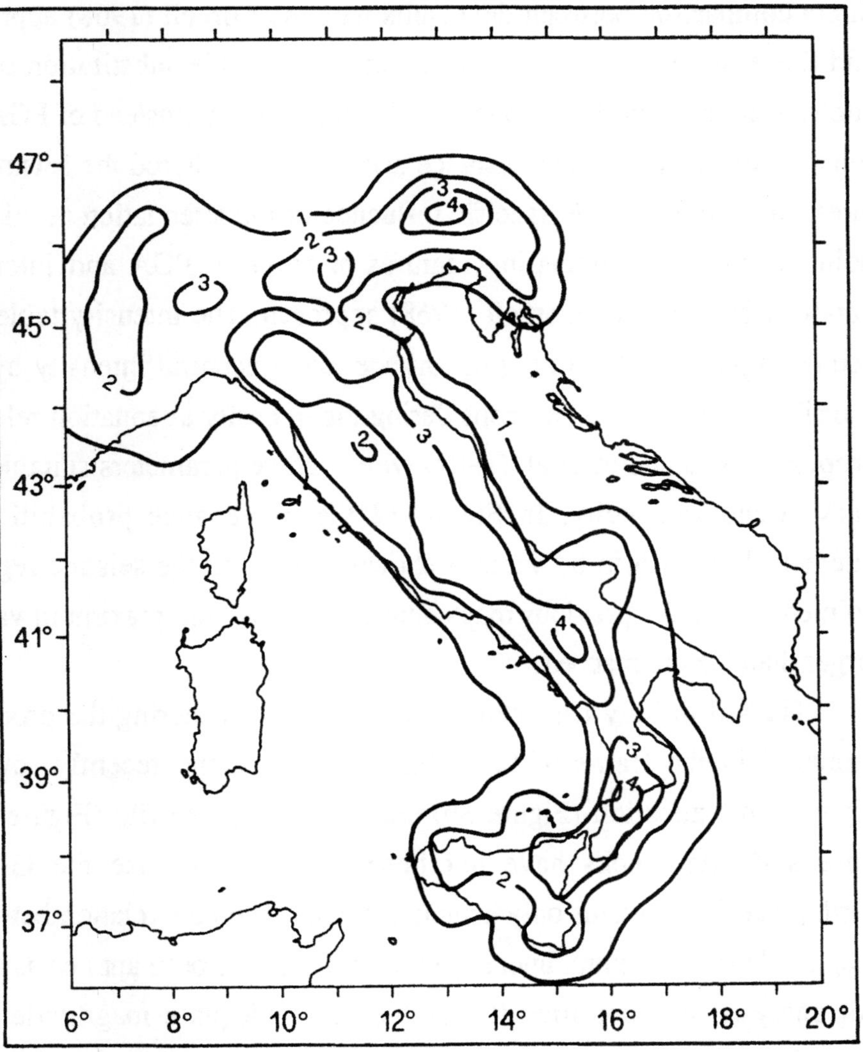

Figure 11. Cornell approach: vertical PGA (in hundreds of g) with 100 year return period, soft boundaries for the SZ, background seismicity added, PFG catalogue used.

half of those computed for horizontal PGA (see Figure 7) and the general pattern of the two figures is similar.

A further elaboration with SEISRISK III (Bender and Perkins, 1987) has been produced in terms of macroseismic intensity for having a direct comparison between the results with the Cornell (1968) approach and those by the mixed one. For doing that a simple substitution of the attenuation table has been done introducing intensity (instead of PGA) as function of magnitude and distance and leaving unaltered the seismicity rates; in such a way the specific influence of the attenuation relation is pointed out when comparing features of maps in PGA and intensity obtained by the same Cornell (1968) approach. The intensity table has been computed transforming magnitude into epicentral intensity by the Karnik (1969) relation and considering the intensity attenuation relation proposed by Grandori et al. (1987) with average parameters suitable for background seismicity. In Figure 12 the exceedance probability of intensity VII MCS in 50 years is shown: almost all the seismic regions pointed out by the previous maps appear also now and maximum values larger than 0.5 are reached.

Hazard in Italy has been computed also considering the products prepared in the frame of the GNDT activities: the recently revised seismogenic zonation (Figure 3b) and the earthquake file (Figure 4b). Fore and aftershocks have not been removed because the GNDT earthquake file contains only principal events. The new relation between magnitude and epicentral intensity (Rebez, 1993) has been applied as well. An analysis has identified 0.3 as the most adequate magnitude step considering the GNDT earthquake file for the quantification of the SZ seismicity rates (Figure 6b). Figures 13, 14, and 15 correspond to Figures 7, 8, and 12 when the first zonation and the PFG catalogue are replaced by the revised zonation and the GNDT earthquake file. The results are now larger in average than the previous ones and this depends mainly on the new magnitude versus epicentral intensity relation. In particular, the forecast related to 100 year return period (Figure 13) shows notable similarities with that obtained before (Figure 7); an average increase of 0.02 g and a wider seismic area in the eastern Alps can be seen, this last evidence derives from the better quantification of the Slovenian seismicity in the GNDT earthquake file. The results for 500 year return period (Figure 14) show again a general increase of about 0.02 g with respect to the forecast made with the PFG catalogue (Figure 8), some seismic areas at the border regions appear and some segments in the Apennines display

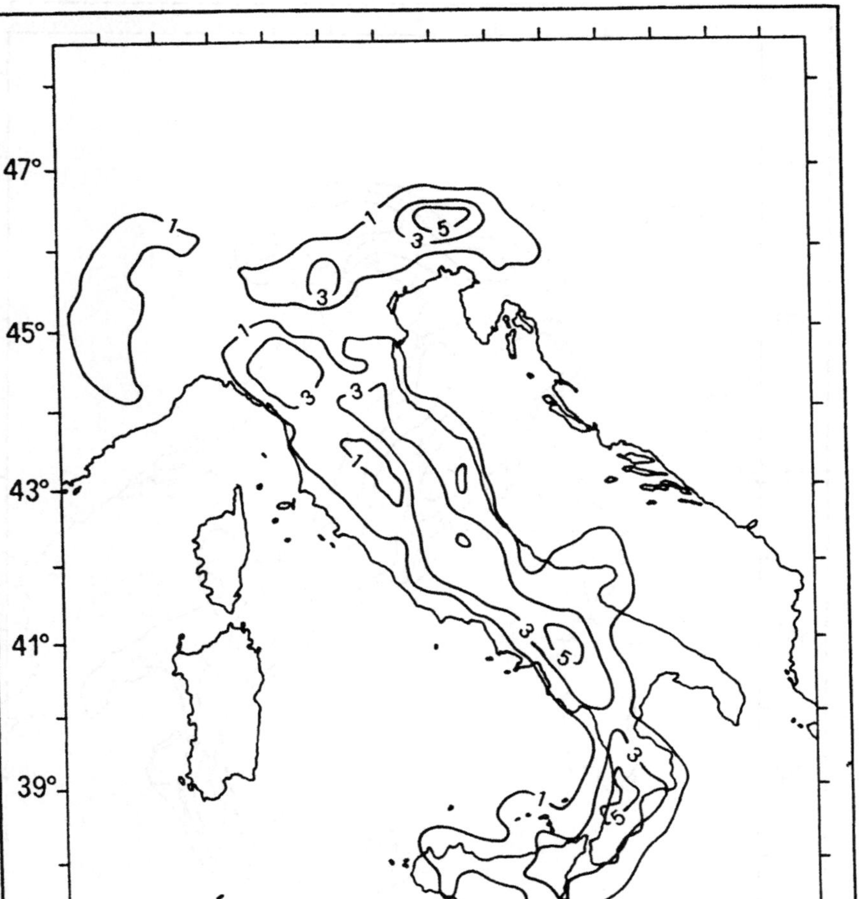

Figure 12. Cornell approach: exceedance probability (in tenths) of intensity VII MCS in 50 years, hard boundaries for the SZ, PFG catalogue used, average attenuation relation for intensity used.

a specific high hazard. Similarly, the forecast in terms of exceedance probability of intensity VII MCS in 50 years (Figure 15) shows higher values of about 10% with respect to those obtained using the PFG catalogue (Figure 12).

As general conclusion it can be said that similarity between the

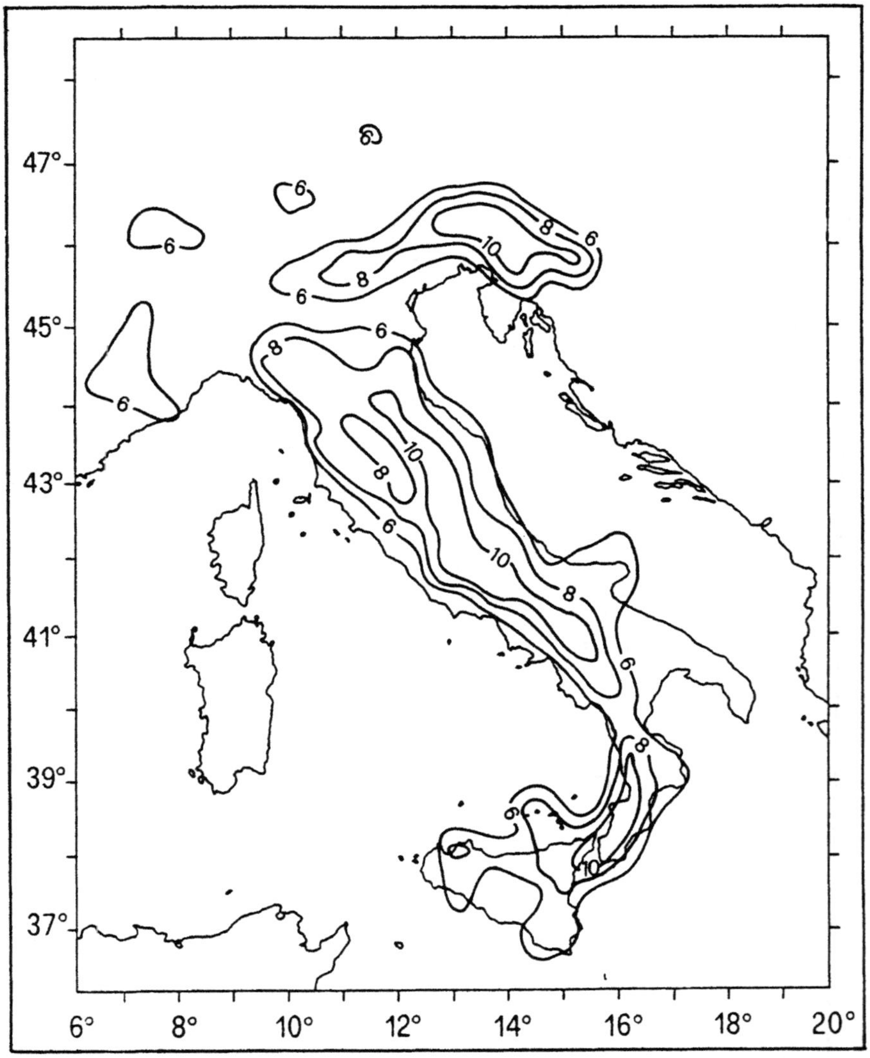

Figure 13. Cornell approach: horizontal PGA (in hundreds of g) with 100 year return period, soft boundaries for the SZ, background seismicity added, GNDT earthquake file used.

results obtained using the PFG and the GNDT catalogues has been found. With the new catalogue, slightly higher results have been obtained, and the eastern Alps display a larger seismic area which includes part of the Dinarides. In addition, the hazard in the western Alps is slightly shifted southwards.

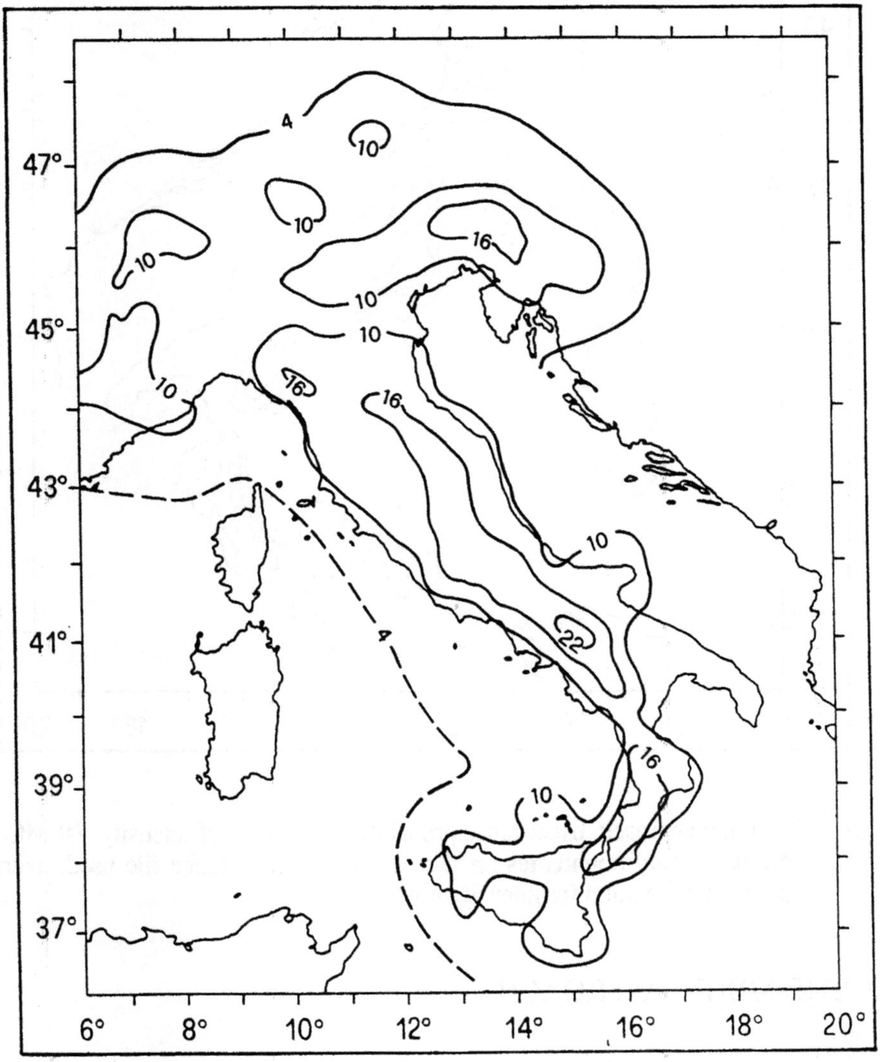

Figure 14. Cornell approach: horizontal PGA (in hundreds of g) with 500 year return period, soft boundaries for the SZ, background seismicity added, GNDT earthquake file used.

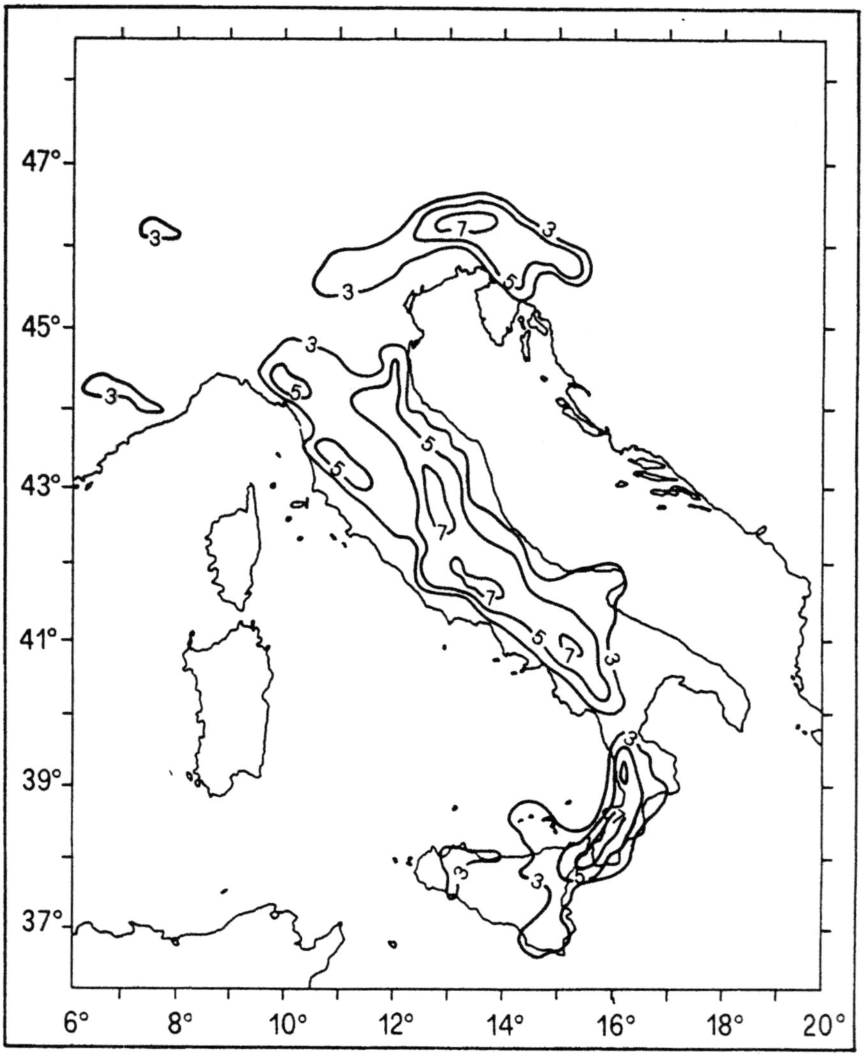

Figure 15. Cornell approach: exceedance probability (in tenths) of intensity VII MCS in 50 years, hard boundaries for the SZ, GNDT earthquake file used, average attenuation relation for intensity used.

5.2. THE MIXED APPROACH

In the application of the mixed method, the PFG catalogue contents have been filtered by a standard space - time window of 30 km and 6 months (3 before and 3 after the main event) regardless of the size of the main event.

Different time periods, according to the reputed completeness of the catalogue, are considered for counting the occurrences; in the present elaboration the quakes with Io > IX MCS since 1000, those with Io equal to VIII and IX MCS since 1680, and those with Io equal to VI and VII since 1780 have been considered. The uncertain intensities, represented in the catalogue as half degrees, have been truncated to the integer value, and the events without epicentral intensity, although with magnitude, have been discarded. The epicentral intensity distribution function, obtained by regression on the catalogue data, can assume one of the following forms: simple exponential, double exponential, or two parameter double exponential. The quality of the regression is remarkably different from one SZ to another, and is strongly conditioned by the quantity of the experimental data. In fact, sometimes the best interpolating function is clearly identified, some others, the poor data do not drive to any objective choice among the functions. In order to provide more robust statistics, a grouping of the SZ's with similar distribution function has been done and the distribution parameters have been recalculated on the global data set. Figure 16 shows the five distribution functions which represent all the SZ's after their grouping: each SZ is, therefore, characterised by its own seismicity rate and by the intensity distribution function of the group to which the SZ belongs. The site intensity distribution function is computed from all SZ contributions attenuated according to their distance from the study site. All quakes with epicentre in the same isoattenuation zone (Figure 5) are supposed to have the same propagation pattern which is modelled by the Grandori et al. (1987) attenuation relation with proper implementations (elliptical and eccentric shape, Drei et al., 1988). The recurrence times of events larger than the threshold intensity at the site have been computed from the epicentral parameters by the Grandori et al. (1987) attenuation relation cited before. As the present application refers to the Poisson condition only, the catalogue events have to be independent

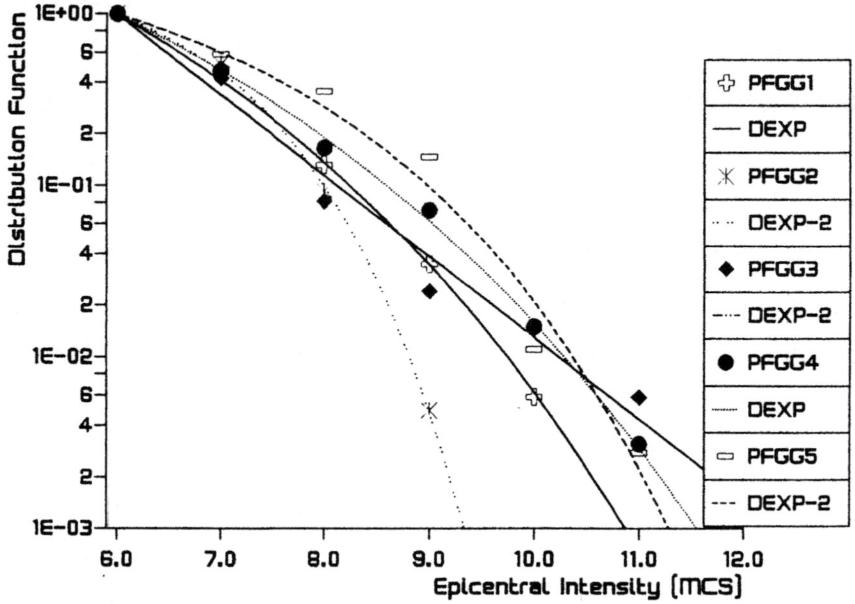

Figure 16. Summary graph of the Io distribution functions in the SZ's after grouping into five groups (PFGG1, ... , PFGG5): DEXP = double exponential, DEXP-2 = two parameter double exponential.

and the interoccurrence time distribution function is restricted to be of the negative exponential form. The earthquakes which occurred in the period 1680 - 1980 with Io > VI MCS have been considered in the calculation. The exceedance probability for a fixed intensity in a certain time is given by the combination of site intensity and interoccurence time distribution functions. A further calculation with a single attenuation relation for all the quakes regardless of their location (the parameters suitable for background seismicity have been chosen) has been performed for having a direct comparison with the results obtained according to the Cornell (1968) approach. Figures 17 and 18 show the exceedance probability of intensity VII MCS in 50 years obtained using different attenuation relations (Figure 17) and the average one (Figure 18). When comparing this latter map

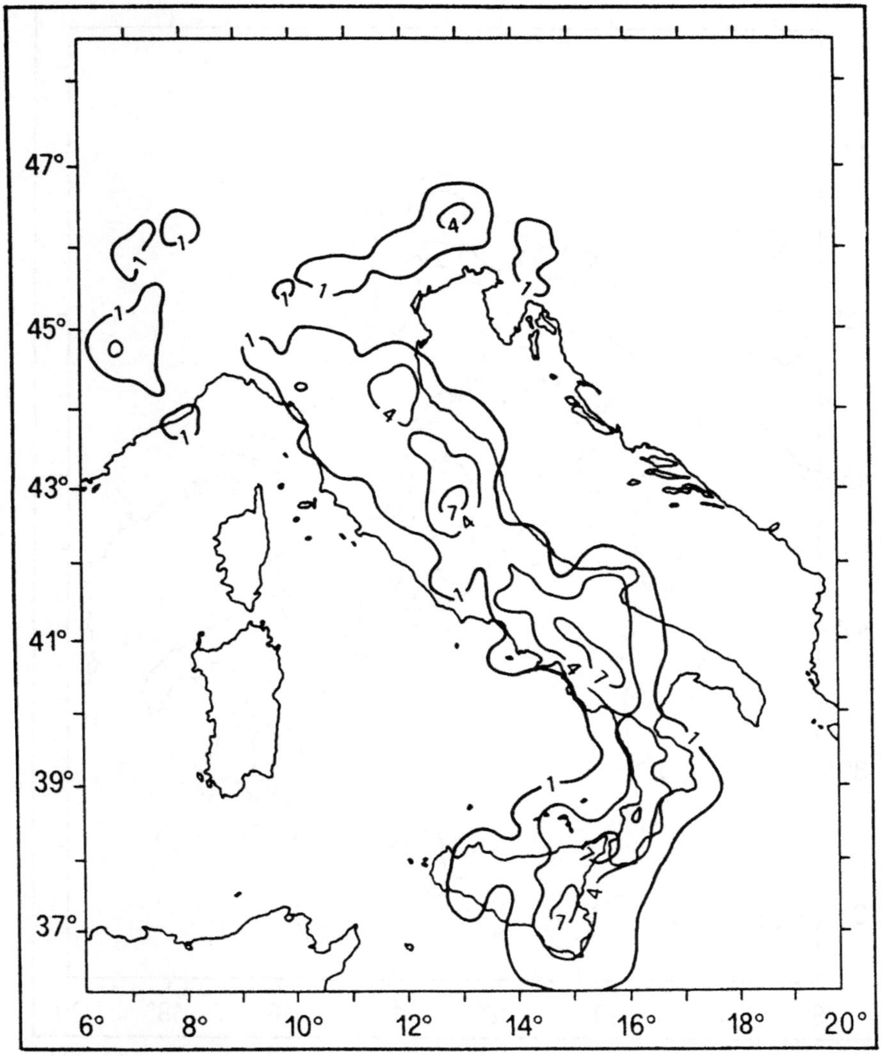

Figure 17. Mixed approach: exceedance probability (in tenths) of intensity VII MCS in 50 years, PFG catalogue used, different anisotropic attenuation relations for intensity used.

(Figure 18) with that obtained by the Cornell (1968) approach (Figure 12) it can be seen that the mixed approach gives higher hazard in eastern Sicily and central Apennines and lower hazard in central and eastern Alps. Similar features to those already described in the previous maps can be

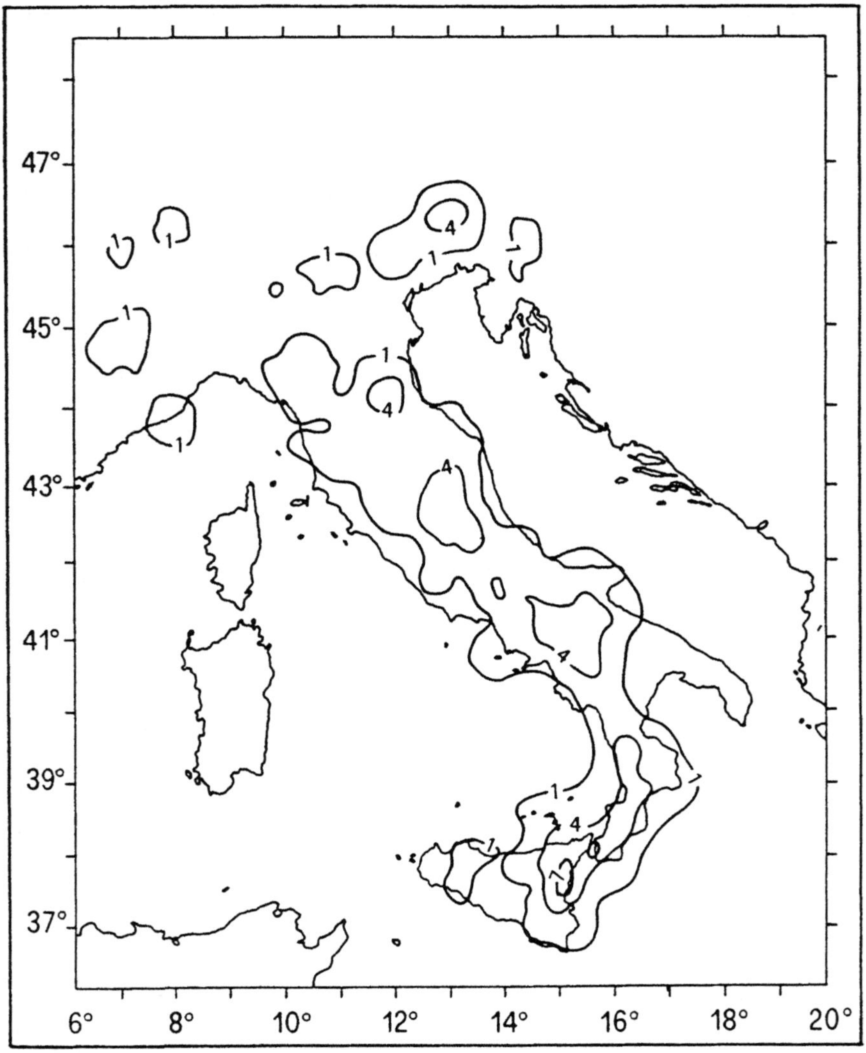

Figure 18. Mixed approach: exceedance probability (in tenths) of intensity VII MCS in 50 years, PFG catalogue used, average attenuation relation for intensity used.

seen also when different anisotropic relations are used (Figure 17), with the exception of some high hazard spots in central and southern Apennines which appear, as expected, when using different attenuation relations.

6. Conclusions

In the present work two probabilistic approaches of hazard assessment have been applied to the Italian territory for reviewing the present seismic zonation. The produced maps can be compared and, therefore, the consistency of the results can be estimated also considering the actual recorded values. The first consideration is that the results obtained using the PFG catalogue are lower than some recorded values (the maximum value is 0.20 g for the 500 year return period, Figure 8): in fact, considering only the last decades, PGA values as high as 0.61 g were recorded during the 1972 Ancona earthquake, as 0.36 g during the 1976 Friuli quake, as 0.34 g during the 1980 Irpinia event, and as 0.23 during the 1984 Val Comino one (Chiaruttini and Siro, 1981; Margottini et al., 1992). These differences can be explained by the fact that the values here presented are expected to be exceeded at about 10% probability in 50 years, and are referred to a standard soil (rock and deep alluvium); they do not take into any account, then, bad geologic and morphologic situations. In addition, the Sabetta and Pugliese (1987) relation attenuates very rapidly in comparison with other relations developed in other regions (see comparisons in Peruzza and Slejko, 1993), but it is in good agreement with the relation found for the cited Friuli earthquake (Chiaruttini and Siro, 1981). If the hazard results (Gruppo di Lavoro Scuotibilità, 1979), on which the zonation of the present seismic code is based, are considered, it can be seen that the maximum intensity with 500 year return period is IX MCS with very limited spots of intensity X MCS. It is not easy to transform this intensity value into a PGA value, as the available relations (Chiaruttini and Siro, 1981; Margottini et al., 1992) are not well constrained for high values, but a PGA between 0.13 g and 0.20 g remains roughly associated with intensity IX MCS. The agreement with the results here presented is, therefore, satisfactory. The map obtained using the revised seismogenic zonation and the GNDT earthquake file repeats in general what shown by the previous elaborations with a general average increase of 0.02 g. The absolute maximum is reached in Irpinia with 0.24 g (0.29 g if the standard deviation of the PGA attenuation relation is accounted for) when the map of maximum PGA with 500 year return period is considered (Figure 14).

Different considerations have to be done regarding the results obtained in terms of exceedance probability for intensity VII MCS in 50 years. The mixed approach is sensitive to the epicentral distribution and the short time period (300 years) used for the assessment of the interoccurrence times decreases the hazard of regions with relevant historical seismicity. The results obtained using differentiated anisotropic attenuation relations are quite similar to those obtained using the Cornell (1968) approach, while they are lower when the general average relation is used. This fact emphasises the importance of the correct quantification of the attenuation relations.

For evaluating the intensity results obtained, the map of the maximum pseudo-observed intensity has been computed. This map represents the local intensity as obtained from epicentral parameters and the Grandori et al. (1987) average intensity attenuation relation (considered previously for constructing Figure 12): it is not, therefore, an actual observed intensity map. Figure 19 shows the maximum pseudo-observed intensity since 1000 considering the data from the PFG catalogue. It can be seen that values larger than IX MCS should have been observed widely along the southern Apennines and the eastern coast of Sicily. Further limited high intensity spots appear in central Apennines, central and eastern Alps. Intensity larger than VII MCS can be seen all over the peninsula with the exception of the western Alps, the Po plain, limited parts of coastal central Italy, and central Sicily. A direct comparison between these values with those referred to the forecasts

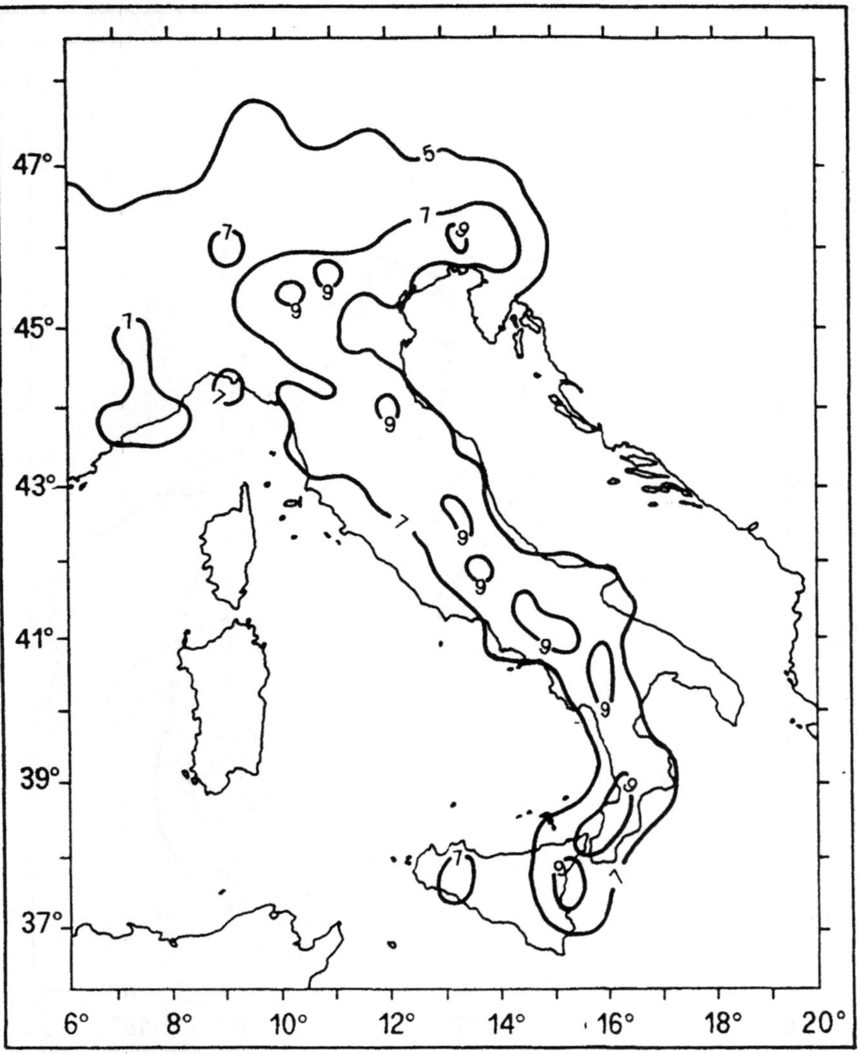

Figure 19. Map of maximum pseudo - observed intensity since 1000, PFG catalogue and average intensity attenuation relation used.

(Figures 12, 17, and 18) is not easy and only the shape of intensity isolines can be considered. The results with the Cornell approach (Figure 12) give more emphasis to the hazard of the eastern Alps and remarkably lower to eastern Sicily, while the mixed method (Figures 17 and 18) shows good similarities with the pseudo - observed values especially

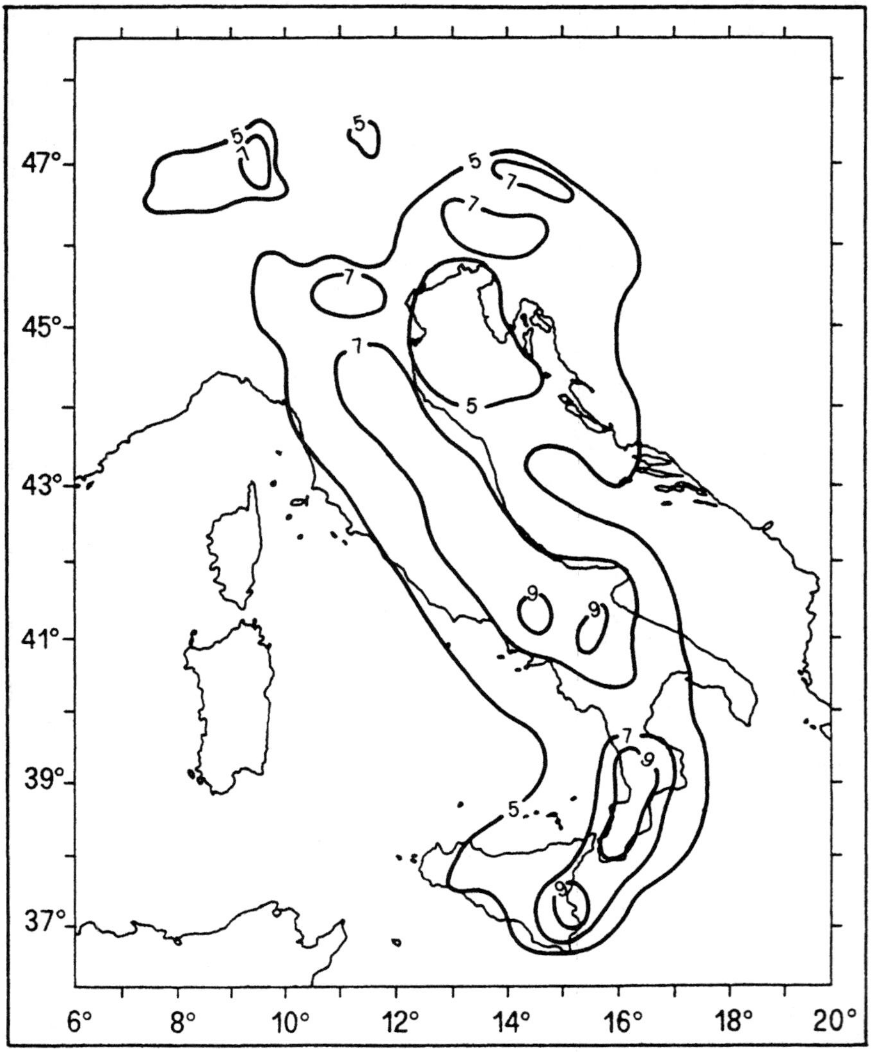

Figure 20. Map of maximum pseudo - observed intensity since 1000, GNDT earthquake file and average intensity attenuation relation used.

when individual attenuation relations are considered (Figure 17). Similarly, the map of maximum pseudo - observed intensity has been computed considering the GNDT earthquake file (Figure 20). The intensity values are now largely lower than those obtained using the PFG catalogue: the difference is slightly larger than one degree MCS. Some areas of intensity larger than IX MCS can be seen in the southern Apennines, along the Calabrian Arc and in eastern Sicily only. Consequently, areas of intensity VII MCS do not cover most of Italy but only surround those of higher intensity. The difference pointed out comparing the two maps (Figures 19 and 20) can be explained by the deep revision of the historical records performed, and justifies, together with the use of the new magnitude - intensity relation (Rebez, 1993), the differences obtained in the forecasts of Figures 12 and 18.

From the global analysis of all the results, the most hazardous areas are the whole Apenninic chain and the eastern Alps. The seismicity displays different characteristics in the different regions: for example the eastern Alps manifest seismicity with short return period and the importance of central Apennines is pointed out mainly by the mixed approach.

Acknowledgements. The present work has been performed in the framework of the activities of the Working Group "Seismic Hazard" of GNDT: the contribution of all the members is here acknowledged and thanked. The elaborations here presented were done by Laura Peruzza and Alessandro Rebez (both at OGS, Trieste). A special thank is due to Ezio Faccioli, Vincenzo Petrini (both at Politecnico, Milano) and Massimiliano Stucchi (CNR IRRS, Milano) who participated with continuity to the preparation of the input data and to the analysis of the results, to Fabio Sabetta (Servizio Sismico, Roma) and David Perkins (USGS, Denver), who commented usefully the manuscript. Many thanks are due to Peter Guidotti, who corrected the English manuscript, and to Corrado Fragiacomo, Manuela Sedmak and Alfonso Salemme (all at OGS, Trieste), who drew the figures.

References

Albini, P. and Barbano, M.S., (eds.) (1991) *GNDT Atti del Convegno 1990 vol. 2 Macrosismica*, Tip. Moderna, Bologna, 343 pp.

Algermissen, S.T., Perkins, D.M., Isherwood, W., Gordon, D., Reagor, G. and Howard, C. (1976) Seismic risk evaluation of the Balkan region, in *Proceedings of the Seminar on seismic zoning maps*, Unesco, Skopje, pp. 172-240.

Bender, B. and Perkins, D.M. (1987) Seisrisk III: a computer program for seimic hazard estimation, U.S. Geological Survey Bulletin 1772, 48 pp.

Chiaruttini, C. and Siro, L.(1981) The correlation of peak ground horizontal acceleration with magnitude, distance, and seismic intensity for Friuli and Ancona, Italy and the Alpide belt, *Bull. Seism. Soc. Am.* **71**, 1993-2009.

Cornell, C.A. (1968) Engineering seismic risk analysis, *Bull. Seism. Soc. Am.* **58**, 1583-1606.

Costa, G., Panza, G.F., Suhadolc, P. and Vaccari, F. (1992) Zoning of the Italian region with synthetic seismograms computed with known structural and source information, in *Proceedings 10th World Conf. Earth. Engineer.*, Balkema, Rotterdam, pp. 435 - 438.

Drei, A., Grandori, G., Molina, C., Mulas, M., Perotti, F. and Petrini, V. (1988) Individuazione delle principali aree sismogenetiche italiane e valutazione, per ciascuna area, della legge frequenza intensità epicentrale e della legge di attenuazione, Professional technical report, Politecnico Milano, 161 pp.

Faccioli, E. (1979) Engineering seismic risk analysis of the Friuli region, *Boll. Geof. Teor. Appl.* **21**, 173 - 190.

Gardner, J.K. and Knopoff, L. (1974) Is the sequence of earthquakes in southern California, with aftershocks removed, Poissonian?, *Boll. Seism. Soc. Am.* **64**, 1363 - 1367.

Giorgetti, F., Nieto, D. and Slejko, D. (1980) Seismic risk of the Friuli - Venezia Giulia region, *Publ. Inst. Geoph. Pol. Acad. Sc.* **A-9 (135)**, 149 - 169.

GNDT (1991) *Atti del Convegno 1990 Vol. 1 Zonazione e riclassificazione sismica*, Tip. Moderna, Bologna, 479 pp.

GNDT Macroseismic W G (coord. Stucchi M.) (1992) The "working catalogue" of the National Group for Protection against Earthquakes (GNDT) of Italy. Problems and results of a preliminary investigation of major earthquakes before 1690, in *23rd General Assembly of the ESC, Activity report 1990 - 1992 and Proceedings. Vol. II*, Geoph. Inst. Czech. AC. Sciences, Prague, pp. 114 - 117.

Grandori, G., Garavaglia, E. and Petrini, V. (1991) Analysis of some of the uncertainties in the statistical elaboration of historical data for local seismic hazard evaluation, in L. Esteva and S. E. Ruiz (eds.), *Proceedings 6th International Conference on Application of Statistics and Probability in Civil Engineering*, Impresiones Profesionales, Mexico City, pp. 512 - 520.

Grandori, G., Guagenti, E.and Petrini, V. (1984) On the Use of Renewal Processes in Seismic Hazard Analysis, in *Proc 8th World Conf, on Earthquake Engineering*, **1**, San Francisco, pp. 287 - 294.

Grandori, G., Perotti, F. and Tagliani, A. (1987) On the attenuation of macroseismic intensity with epicentral distance, in A.S. Cakmak (ed.), *Ground motion and engineering seismology*, Developments in Geotechnical Engineering 44, Elsevier, Amsterdam, pp. 581-594.

Gruppo di Lavoro Scuotibilita' (1979) Carte preliminari di scuotibilita' del territorio nazionale, C.N.R. P.F. Geodinamica pubbl. 227, ESA, Roma, 25 pp.

Karnik, V. (1969) *Seismicity of the European area II*. Reidel Publ. Co. Dordrecht, 364 pp..

Margottini, C., Ambraseys, N.N. e Screpanti, A. (1993) La magnitudo dei terremoti italiani del XX secolo, Technical Report, ENEA, Roma, 57 pp.

Margottini, C., Molin, D. and Serva, L. (1992) Intensity versus ground motion: a new approach using Italian data, *Engineering Geology* **33**, 45 - 58.

McGuire, R.K. (1976) Fortran computer program for seismic risk analysis, U.S.G.S. Open File Report 76-67, 92 pp.

Monachesi, G., Peruzza, L., Slejko, D., and Stucchi, M. (1994) Seismic hazard assessment using intensity point data, *Soil Dynamics and Earthquake Engineering* **13**, 219 - 226.

Peruzza, L. and Slejko, D. (1993) Comparison of different approaches to seismic hazard assessment, *Natural Hazards* **7**, 133 - 153.

Petrini, V. (ed.) (1980) Proposta di riclassificazione sismica del territorio nazionale, C.N.R. P.F. Geodinamica pubbl. 361, ESA, Roma, 83 pp.

Petrini, V. (1991) Il rischio sismico in Italia, in E. Boschi (a cura di), *Il rischio sismico*, Le Scienze quaderni 59, Le Scienze S.p.A., Milano, pp. 3 - 5.

Petrini, V., Bosi, C., Bigi, G., Eva, C., Grandori, G., Iaccarino, E., Luongo, G., Postpischl, D., Praturlon, A., Riuscetti, M., Scandone, P., Scarpa, R., Stucchi, M., Vezzani, L. (1981) Carta della pericolosità sismica d'Italia, C.N.R. P.F. Geodinamica, Pubbl. 442.

Petrini, V., Stucchi, M. and Tento, A. (1987) Classificazione sismica e revisione del catalogo sismico: analisi dell'influenza del catalogo sulla carta di pericolosità sismica ed alcune ipotesi per la revisione di quest'ultimo, in E. Boschi and M. Dragoni (a cura di), *Aree sismogenetiche e rischio sismico in Italia. I*, Galilei Ed., Lausanne, pp. 147 - 165.

Postpischl, D. (ed.) (1985a) Atlas of isoseismal maps of Italian earthquakes, C.N.R. P.F. Geodinamica, Graficoop, Bologna, 164 pp.

Postpischl, D. (ed) (1985b) Catalogo dei terremoti italiani dall'anno 1000 al 1980, C.N.R. P.F. Geodinamica, Graficoop, Bologna, 239 pp.

Pugliese, A. e Sabetta, F. (1993) Attenuazione della componente verticale del moto del terreno, in *Atti 6° Convegno Ingegneria Sismica in Italia*, **1**, pp. 75 - 84.

Rebez, A. (1993).Ricerca di una relazione fra intensità macrosismica e magnitudo per l'utilizzo del catalogo GNDT ai fini di pericolosità sismica, in *GNGTS 12° Convegno Nazionale Riassunti delle comunicazioni*, pp. 25.

Sabetta, F. and Pugliese, A. (1987) Attenuation of peak horizontal acceleration and velocity from Italian strong-motion records, *Bull. Seism. Soc. Am.* **77**, 1491-1513.

Scandone, P., Patacca, E., Meletti, C., Bellatalla, M., Perilli, N. e Santini, U. (1991) Struttura geologica, evoluzione cinematica e schema sismotettonico della penisola italiana, in GNDT (ed.), *Atti del Convegno 1990 Vol. 1 Zonazione e riclassificazione sismica*, Tip. Moderna, Bologna, pp. 119 - 133.

Servizio Sismico del Consiglio Superiore dei Lavori Pubblici (1986) Atlante della classificazione sismica nazionale, Ist.Poligrafico e Zecca dello Stato, Roma, 209 pp.

Slejko, D. (1992) Seismic hazard assessment for the Italian seismic code, in V. Schenk and D. Mayer-Rosa (eds.), *ESC 23rd General Assembly Activity report 1990 - 1992 and Proceedings*, Czech. Ac. Sciences, Praha, pp. 383 - 386.

Stucchi, M., Camassi, R. e Monachesi, G. (1993) Il catalogo di lavoro del GNDT,

CNR GNDT GdL "Macrosismica", Rapporto interno, GNDT, Milano, 89 pp.
Westaway, R. (1992) Seismic moment summation for historical earthquakes in Italy: tectonic implications, *J. Geoph. Res.* **97**, 15437 - 15464.

EARTHQUAKE HAZARD ASSESSMENT FOR THE CZECH REPUBLIC AND ADJACENT AREA

V. SCHENK, Z. SCHENKOVÁ, P. KOTTNAUER
Institute of Rock Structure and Mechanics, Academy of Sciences,
182 09 Praha 8 - Libeň, Czech Republic

1. Summary

This contribution presents the earthquake hazard calculations that have been realized for the territory of the Czech Republic and its adjacent area. Both a review of the calculations obtained and an outlook of possible new methodological aspects in earthquake hazard calculations are summarized. The hazard calculations have been made mainly for the region of 48.5°-51.2°N, 12°-19°E, which belongs to an area of relatively low seismicity, and which is affected from time to time by moderate and strong earthquakes from surrounding seismogenic zones. These zones are located either in the Alpine-Carpathian orogenic system or on the boundaries of obviously rigid geologic blocks (e.g. the Bohemian Massif), or they are connected with neotectonic Graben systems. In order to make the earthquake hazard calculations, we delineated a large number of seismogenic zones. Influences of several factors (e.g., classification of dependent and independent events, normalization of the earthquake regime per period of observation and/or the size of the area under study, and application of different attenuation laws, effects of near-surface sedimentary deposits) upon the final hazard values have been taken into account. The earthquake hazard for the swarm pleistoseismal area of Western Bohemia / Vogtland has also been evaluated. Application of artificial intelligence techniques suggests that, in the vicinity of a resistant block of the Bohemian Massif, an earthquake-prone zone has been identified. A discussion of further steps regarding the ways in which a large amount of input data can be applied, and of an assessment of their reliability in earthquake hazard calculations is presented.

2. History of Earthquake Hazard Assessment in the Czech Republic

The first attempt to assess earthquake hazard Czechoslovakia is reflected in the 1973 in Building Code ČSN 73 0036 (Seismic Loads of Buildings). Since 1978 studies have been conducted to define the earthquake hazard for proposed nuclear power plant sites. The assessments applied in calculations for an earthquake-resistant design were worked out using a seismostatistical approach that combined Cornell's algorithm (Cornell, 1968, 1971; Cornell and Merz, 1974) with program SERIAL (Schenková *et*

V. Schenk (ed.), Earthquake Hazard and Risk, 125–140.

al., 1981; Schenk *et al.*, 1981). In program SERIAL not only seismological aspects but also the geological structure of the whole area including all attenuation features are taken into account.

Program EQRISK (McGuire, 1976) was used for a rough estimate of earthquake hazard values in the whole Bohemian Massif (Schenk *et al.*, 1981, 1989a, 1993). This program applies a mean intensity attenuation with distance.

Earthquake hazard calculations performed for the pleistoseismal area of the Western Bohemia / Vogtland seismogenic region in which shallow earthquake swarms occur (Schenk *et al.*, 1989b) showed that the probabilistic approach mentioned above could be applied without any changes, too. Only the attenuation law must be eliminated from the calculation because of the pleistoseismal area. The earthquake hazard assessment seems to be realistic provided that strictly physically independent events were used. A statistical separation of dependent and independent events remains still a special problem.

In the eighties, a new method of determining maximum expected macroseismic effects was developed (Schenk 1984; Schenk and Mantlík, 1985). An algorithm allowing conversion of the observed isoseismal field into higher expected levels was used. The results obtained were taken into account in the compilation of the revised seismic zoning map of Czechoslovakia (Kárník *et al.*, 1988).

In the latter half of the eighties, expert system GEO-1.2 and other methods of artificial intelligence techniques were introduced to estimate maximum possible earthquakes of the Bohemian Massif regardless of their time occurrence. This approach allows active seismogenic zones and possible earthquake prone zones to be delineated (Schenk *et al.*, 1991).

At the beginning of the nineties, a calculation of earthquake hazard was included into a governmental project assessing natural and industrial environmental hazards of the North Bohemian Brown-Coal Basin (Schenk *et al.*, 1994a). In this case, the standard approach of the hazard calculation was extended by involving the influence of near-surface sedimentary deposits of natural (alluvium, loesses, gravel-sands, etc.) and man-made (mining refuse dumps, surface fills, etc.) origins (Schenk *et al.*, 1996a). Corrections were defined with respect to seismic wave transfer functions of a surface sedimentary layer.

Since instrumental data on strong motions are lacking for the territory of the Czech Republic, and the majority of event strength data are originally in terms of macroseismic intensities, this parameter was used in quantifying the level of the earthquake hazard. In addition, on the territory of the Czech Republic, users, designers, land and urban planners have to follow so far valid Building Code ČSN 73 0036, in which the macroseismic intensity is a leading parameter.

3. Input Data

Earthquake activity in Central Europe is determined by its geologic history and tectonics. The location of earthquake foci in the crust indicate weakened zones where the stress accumulated by recent tectonic processes is released. The situation in the Czech Republic and its surroundings is demonstrated in the latest version of the map of

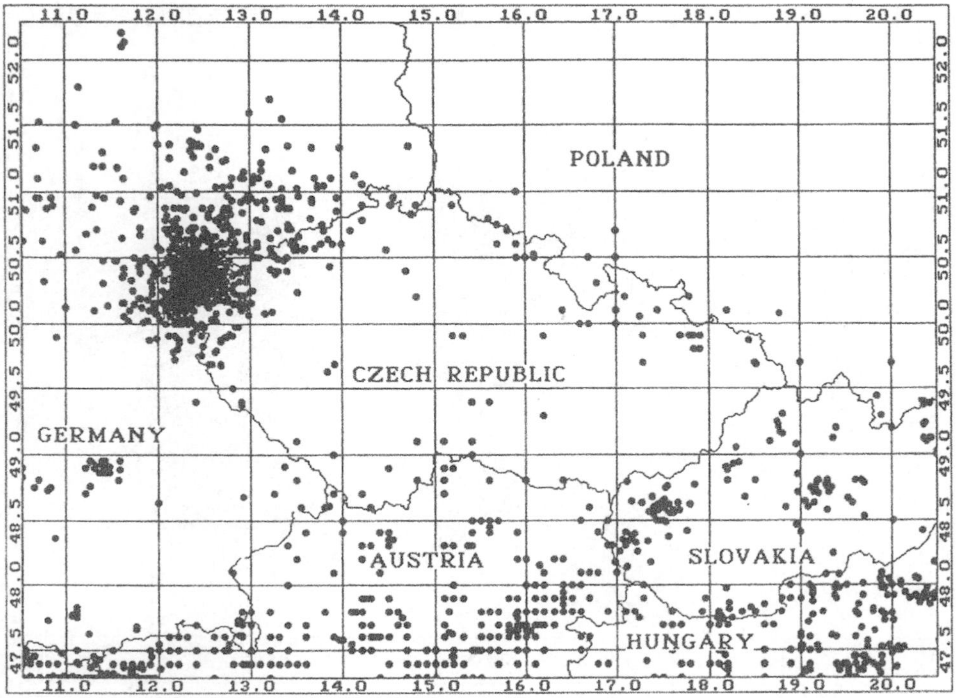

Figure 1. Earthquake epicentre map (Schenková *et al.*, 1993)

earthquake epicentres (Fig. 1, Schenková *et al.*, 1993). The general seismic pattern does not change substantially in a short time period, e.g. one or two decades. So far, the only way of obtaining new knowledge on focal zones is the observation of microearthquakes.

The earthquakes in the Bohemian Massif have taken place mainly in its peripheral areas. The macroseismic intensities of the strongest events were 7.5° MSK (6 March 1872, near Gera in Saxony), 7° MSK (10 January 1901, in the vicinity of Trutnov in the Sudeten, jointed with the Hronov-Poříčí fault). The intensities of the strongest shocks having occurred in earthquake swarms in the western part of the Bohemian Massif near the towns of Kraslice and Aš not exceed 7° MSK. The epicentres between towns Těšín and Cracow (I_0 = 7°-8°) belong to the seismically active belt stretching along the Peripienian lineament of the West Carpathians. As a rule, earthquake foci occur within the upper part of the Earth's crust, i.e. $h \leq 15$ km. Exceptions are in the Beskydy Mts ($h = 20\text{-}40$ km).

Macroseismic effects that have been observed in the Czech Republic, however, are not confined to the Bohemian Massif only. Figure 1 clearly shows that also earthquakes in the Pre-Alps, the Eastern and Southern Alps, the Franconian and the Swabian Jura, the Brabant Massif, Saxony, Southern Poland, Central and Western Slovakia and even in Slovenia and Vrancea affect the territory of the Czech Republic.

The computerized file of defined localities of the Czech Republic shaken by individual earthquakes with foci located on the territory of the Czech Republic, as well

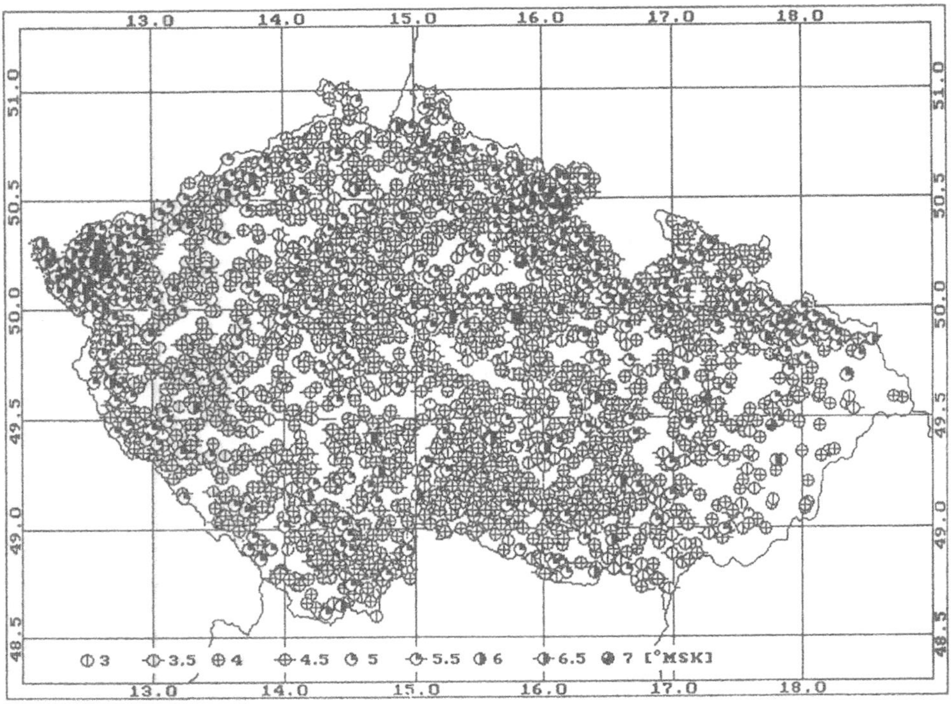

Figure 2. Maximum observed intensity point map (Schenková *et al.*, 1996)

as outside of it, was compiled and the geographical coordinates were added. It consists of about 4,000 macroseismic intensity data points and about 14,000 observations altogether. The resulting maximum observed intensity point map for the Czech Republic is given in Figure 2 (Schenková *et al.*, 1996). This computerized macroseismic file allows the maximum observed intensities with digitized geological information, e.g. active faults, Quaternary sediments, water-bearing discharges, etc., to be compared. Also local amplifications of macroseismic effects can be studied.

For evaluation and correction of the data for the most reliable calculation of earthquake hazard, a new statistical procedure of time-normalization of earthquake occurrences has been developed (Schenk, 1983) and successfully applied in input data homogenization.

4. Seismic zoning map and map of the maximum expected macroseismic intensity of the Czech Republic

The seismic zoning map is the simplest version of an earthquake hazard map. It summarizes and generalizes the extreme macroseismic effects observed during the historical period. Stationarity in the positions of epicentres and in the earthquake regime is assumed. This information invariably constitutes a part of building codes as in the case of the existing Building Code ČSN 73 0036. The text of ČSN contains a brief commentary on earthquakes in Czechoslovakia and a map of seismic zones, originally established by Dvořák, which delineates for the Czech Republic the regions of expected maximum intensity ranging from 6° to 7° MCS.

In the last two decades, new earthquake data appeared, additional isoseismal maps were constructed, earthquake parameters calculated and revised maps of earthquake epicentres, new maps of observed intensities as well as maps of expected maximum intensities were compiled.

Macroseismic fields were analyzed and theoretical models of isoseismals were constructed for the individual focal regions. These theoretical estimates of the pattern of isoseismals enable us to infer the likely pattern of the individual isoseismals of the earthquakes for which data are scanty or missing (Procházková and Brouček, 1986; Schenk and Mantlík, 1985).

One of the approaches, which was adopted to supplementing the observations in this way is a statistical method of conversion of observed macroseismic fields up to a higher intensity level by generalized circular isoseismals (Schenk, 1984). The transformation allows a greater number of observed macroseismic maps to be modified to the level of maximum observed intensity documented by at least one isoseismal map. It Further allows an estimate of a possible scatter of macroseismic effects caused by regional geological structures and of a general relation of isoseismal distances to the epicentre to be defined. For some of our regions (southern corner of the Bohemian Massif, the Lesser Danube basin and the adjacent region, the Eastern Alps and the Franconian Jura) the observed isoseismals were adjusted to higher epicentral intensities

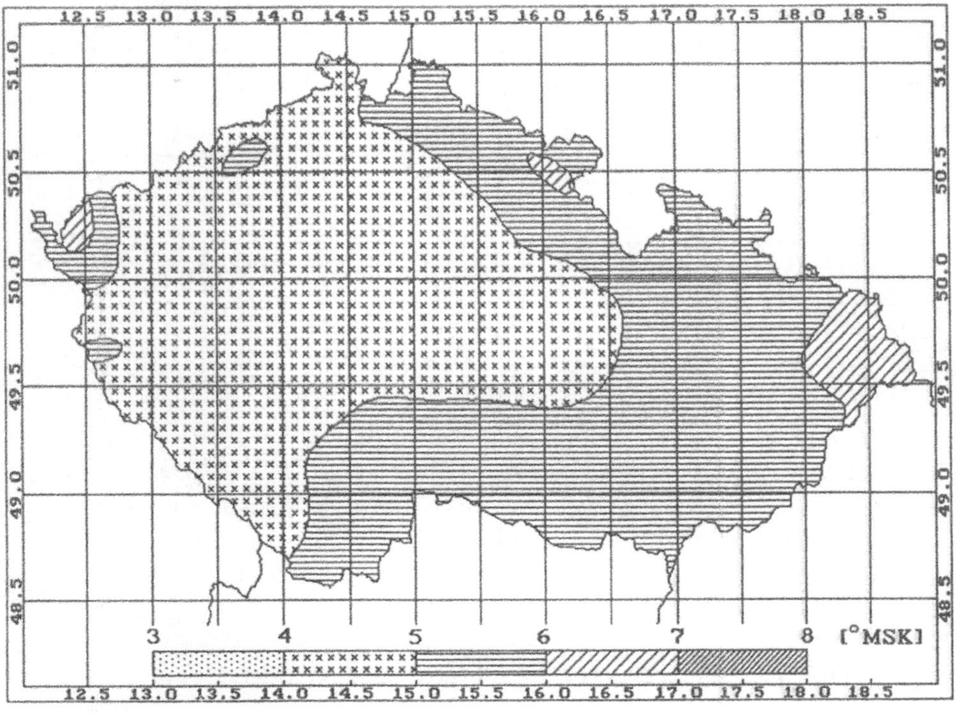

Figure 3. Seismic zoning map (Kárník *et al.*, 1988)

(by ½° to 1°). In this simple way, the map of maximum expected macroseismic intensity in Czechoslovakia (Fig. 3) was compiled (Schenk and Mantlík, 1985).

On the basis of the material mentioned above, a new version of the seismic zoning map of Czechoslovakia was constructed (Kárník *et al.*, 1988), primarily designed for revised Building Code ČSN 73 0036. The part of this seismic zoning map for the Czech Republic is given in Figure 4. The resulting isolines form an envelope of observed as well as of theoretical zones in the recent maps of the authors (Procházková and Brouček, 1981, 1986; Kárník *et al.*, 1984; Schenk and Mantlík, 1985) relating to the topics. In the Bohemian Massif the zones of intensity 7° correspond to observations. The same statement is valid for isolated zones of intensity 6° in Western Bohemia. However, a more extent zone of intensity 6° in south Bohemia and Moravia was estimated by assumed seismic shaking from the largest historical events in the Eastern Alps or in NE Moravia. The theoretical models supply the scattered observations of intensity 6° in that area and this approach results in the map corresponding to the worst ground conditions.

The zoning map in Figure 4 corresponds to observations made over an interval of five to six hundred years up to the higher intensities of 7°. Due to the unknown recurrence rate of the extreme events, it is not feasible to assess the probabilities of their occurrence in the map. It must be emphasized that, the present zoning map displays intensities closer to an upper threshold than to certain average ground conditions. As compared to the map compiled by Dvořák, which is attached to the existing Building Code ČSN 73 0036, the modifications were made in NE Moravia, in

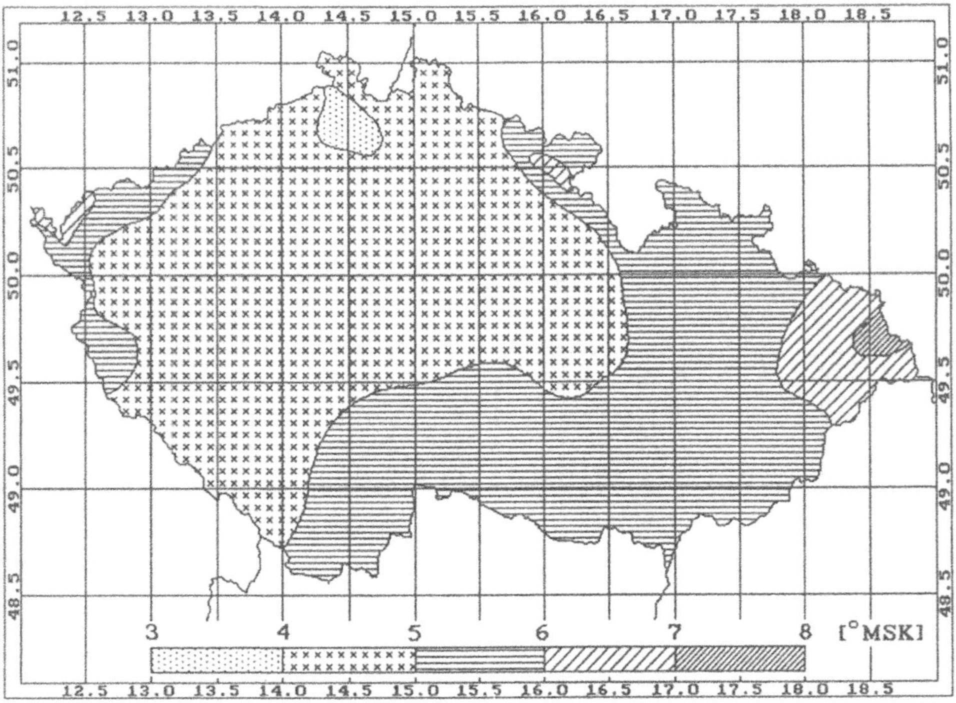

Figure 4. Map of the maximum expected macroseismic intensity (Schenk and Mantlík, 1985)

the area of Cheb and Sokolov (Western Bohemia) and in the region of Most-Duchcov (NE Bohemia). Unfortunately, the revision of Building Code 73 0036 is still under processing.

5. Probabilistic estimation of earthquake hazard

In general, the probabilistic calculation of earthquake hazard follows the scheme:
(1) delineation of earthquake source regions in terms of their boundaries, level of activity and the upper intensity threshold,
(2) determination of the macroseismic intensity changes with distance, magnitude, focal depth and ground conditions, particularly in the direction from the source region to the site,
(3) application of a theoretical model to the calculation of the seismic hazard at the site under study.

In order to obtain the most reliable earthquake hazard assessments, great attention must be given to the preparation of input data, especially to those which extrapolate and thus predict future values.

Maps of epicentres serve as a basis for delineation of earthquake source zones for which their regional geological structures, positions of geophysical anomalies and information on recent crustal movements associated with the origin of earthquakes have to be taken into account. Schemes of the source regions were compiled for Central Europe using ways from a simple separation of large seismotectonic units up to contouring isolated zones or clusters of epicentres. Therefore, there are several schemes of earthquake source regions within the radius of 300-400 km round the Czech Republic constructed for different purposes (e.g. Schenková *et al.*, 1981; Kárník *et al.*, 1984; Schenk *et al.*, 1989a). It is a necessary assumption that each earthquake source region should contain a sufficient number of shocks in order to determine its seismogenic regime statistically and to express it by the cumulative annual intensity-frequency relation (Schenk, 1983).

Threshold values of M_{max} or I_{max} for individual source regions in Central Europe often correspond to the maximum values observed in each region during a period of approximately 500 years. Statistical estimations of threshold values of I_{max} or M_{max} (the truncated Gutenberg-Richter relation, the theory of the extreme values (III distribution, Gumbel, 1958; Schenk *et al.*, 1989a; Schenk and Kottnauer, 1991)) are frequent. They are effective, however, only in the case of numerous and homogeneous sets of observations. As a rule, these procedures as a rule do not take into account the fluctuation of earthquake activity in the future or activation of new foci.

The value of M_{max} or I_{max} can be determined on the basis of relations between the magnitude of the earthquake and fault dimensions (Schenk and Schenková, 1983) and between a magnitude and recent crustal movement data (Schenk *et al.*, 1982). For an approximate estimation of the earthquake magnitude Schenk and Schenková (1983) derived for the Bohemian Massif the relation

$$M = 1.85 \cdot \log L + 2.0 \,, \tag{1}$$

where L is the length of the seismoactive part of the fault in km, which depends entirely on the reliability of determining the active fault length.

For all Central European seismogenic zones it applies that a relatively low amount of seismic information does not allow a reliable prediction and determination of their earthquake activity to be provided. Therefore, the seismic data analysis must be complemented with an analysis of all available information associated with the earthquake origin. It includes, for example, data on geological structures of the Earth's crust, on fault mobility, on geophysical fields, etc.

In most cases, the prediction of earthquake activity assumes the application of a certain type of extrapolation, i.e. in principle, adoption of the results which were obtained by studying a partial solution of any process and then applied to the whole process. It is evident that the successful geologico-geophysical prognosis has to be based on the prognostic features of physical characteristics of geologic media with respect to their regional and/or local structure. Such an approach represents a rather complicated task of finding mathematical relations or statistical dependences of different data types under multidimensional conditions. The above mentioned solution can only be obtained by applications of computer systems and artificial intelligence techniques (Schenk *et al.*, 1992).

The level of earthquake hazard for a particular site depends much on the macroseismic intensity (seismic vibrations) attenuation with distance from all surrounding potential source regions. In general, the decrease depends on the depth of the focus, on the earthquake mechanism and on geological conditions along the path. For the application of computer program SERIAL individual attenuation curves are approximated by two linear branches. The first branch corresponds to the pleistoseismal area of radius r_0 within which a very small or no intensity decrease occurs. The other branch reflects the simplified attenuation process and can be expressed by a simple formula (Schenková *et al.*, 1981)

$$\Delta I = I - I_0 = c_1 - c_2 \ln(r + r_0) , \tag{2}$$

where I_0 [°MSK] is the epicentral intensity and I [°MSK] is the intensity at distance r [km]. Program EQRISK adopts an average attenuation curve (Schenk *et al.*, 1981; Schenk *et al.*, 1989a). After a critical evaluation of available macroseismic values (mostly Procházková and Kárník, 1980), the attenuation dependence for the Bohemian Massif was found in the form

$$\text{for } r \le 6.9 \text{ km:} \quad \Delta I = 0 , \tag{3a}$$

$$\text{for } r > 6.9 \text{ km:} \quad \Delta I = 3.512 - 1.479 \cdot \ln(r + 3.844) . \tag{3b}$$

In principle, both seismostatistical approaches applied in the Czech Republic (program EQRISK and its modified version - program SERIAL) are based on the algorithms proposed by Cornell (1968, 1971) and Cornell and Merz (1974). The basic assumptions of these algorithms are the validity of the Poisson distribution for the occurrence of earthquakes with time, and the exponential epicentral intensity-frequency relation. In order to estimate the seismic hazard at any point within the area under investigation, we have to consider all possible earthquake effects from the surrounding source regions if they exceed a certain intensity level at this point.

The probabilistic earthquake hazard for the Czech Republic was recalculated several times during the last fifteen years (Schenk *et al.*, 1981, 1989a; Schenk 1993). Each new issued version improved the previous one owing to the implementation of additional and corrected input data or new methodological aspects of calculation. In this paper the results of the latest two versions are presented.

5.1. EARTHQUAKE HAZARD MAP IN THE CZECH REPUBLIC - VERSION 1993

In 1993 the earthquake hazard calculations for the area of the Czech Republic (Schenk *et al.*, 1989a) were modified by introducing a new delineation of source zones in the north-eastern and the eastern parts of the studied area. At the moment, altogether more than 70 zones have been delineated which could seismically affect the territory of the Czech Republic. The probabilistic hazard calculation was realized by programme SERIAL, especially by its modification which calculates hazard values over the whole defined area with one attenuation relationship (3). Figures 5a and 5b present maps of the earthquake hazard for return periods 500 and 2000 years which reflect a regional pattern of earthquake hazard only (Schenk, 1993).

Comparing these maps with the previous calculations (Schenk *et al.*, 1989a) we can observe negligible differences for small return periods (100 years). For greater return periods (e.g. 1.000 years) some tendencies to decreasing the hazard values can be found in some areas of the Czech Republic, and for rather great extrapolations to the future (return periods of 10.000 years) one can see pronounced tendencies towards higher hazard values, especially in the north-eastern part of the Czech Republic. It can be explained by a reassessment of the input data applied for this part of the Republic

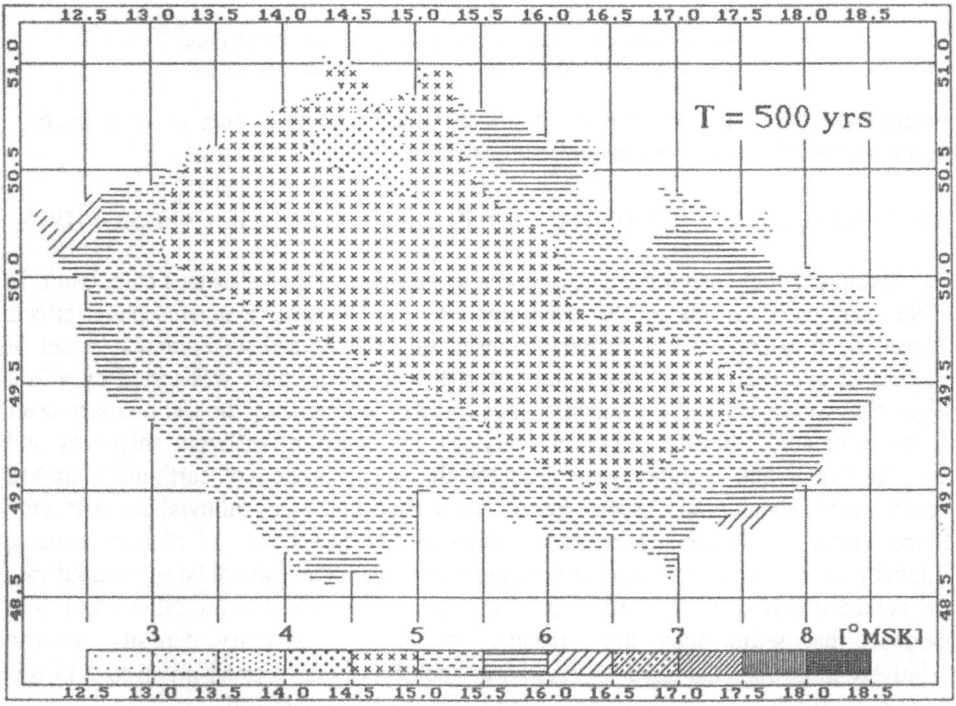

Figure 5a. Earthquake hazard map for the return period of 500 years without expected near-surface geology effects (Schenk *et al.*, 1993)

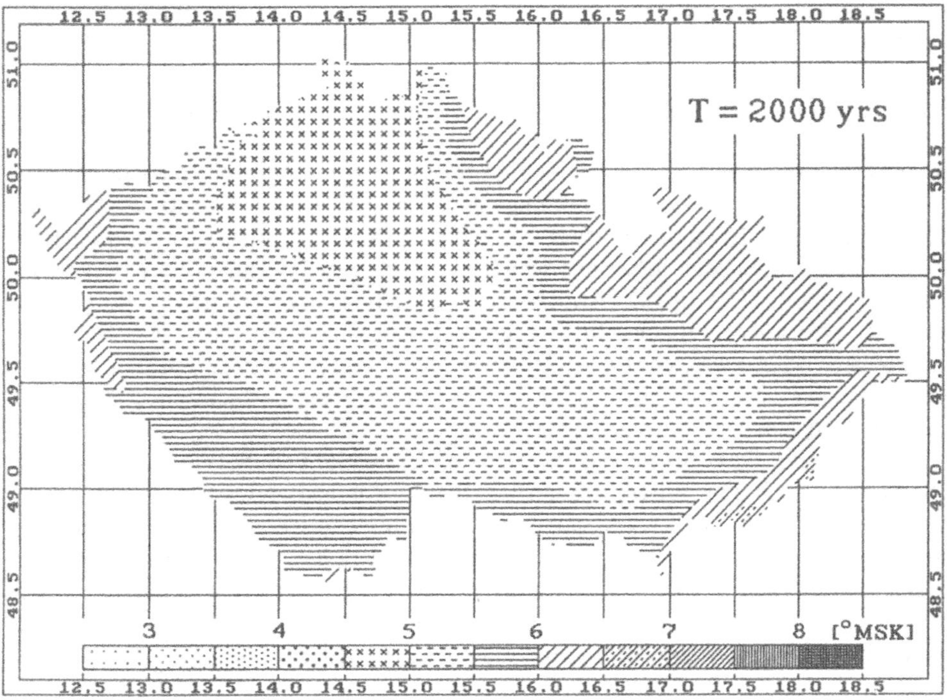

Figure 5b. Earthquake hazard map for the return period of 2000 years without expected near-surface geology effects (Schenk *et al.*, 1993)

where previously expected seismic regimes of individual source zones were analyzed and reestimeted with respect to new results obtained in recent years.

5.2. EARTHQUAKE HAZARD MAP IN THE CZECH REPUBLIC - VERSION 1995

As mentioned above, simultaneously to the standard earthquake hazard calculation in 1993 there were carried out also investigations how to involve macroseismic effects caused by near-surface sedimentary deposits into the hazard assessments. A set of models of one sedimentary layer situated at the rock basement was tested by a 1-D algorithm which allowed the seismic wave transfer functions of the sedimentary layer to be calculated. Transfer functions obtained for layers of different thickness and velocities of P and S waves were used for the determination of earthquake hazard corrections (Schenk *et al.*, 1996a). For example, for dumps, fills, alluvial deposits, etc., it was found that differences in hazard values could attain 1° to 1½° of macroseismic intensity. It is clear that such local geologic inhomogeneities cannot be introduced into the standard earthquake hazard map for the Czech Republic. On the other hand, it is evident, that sedimentary units (basins, foredeeps, Quaternary deposits) with a relatively large regional cover of crystalline rock blocks have to be involved into this hazard calculation if a more realistic hazard assessment is obtained.

For that purpose, the sedimentary units of Cretaceous, Paleogene, Neogene and Quaternary periods were computerized (Fig. 6). Having a regional distribution of these

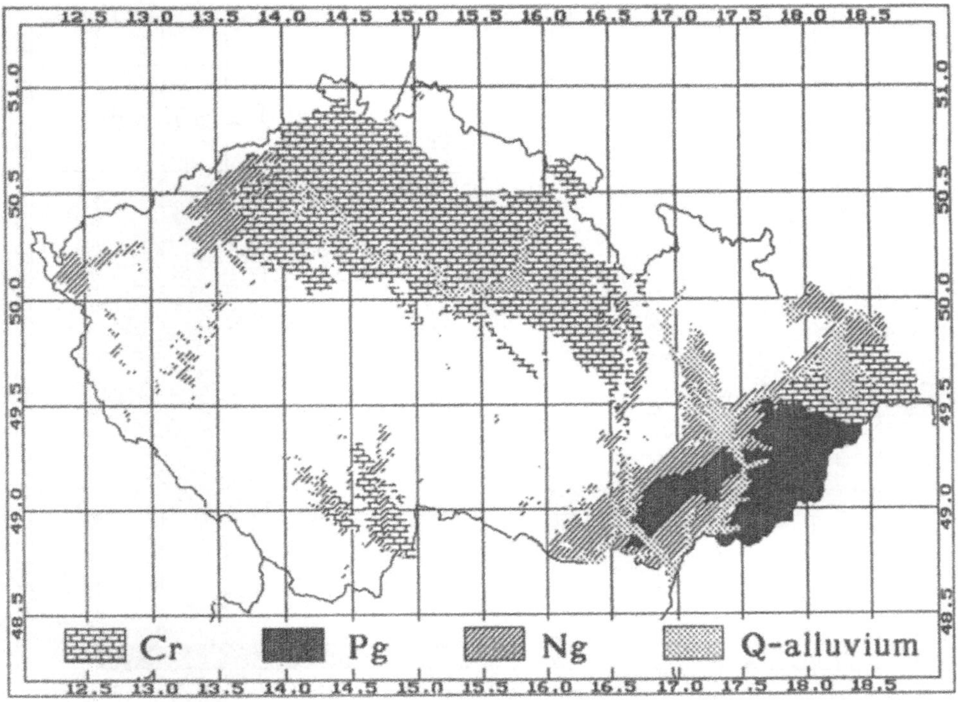

Figure 6. Sedimentary units of Cretaceous (Cr), Paleogene (Pg), Neogene (Ng) and Quaternary (Q-alluvium) periods involved to the earthquake hazard calculations (Schenk *et al.*, 1996b)

four sedimentary units for the Czech Republic, we introduced corrections of the macroseismic intensity defined with respect to the physical properties of sediments (bulk density, P- and S-wave velocities) (Schenk *et al.*, 1996b; Figs 7a and 7b). In principle, sediments and unconsolidated covers increase seismic vibrations, i.e. macroseismic effects have to be increased, too. A comparison of Figures 7a and 5a, 7b and 5b respectively, shows how great differences in earthquake hazard values appear for certain regions if the sedimentary covers are involved into input data for the hazard calculation. One can observe that the pattern of the regional hazard values changes according to the geographic positions of covers (Fig. 6): areas of higher macroseismic effects extend to the detriment of lower effects. The maximum differences of hazard values attain ½° to ¾° MSK for Quaternary alluvial covers.

6. Map of the maximum possible earthquakes $I_{0(max)}$ for the Czech Republic

It is common to all epicentral areas that the seismic information they yield does not enable us to make a reliable prediction and determination of their earthquake activities. That is why the seismic data analysis must be complemented with the analysis of all available information associated with the earthquake origin. It includes, for example, data on the geodynamics and structure of the Earth's crust, on the positions and dimensions of recently active faults, on the boundaries and sizes of individual

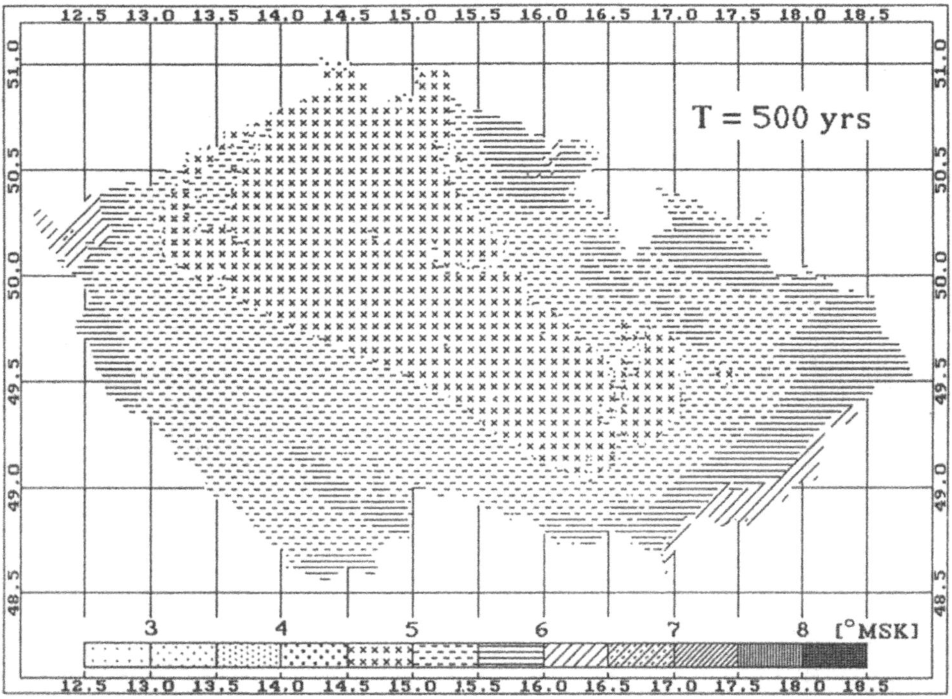

Figure 7a. Earthquake hazard map for the return period of 500 years in which expected near-surface geology effects are included (Schenk *et al.*, 1996b)

geological blocks, on the structural depth discontinuities as well as on the relations between geophysical fields and seismic activity.

A geophysical prognostic model of the maximum possible earthquake occurrence for the outer-Alpine part of the central European territory with the Bohemian Massif in its centre was defined (Schenk *et al.*, 1991) by the GEO-1.2 expert system (Gitis *et al.*, 1991; Schenk *et al.*, 1992). This system analyzes geo-data of an area with respect to expert estimates, synthetizes the prognostic model according to the prognostic functions between the earthquake size expressed by the epicentral intensity I_0 and geo-data defined by statistical regression and pattern recognition techniques (Schenk *et al.*, 1994b) and checks the obtained results. Its fundamental principles and the algorithm of calculation together with the prognostic functions determined for the area belonging only to the West European platform have been discussed in detail in the paper by Schenk *et al.* (1991).

The following initial data available in a digital form, specifying the calculation of the maximum earthquake $I_{0(max)}$ prognostic field, were used: an earthquake catalogue for the period 1300 - 1988, a schematic map of Bouguer anomalies, the MOHO discontinuity, the topographic relief of the Earth's surface and heat flow data. Since the expert system also allows additional fields of each initial quantity (modulus of gradients, local/residual anomalies, different functions of these quantities, etc.) to be calculated, approximately 60 geophysical and geodetical input fields of features were introduced into the calculation of the maximum earthquake $I_{0(max)}$ prognostic field. It

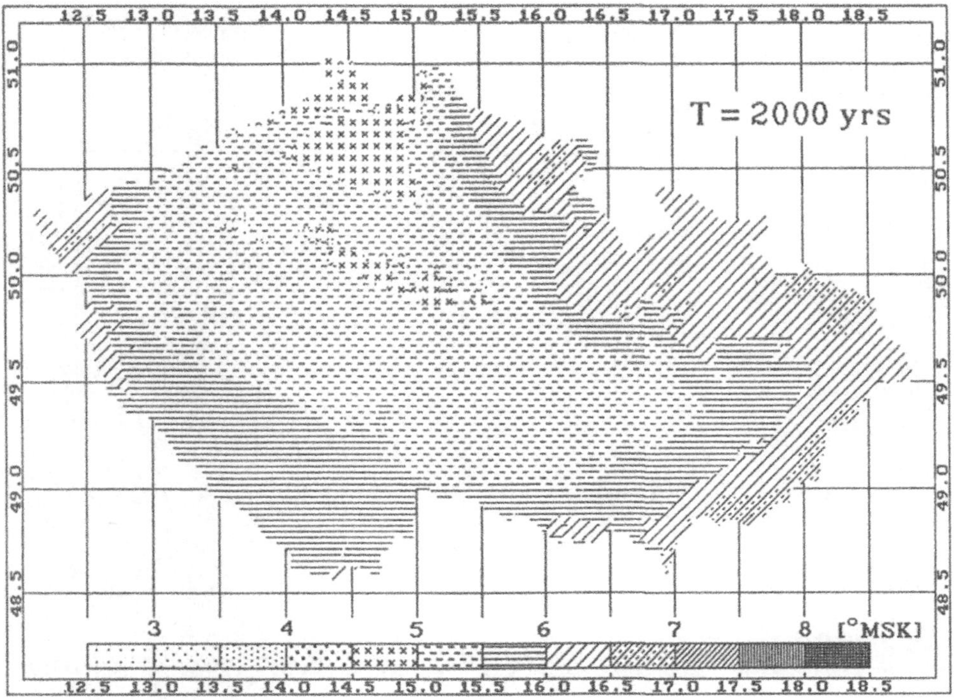

Figure 7b. Earthquake hazard map for the return period of 2000 years in which expected near-surface geology effects are included (Schenk *et al.*, 1996b)

was defined that the greatest contribution to the prognostic field is made by the relief of the MOHO discontinuity. For example, for the crust thickness of 28 km its contribution attains about 2¼° - 2½° MSK.

The prognostic model of the $I_{0(max)}$ field (Schenk *et al.*, 1991; Fig. 8) delineates regions in which $I_{0(max)}$ values are defined by the maximum earthquake intensity thresholds of all retrospective foci. If the regions are correlated with observed earthquake epicentres some of them exhibit good agreement, however, some do not evidence any significant correlation. Such regions can be classified as possible "earthquake prone zones". One such earthquake prone zone was identified on the northern margin of the West Sudeten near Jelenia Góra (round 51°N, 15.5°E). On the territory of the Czech Republic regions up to $I_{0(max)} = 7°$ are predicted.

Since there are some open questions which have to be clarified by other investigations, this map of $I_{0(max)}$ can be understood as a preliminary one. On the other hand, the knowledge concerning an identification of possible earthquake prone zones and a qualitative classification of seismogenic zones will be involved in the next versions of the earthquake hazard map of the Czech Republic.

Another result of interest is connected with an explanation of the seismogenic character of the earthquake zones which are known on the contact of the Bohemian Massif and the West Carpathians. Although only the area belonging to the West European platform was analyzed to find prognostic functions, the functions were also applied to the assessment of the maximum earthquake $I_{0(max)}$ field of the entire

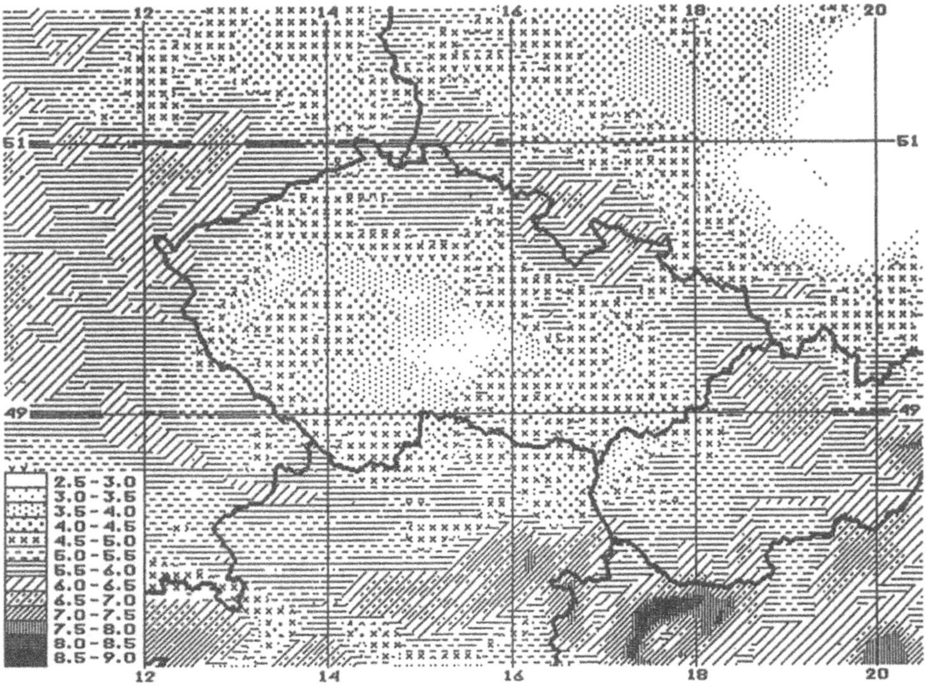

Figure 8. Map of the maximum possible earthquake (Schenk *et al.*, 1991)

geographic territory, i.e. to the marginal parts of the West Carpathians. The obtained results show a close correlation of the seismogenic activity of the Žilina focal zone (49.23°N, 18.74°E) with the West European structures of the Sudetic direction. This zone is located in the Inner Carpathians and the focal depth of the 1858 Žilina earthquake is known to have been about 20 km. From these two aspects, one can infer that seismogenic processes occurring in these foci are most likely linked with the West European structures, located below the Carpathians, and that they have no direct relations to the Carpathian tectonics.

7. Conclusions

It is very probable that serious earthquake damage cannot occur within the Bohemian Massif and antiseismic protection is only needed for very vulnerable or important structures. The earthquake hazard assessments and other results presented in this paper clearly show that the hazard calculations for the Czech Republic are still topical and new substantial improvements can be introduced. For example, the introduction of the problem of variable attenuations of seismic waves (macroseismic effects) in different geographical azimuths of real strong ground motions including their frequency contents and their durations, an involving prone seismogenic zones, an evaluation of their seismic activities, etc.

In case the hazard assessments are used for seismic risk evaluations, existing so far mainly for limited areas only, then rather sophisticated corrections of macroseismic intensities and strong ground motions will have to be introduced.

A revision of Building Code ČSN 73 0036, which will involve the earthquake hazard map for the return period of 475 years, is being performed according to recommendations of the European Association for Earthquake Engineering (see the last version of EUROCODE 8) and with respect to other national codes. Recommendations concerning the selection of sites for important earthquake-resistant industrial and civic structures (power stations, chemical plants, dams, bridges, high-rise buildings, TV towers, etc.) and the location of radioactive waste disposals are made.

8. References

Cornell, C.A. (1968) Engineering seismic risk analysis, *Bull. Seis. Soc. Am.* **58**, 1583-1606.

Cornell, C.A. (1971) Probabilistic analysis of damage to structures under seismic load, in D.A. Howell, I.P. Haigh and C. Taylor (eds.), *Dynamic Waves in Civil Engineering*, Interscience, London, pp. 473-488.

Cornell, C.A. and Merz, H.A. (1974) Seismic risk analysis of Boston, *J. Struc. Div., Proc. Am. Soc. Civil Eng.* **101** (ST 10), 2027-2043.

ČSN 73 0036 (1973) Seismic loads of buildings, in A. Dvořák (ed.), *ČSN Codes*, Úřad pro normalizaci a měření, Praha, 45 p.

Gitis, V.G., Yurkov, E.F. and Osher, B.V. (1991) GEO - An expert system for geological and geophysical prognosis, *Proc. of the 4th Intern. Conf. on Seismic Zonation*, v. III, Earthq. Eng. Res. Inst., Stanford, 59-64.

Gumbel, E.J. (1958) *Statistics of Extremes*, Columbia Univ. Press.

Kárník, V., Schenková, Z. and Schenk V. (1984) Earthquake activity in the Bohemian Massif and in the Western Carpathians, *Trav. Géophys.* **29**, 9-33.

Kárník, V., Procházková, D., Schenk, V., Schenková, Z. and Brouček, I. (1988) Seismic zoning map of Czechoslovakia - version 1987, *Stud. geoph. et geod.* **32**, 144-150.

McGuire, R. (1976) Fortran computer program for seismic risk analysis, *U.S. Geol. Survey Open-File Rept.* **76-67**, 90 p.

Procházková, D. and Kárník, V. (eds.) (1980) *Atlas of Isoseismal Maps for Central and Eastern Europe*, Geoph. Inst., Czechosl. Acad. Sci., Prague.

Procházková, D. and Brouček, I. (1981) Map of maximum observed intensity of Czechoslovakia, in V. Kárník, D. Procházková and Z. Schenková (eds.), *Atlas of Seismological Maps of Central and Eastern Europe*, Geoph. Inst., Czechosl. Acad. Sci., Prague.

Procházková, D. and Brouček, I. (1986) Map of maximum intensities on the territory of Czechoslovakia, *Contr. Geophys. Inst. Slovak Acad. Sci., Veda, Bratislava* **16**, 25-32.

Schenk, V. (1983) On the problem of time-normalization of the magnitude-frequency relation, *Ann. Geoph.* **1**, 439-442.

Schenk, V. (1984) Conversion of observed macroseismic field up to higher intensity level by generalized circular isoseismals, *Eng. Geology* **20**, 153-160.

Schenk, V. (1993) Seismic hazard in the Czech Republic and Slovakia, in R.K. McGuire (ed.), *The Practice of Earthquake Hazard Assessment*, IASPEI and ESC, pp. 87-91.

Schenk, V. and Kottnauer, P. (1991) Probabilistic methods in maximum earthquake assessment, in G.W. Borm (ed.), *Computational Mechanics Publications* (Proc. 5th Intern. Conf. on Soil Dynamics and Earthquake Engineering) **110**, pp. 1-7.

Schenk, V. and Mantlík, F. (1985) Map of maximum expected macroseismic intensity in Czechoslovakia, *Proc. of the 3rd Intern. Symp. on the Analysis of Seismicity and on Seismic Risk*, Geoph. Inst., Czechosl. Acad. Sci., Prague, 440-445.

Schenk, V. and Schenková, Z. (1983) Seismic hazard of a selected site from the recent knowledge viewpoint (in Czech), in *Bezpečnost a Provozní Spolehlivost Reaktoru VVER - Seizmické Zatížení*. ÚISJP, Prague, 7-12.

Schenk, V., Schenková, Z. and Kárník, V. (1981) Seismic hazard assessment in the Bohemian Massif, *Proc. of the 2nd Intern. Symp. on the Analysis of Seismicity and on Seismic Risk*, Geoph. Inst., Czechosl. Acad. Sci.,

Prague, 531-544.
Schenk, V., Kárník, V. and Schenková, Z. (1982) Seismotectonic scheme of Central and Eastern Europe, *Studia geoph. et geod.* **26**, 132-144.
Schenk, V., Schenková, Z., Mantlík, F., Kottnauer, P. and Grünthal, G. (1989a) Seismic hazard assessment for Central Europe, *Proc. of the 4th Intern. Symp. on the Analysis of Seismicity and on Seismic Risk*, Geoph. Inst., Czechosl. Acad. Sci., Prague, 494-503.
Schenk, V., Schenková, Z. and Grünthal, G. (1989b) Problems of seismic hazard assessment for the pleistoseismal area of swarm earthquakes in Western Bohemia and Vogtland, *Tectonophysics* **167**, 217-222.
Schenk, V., Gitis, V.G., Schenková, Z., Mantlík, F., Kottnauer, P., Yurkov, E.F. and Shchukin, Yu.K. (1991) Maximum earthquake prediction in Central Europe given by the GEO-1.2 expert system, *Proc. of the 4th Intern. Conf. on Seismic Zonation*, v. III, Earthq. Eng. Res. Inst., Stanford, 83-91.
Schenk, V., Gitis, V.G. and. Schenková, Z. (1992) Aspects of artificial intelligence techniques in earthquake hazard assessment and seismic risk mitigation, *Cahiers du Centre Européen de Géodynamique et de Séismologie* **6**, 145-152.
Schenk, V., Schenková, Z. and Kottnauer, P. (1993) Earthquake Hazard Map for Central Europe - Version 1993, Geoph. Inst., Acad. Sci. Czech Rep., Prague, unpublished.
Schenk, V., Kottnauer, P. and Schenková, Z. (1994a) *Earthquake Hazard Map of Czech Republic (T=500 yrs); Region: North Bohemian Brown-Coal Basin*, Set of 21 maps in scale 1:25 000 and Explanation (in Czech), Geoph. Inst., Acad. Sci. Czech Rep., Prague.
Schenk, V., Schenková, Z. and Gitis, V.G. (1994b) Features of geonomic prognostic functions for the maximum possible earthquake, *Natural Hazards* **10**, 97-115.
Schenk, V., Schenková, Z. and Kottnauer, P. (1996a) Earthquake hazard correction for effects of near-surface sediments: An example of the Bílina district, Czech Republic, *Annali di Geofisica*, under preparation.
Schenk, V., Schenková, Z. and Kottnauer, P. (1996b) Earthquake hazard assessment for the Czech Republic, *Natural Hazard*, under preparation.
Schenková, Z., Schenk, V. and Kottnauer, P. (1993) *Earthquake Catalogue for the Czech Republic and Adjacent Areas*, Geoph. Inst., Acad. Sci. Czech Rep., Prague, unpublished.
Schenková, Z., Schenk, V. and Kárník, V. (1981) Seismic hazard estimate for a low seismicity region - example of Bohemia, *Pure Appl. Geoph.* **119**, 1077-1092.
Schenková, Z., Kottnauer, P. and Schenk, V. (1996) Maximum observed intensity point map for the Czech Republic - version 1993, *Natural Hazard*, under preparation.

Sensitivity of seismic hazard estimates to the use of historical site data

Mucciarelli M.*, Albarello D.** and Stucchi M.***
* *ISMES, Bergamo, Italy*
** *Dip. Scienze della Terra, Un. Siena, Italy*
*** *IRRS-CNR, Milano, Italy*

Abstract

At 4 sites in Italy, a set of comparative seismic hazard estimates was performed with the aim to assess the sensitivity of each estimate to the use of different data sets. Estimates of expected average return periods for strong earthquakes have been carried out using a new approach based on probabilistic counting and logistic-type attenuation models. The different data sets used were:

- the National Catalogue of Italian Earthquakes of epicentral data from 1000 until 1980 (Postpischl, 1985);
- a new catalogue prepared considering all available results of recent researches on historical seismicity carried out in the framework of GNDT (National Protection Against Earthquakes of the National Council of Research)
- data base of observed intensities compiled by the same research group.

Significant differences were found which, in particular, show the importance of observed site intensities for a reliable hazard assessment.

1. Introduction

The assessment of site specific seismic hazard, i.e. the estimated probability that a given site will be affected within a given time interval by a seismic ground shaking exceeding a given threshold, represents one of the most important tasks of the applied seismological research. When knowledge about local seismogenic processes is lacking, phenomenological probabilistic models are adopted to perform the requested hazard estimates (see, e.g., Epstein and Lomnitz, 1966; Cornell, 1968; McGuire, 1976; Grandori et al., 1984) . These models are defined and parametrized on the basis of available data about the seismic activity in the area of interest. Thus, hazard estimates obtained following this kind of approach are strongly dependant to the level of available knowledge about the seismic history of the site under study.

In this work, seismic hazard estimates have been performed at four sites in Italy (Ancona, Pisa, Bagnara Calabra and Reggio Calabria, see fig. 1) on the basis of

V. Schenk (ed.), Earthquake Hazard and Risk, 141–151.

different data sets concerning the relevant seismic history at each site, expressed in terms of intensity.

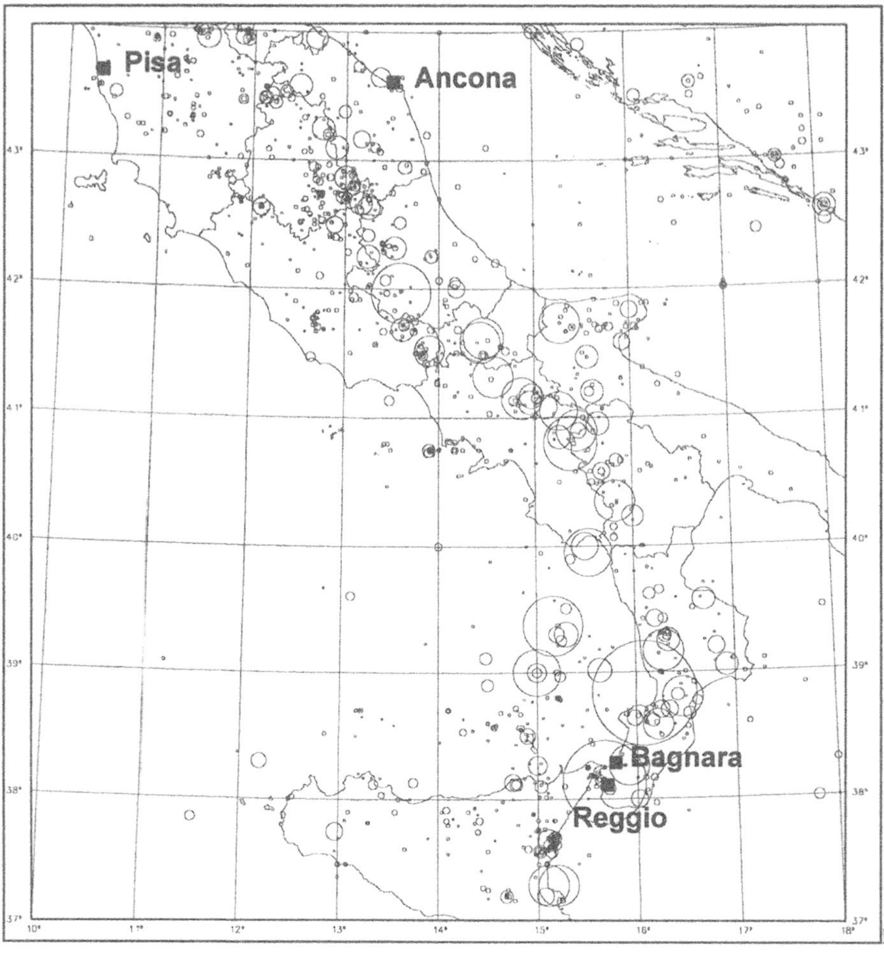

Fig. 1: Seismicity of Central and Southern Italy and location of the four sites studied.

Macroseismic intensity describes the effect of ground shaking on people, bulding and the environment. For Italy, it is here condidered to be a good representation of hazard, for two reasons: the careful reassessment of intensity performed by GNDT (see paragraph 2 and 3 of the chapter "Data set") that gave homogeneous estimates all over the territory, and the fact that the main concern in Italy is for to non-engineered structures that are used in the definition of macroseismic scales and that represent the majority of Italian buildings.

With this study, we also want to stress the importance of intensity data base. In the usual catalogues, all the information contained in the macroseismic intensity map was compressed in a few parameters (epicenter, maximum or epicentral intensity). Now the easily available computer power can be coupled with suitable statistical techniques in

order to fully exploit all the information given for each point. For example, if a town shows ground motion amplification for every earthquake occurring, the use of site observed intensities will greatly improve hazard definition that may be otherwise underestimated. If a large number of observed intensity are available, this will reduce the uncertainty in hazard estimate due to the attenuation relationship.
Site dependent hazard estimates may then be used as a checking tool for the reliability of hazard estimates performed over a wide territory with methods involving the use of seismotectonic zoning or active faults.
The aim of this comparative analysis is to study the sensitivity of hazard estimates with respect to the available basis of knowledge. To this purpose, for each site, the same approach to hazard estimate has been adopted and reparametrized on the basis of the different data sets. The adopted approach is shortly described in the following section.

2. Seismic Hazard Assessment

The definition of phenomenological models suitable for hazard estimates presents a number of problems. One of the most important is the fact that a large part of available knowledge about seismicity has to be obtained by the analysis of documents. Thus it has a non-instrumental character and cannot be expressed in terms of real-type numerical values, but only using integer numbers belonging to a non-metric, ordinal scale (The original use of Roman figures emphasises this characteristic). Furthermore, this kind of data is mostly available in epicentral areas where the effects of the earthquake were stronger and are generally lacking far from epicentres as well as in inhabited areas.
These specific features of "historic" earthquake data have to be taken into account when probabilistic models for seismic hazard assessment are defined. In particular, the model assumptions should be compatible with the use of ordinal/discrete scales of ground shaking and damage (Macroseismic Intensity Scales) and should include some treatment of the appropriate "experimental" uncertainty. The model should include some suitable "attenuation relationship" that, on the basis of available data (away from the site), allows estimates of the expected site intensity when observed data are not available. Furthermore, it is needed, in the framework of a unique and coherent methodology, to use both the data directly observed at the site and those estimated on the basis of attenuation relationships.
In the following analysis, a probabilistic approach fulfilling these requirements has been adopted. It allows reliable seismic hazard estimates on the basis of ill-defined site specific seismic data. This formulation (Magri et al., 1994) is based on the use of a discrete distribution function $p_S(I)$ describing, for each earthquake, the probability that site effects can be described by each possible intensity value I. By definition it holds that

$$\Sigma_I\, p_S(I)=1$$

where the sum is extended over the discrete interval [Imin, Imax] with Imin and Imax representing respectively the lower and upper limits of the adopted macroseismic scale. The probability function $p_S(I)$ can be defined using available macroseismic data at the site on the basis of expert seismological judgement or other more or less formalised procedures. As an example, if uncertainty is assumed to exist only between the two possible MCS intensities VIII and IX, the distribution function $p_S(I)$ could be described by the twelve elements array

$$p_S(I) = [0.,0.,0.,0.,0.,0.,0.,0.5,0.5,0.,0.,0.]$$

for I in the range [I, XII] degrees MCS.

When observed data are lacking, the dependence of this distribution function with the distance r from the epicentre and with epicentral intensity is modelled by a probability function R(I|r,J). It represents the probability that the macroseismic intensity expected at a site will be greater or equal to I given the distance r from the site where the maximum intensity J has been experienced for the earthquake under examination. The formula

$$P_S(I)=\Sigma_K \, p_S(K) \qquad [1]$$

is adopted where the sum extends over the discrete interval [I, I_{max}] and $P_S(I)$ is the probability that, at the site, seismic intensity is greater or equal to I. It holds that

$$P_S(I)=\Sigma_J \, p_e(J) \, R(I \mid r,J) \qquad [2]$$

where $p_e(J)$ is the probability that the maximum observed intensity for that earthquake is J. The sum is extended from I_{min} up to I_{max}. For the Italian region, the probabilistic attenuation function R has been determined using more than 10.000 single intensity observations (Magri et al., 1994) in the logistic form

$$R = \exp(a + b \log(r)) \, / \, (1 + \exp(a + b \log(r))) \qquad [3]$$

where

$$a = 1.00 + 1.95 \, (J - I)$$
$$b = -1.15 - 0.16 \, (J - I)$$

and log is the natural logarithm.

In this formalization, site seismic history results in an array of distribution functions $P_S(I)_i$ representing, for each earthquake i, the probability that at the site the intensity is I. The most significant advantages of this formulation with respect to the more usual ones have been extensively described in the reference work (Magri et al., 1994) and will not be reported here.

In this framework, some relevant statistics can be obtained (Magri et al., 1994). In particular, the average return period T(I) and the number N(I) of earthquakes can be estimated for expected site intensity greater or equal to I and respectively are given by

$$T(I) = \Delta T / \Sigma_i P_S(I)_i \qquad [4]$$

$$N(I) = \Sigma_i P_S(I)_i \qquad [5]$$

where $P_S(I)_i$ can be obtained by the equation [3] or directly by the local $p_S(I)$ distribution via equation [1]. The sums in [4] and [5] are extended over the whole number of earthquakes occurred in the time span ΔT. Suitable evaluations of the standard deviation associated to the estimator T(I) can also be supplied (see Magri et al., 1994).

3. Data Set

The analysis was carried out on the basis of three data sets, which are briefly explained in the following.

1. Catalogue of Italian earthquakes (Postpischl, ed., 1985), hereafter called "PFG catalogue", has been produced in the frame of the Italian Geodynamics Project (Progetto Finalizzato Geodinamica). It includes epicentral data of about 37,000 earthquakes which occurred in the Italian area in the period 1000-1980 and represents a fairly updated version of an unpublished earthquake data file by the Electric Agency of Italy (ENEL, 1978) which consisted of about 20,000 earthquakes. The updating was made by merging together the ENEL file with other parametric earthquake catalogues covering smaller portions of Italy, which were supposed to be more reliable, for their area, of the ENEL catalogue, and then sorting out multiple determinations automatically. For some tens of major earthquakes the parameters were derived from the new intensity data sets produced within the frame of the same PFG project after intensive investigation (Postpischl, ed., 1985b).

2. The "NT" earthquake catalogue (Stucchi et al., 1993) is a file covering the Italian and surrounding's areas containing only epicentral information about main events selected after a declustering procedure based on a space-time-window of +/- 90gg., +/- 30 km. The reason for such a choice is of operative nature only: actually, here it is considered that a main event in such a space-time-window is representative for hazard evaluation because it is the most damaging with respect to its aftershock. Moreover, the majority of hazard models requires aftershocks to be removed prior to the analysis. In such a way, the file contains "only" about 4000 entries (lower threshold: Io = 5/6, Ms = 4); one third of them are compiled from data sets produced after intensive historical investigation; two thirds of them are taken from other parametric catalogues (including the PFG one), without revision so far.

Both PFG and NT catalogues are used in combination with the attenuation relationship described by equation [3].

3. The third data set (hereafter "HNT" catalogue) is constituted by epicentral data from the "NT" catalogue plus a set of the observed intensities at the sites under study, obtained by published studies or retrieved in the frame of the investigation mentioned above. This data set adds to the almost 4000 entries of the "NT" catalogue about 30000 observations to date . For the four sites under study, the data used are summarized in table 1. Mixing the documented site intensity data and the ones computed from epicentral data via eq.[2] allowed the compilation of site catalogues of "expected" felt intensities.

The number of documented site intensities represent a significant percentage of the total number of expected intensities estimated from using the "NT" catalogue and eqq. [2] and [5]). Fig. 2 reports the cumulative numbers of events (site inensities) as a function of intensity both for the observed and the theoretical intensities for the town of Bagnara Calabra.

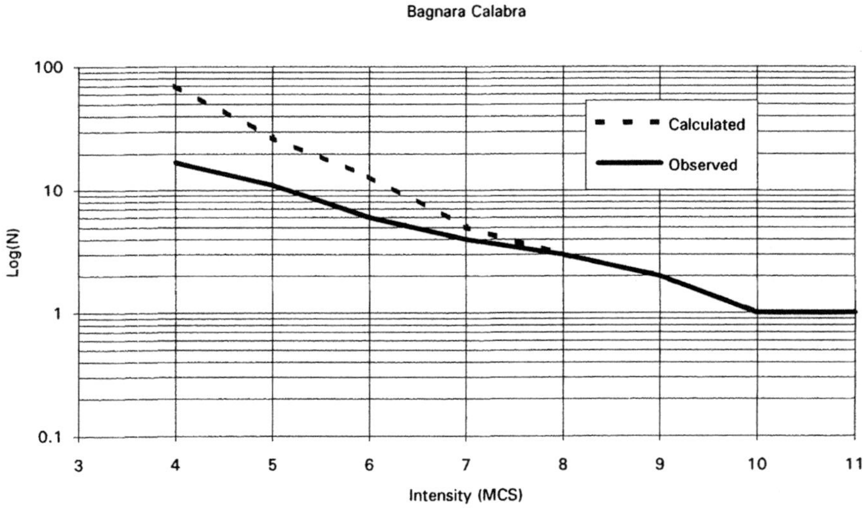

Fig. 2 Cumulative number of available data as a function of intensity for a sample site.

It can be seen that the ratio is 1:1 for the higher intensities, and remains above 1:2 down to intensity 5. This means that for the more damaging earthquakes the observed intensities are available and can be used instead of the computed ones in hazard estimate. Similar considerations apply to the other three sites. The difference in the low intensity range could be attributed to three factors: a bias in intensity assessment, an inappropriate attenuation function or an incompleteness of historical data. The first possibility is excluded by the homogeneity of the used database. As mentioned in the previous chapter, the attenuation law was derived from more than 10,000 observations and no significant bias was observed with the increasing of attenuation. The most realistic possibility is thus the fact that the information concerning low site intensity is not as complete as the one regarding stronger intensities.

4. Main Results

Using the three catalogues described above ("PFG", "NT" and "HNT"), expected average return periods (ART) have been estimated in correspondence of different intensity thresholds at the four sites under study: Pisa and Ancona in Central Italy and Reggio Calabria and Bagnara Calabra in Southern Italy (Fig. 1). The time spans considered for probabilistic analysis (*s.l.*, the "completeness" periods) for each intensity class (see Tab. 1) have been estimated using the speditive approach proposed by Mulargia et al. (1987).

Table 1: Completeness period and summary of data available for the four sites studied.

Site	Ancona	Pisa	Reggio C.	Bagnara C.
Number of observations	37	21	49	20
I_{max} (MCS)	8	6/7	11	11
First observed intensity (year)	1269	1542	1783	1783
Completeness for I = 2 (starting year)	1872	1888	1839	1783
3	1690	1805	1839	1783
4	1690	1739	1783	1596
5	1690	1168	1783	1626
6	1690	1168	1783	1509
7	1358	1168	1783	1509
8	1358	1168	1509	1509
9	1358	1168	1509	1509
10	1358	-	1509	1509
11	-	-	1509	1509

Fig. 3 reports the results obtained for the city of Pisa. It can be seen that using the PFG catalogue the average return times (ART) are always lower than the ones obtained using NT file. This can be due to the fact that the historical seismicity studies used for preparing the NT catalogue led to the conclusion that most of the intensities attributed in the PFG catalogue were overestimated, and only few appeared to be underestimated. Another interesting feature of the results obtained using PFG values is a sort of bilinear trend with different slopes in the ranges 2-5 and 6-10 degree MCS. This effect is probably a consequence of the fact that the NT and HNT catalogues discard aftershocks which significantly affects expected ART values in correspondence of lowest intensities.

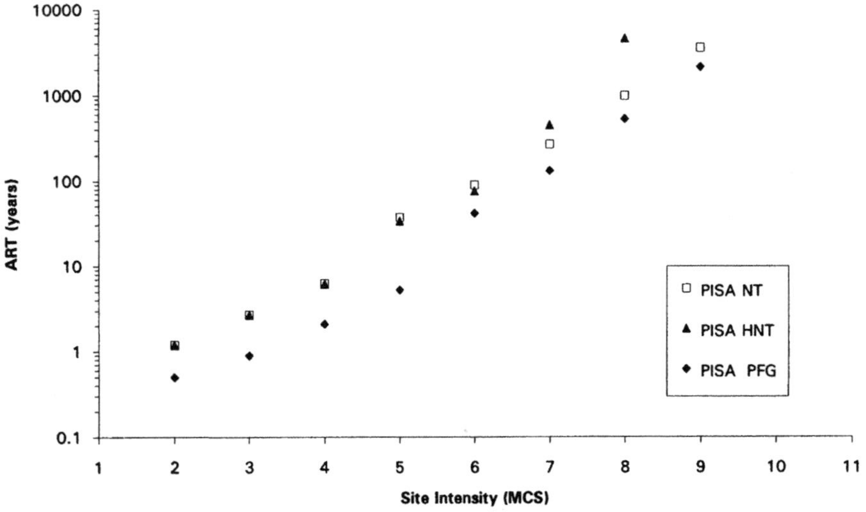

Fig. 3 Average return times for the site of Pisa

The comparison between the ART values obtained using NT and HNT catalogues shows that almost identical estimates are obtained for the lowest intensity range (II-VII) while a significant divergence can is observed for highest intensity values. This fact could imply that the logistic attenuation model adopted is reliable "on average", as confirmed when a large number of historical data is available. When the frequency of events decreases, even a single, small uncertainty in site intensity estimate can strongly affect the ART producing relatively large fluctuations. Moreover, results in figure 3 show that the inclusion of observed values (HNT catalogue) may act as a natural upper bound for expected intensities. This results in a quasi-asymptotic behaviour that increases by more than ten times the ART for intensity 8 and does not allow any estimate for intensity 9 MCS.

A very similar pattern has been obtained for ART estimates performed at the city of Ancona (Fig. 4) on the basis of data from the three catalogues considered.

As concerns the two last sites (Bagnara Calabra and Reggio Calabria), only information contained in the HNT catalogue has been considered. The ART estimates performed at these sites are shown in fig. 5. Despite the relatively low distance between these sites (few tens of Km) very different patterns have been obtained. When the highest intensities (>VII) are considered, ART values at Bagnara Calabra and Reggio Calabria show very similar patterns. However, the ART values estimated for the lowest intensities range, diverges significantly. In particular, ART values at Bagnara Calabra follow a log-linear trend common to both highest and lowest intensity classes. On the contrary, ART estimates at Reggio Calabria seems to imply two different patterns in correspondence of expected intensities lower or higher than VII MCS.

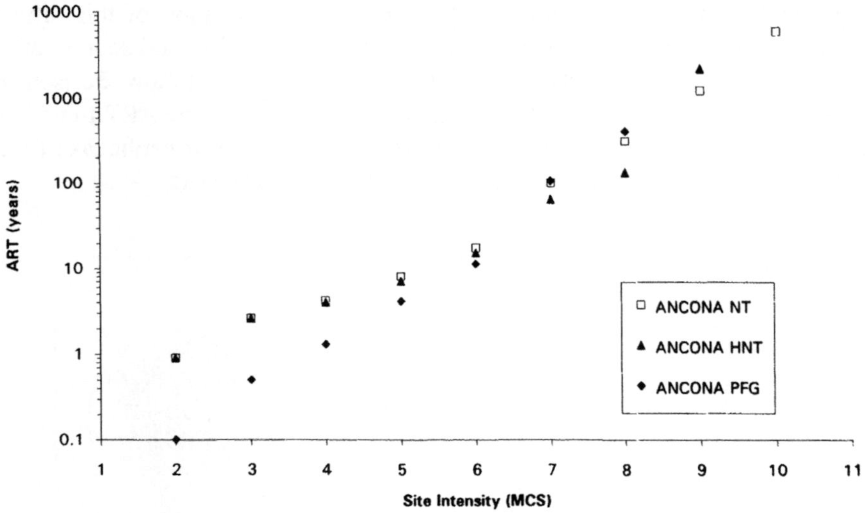

Fig. 4 Average return times for the site of Ancona

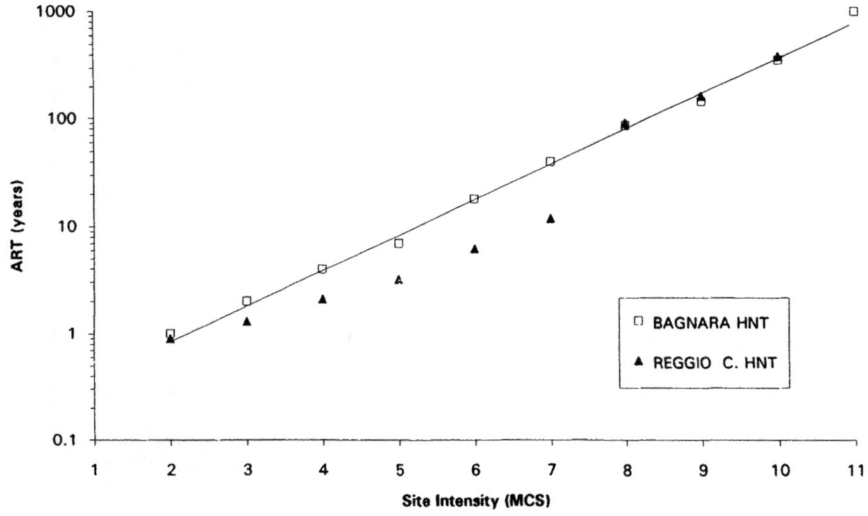

Fig. 5 Average return times for the sites of Bagnara Calabra and Reggio Calabria

This significant difference could be interpreted in the framework of the local seismotectonic setting (Fig. 6). The city of Reggio Calabria is located nearby one important seismogenic fault which presumably determines alone the local seismic hazard. As concerns the town of Bagnara Calabra, local seismic hazard is connected to the presence of three different seismogenic sources not located nearby the site.

Taking into account this different setting, a possible interpretation of this different pattern can be given as follows. The sum of the contribution of several sources affects the trend of ART vs. intensity so that the inter-event time follow Poisson and Gutenberg-Richter distributions. When a single fault is involved, the ART trend is not linear, and this may be due to two factors: the recurrence rule of earthquake (quasi-periodic or characteristic) and the enhanced weight of the aftershocks.

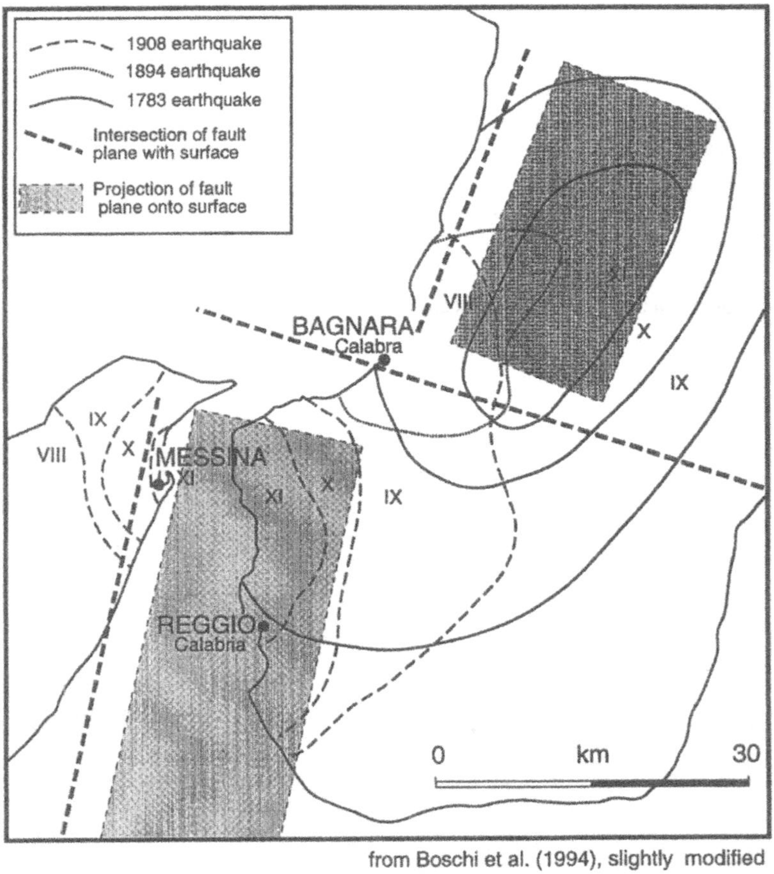

Fig. 6 Seismotectonic setting for the sites of Bagnara Calabra and Reggio Calabria

5. Conclusions

The intensity observed at four sites were used to establish if and how much their inclusion in hazard estimates could affect the results.

The main results of the adopted methodology for the four considered sites can be summarised as follows:

1. The effect of source aftershock removal is negligible for site hazard estimate if the interest is limited to the damaging part of the intensity scale (I>=6 MCS).

2. The use of observed intensities gives results that suggest the existence of an upper bound in the maximum expected intensity.
3. The usual assumptions on earthquake distribution (Gutenberg-Richter, Poisson) seem to be valid for site hazard estimate only if there is a large enough number of sources affecting the site.
4. The availability of a good set of observed site intensity is able to induce significant changes in seismic hazard estimates with respect to the one performed using only an earthquake catalogue and attenuation function.

Acknowledgements

The financial support of CNR-GNDT is gratefully acknowledged.
The comment of Prof. Shapira and of an anonymous reviewer helped improve the manuscript.

References

Boschi E., Pantosti D. and Valensise G., 1994, L' identificazione geologica delle faglie sismogenetiche, Le Scienze, 310, 36-49

Cornell C.A., 1968. Engineering seismic risk analysis. Bull. Seis. Soc. Am., 58, 1583-1606

ENEL, 1978, Catalogue of Italian Earthquakes, Open File

Epstein B. and Lomnitz C., 1966. A model for the occurrence of large earthquakes. Nature, 211, 954-956

Grandori G., Guagenti E. and Petrini V., 1984. On the use of Renewal processes in seismic hazard analysis. Proc. 8th World conf. on Earthquake Engineering, 1, S.Francisco, 287-294

Magri L., Mucciarelli M. and Albarello D., 1994. Estimates of site seismicity rates using ill-defined macroseismic data. PAGEOPH, 135, 4, 617-632

McGuire R.K., 1976. Fortran computer program for seismic risk analysis. U.S.G.S. open file Report 76-67, 92 pp.

Mulargia F., Gasperini P. and Tinti S., 1987. A procedure to identify objectively active seismotectonic structures. Boll.Geofis.Teor. Appl., XXIX, 114, 147-164

Postpischl D., ed., 1985, Catalogo dei terremoti italiani dall'anno 1000 al 1980, CNR-PFG

Postpischl D., ed., 1985 b, Atlas of isoseismal maps of the Italian area, CNR-PFG.

Stucchi M., Camassi R. and Monachesi G,.1993, Il catalogo di lavoro del GNDT. CNR GNDT GdL "Macrosismica", Rapporto interno, Milano, 89 pp.

RECENT TRENDS IN REGIONAL AND GLOBAL SEISMIC HAZARD ASSESSMENT

D. GIARDINI[1),2)], E. BOSCHI[2)]
[1)] *Dept. of Geology, III University of Rome*
[2)] *Istituto Nazionale di Geofisica*

1. Abstract

A recent trend in the global practice of seismic hazard assessment is the establishment of regional and global programs based on the multidisciplinary cooperation of seismologists, geologists and engineers. Here we review problems and limitations in the practice of seismic hazard assessment, the use of different probabilistic approaches in different tectonic areas and the implementation and significance of geological input in seismic hazard assessment. We also present a survey of global and regional cooperative programs in multinational and multidisciplinary seismic hazard assessment and we review the status and plan of the UN/IDNDR Global Seismic Hazard Assessment Program.

2. Introduction

Earthquakes are the expression of the continuing evolution of the planet Earth and of the deformation of its crust and occur worldwide; while the largest events ($M>7.5$) concentrate on plate boundary areas and active plate interiors, large and moderate earthquakes ($6<M<7.5$) may take place, if rarely, in all continental areas (Fig. 1). The notion that the primary targets for seismic hazard assessment should be the areas of high seismicity is incorrect. Events of moderate and small dimensions ($5<M<6.5$) occur virtually everywhere and may turn catastrophic in earthquake prone areas with poor building construction practice, as tragically shown by the 1993 event in central India (M=6.3, 14.000 casualties), the 1960 event in Morocco (M=5.8; 12.000 casualties) and the 1972 event in Nicaragua (M=6.2; 5.000 casualties; damages for 40% of national budget).

Vulnerability to disaster is increasing as urbanisation and developments occupy more areas that are prone to the effects of significant earthquakes. In order to minimize the loss of life, property damage and social and economic disruption caused by earthquakes, it is essential that reliable estimates of seismic hazard be available to

V. Schenk (ed.), Earthquake Hazard and Risk, 153–168.

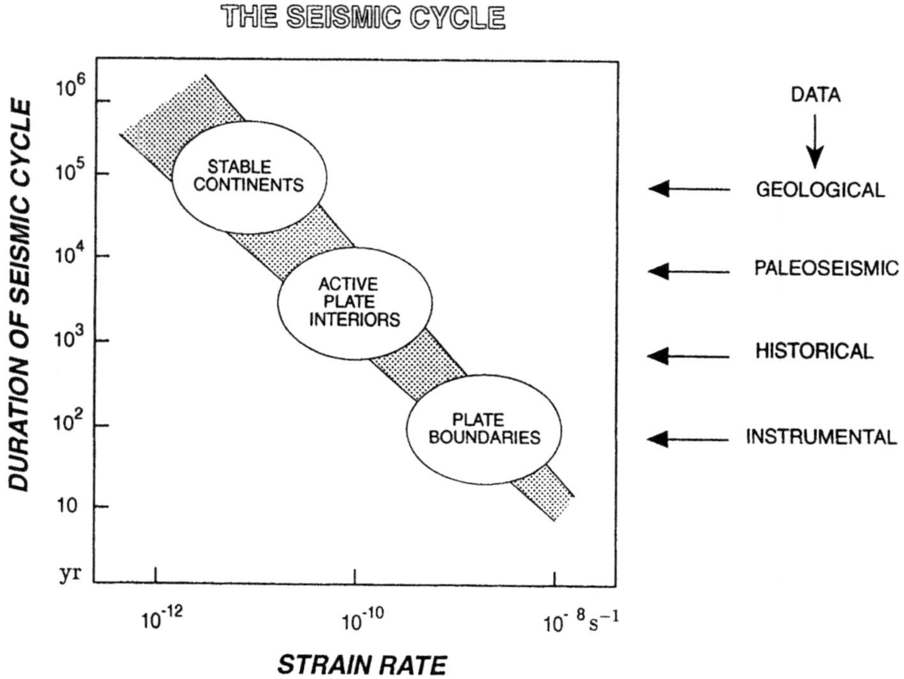

Figure 1. Strain rate and duration of characteristic seismic cycle for different seismotectonic provinces (modified after Burton, personal communication, 1993). The input data required to control earthquake recurrence at different time scales are indicated on the right.

national decision makers and engineers for land use planning and improved building design and construction.

The seismic hazard is defined as the probabilistic measure of ground shaking associated to the recurrence of earthquakes. The assessment of seismic hazard is the first step in the evaluation of the seismic risk, obtained by convolving the seismic hazard with local site amplification effects tied to soil conditions and with the intrinsic value and vulnerability of the existing buildings and infrastructures. Frequent, large events in remote areas result in high seismic hazard but pose no risk; on the contrary, moderate events in densely populated areas entail small hazard but high risk.

Concrete measures for seismic risk mitigation include the definition of the seismic zonation of the territory, with established safety coefficients used in the design and construction of private and public buildings, a strategy for the defense of the historic and cultural heritage, the protection of the large-scale addition of man to the habitat (industrial and chemical complexes, nuclear and conventional power plants, technological concentrations, large dams, major communication systems), a long-term strategy for land-use planning in earthquake-prone areas taking into account the long-term effects of catastrophic earthquakes (the disruption of the economic chain, the human resettlement, the reconstruction to modern standards), the preparation of emergency plans in areas of high risk well in advance of the earthquake, and an improved practice of earthquake surveillance. All of these measures rely on accurate seismic hazard assessment.

Two main limitations affect the global practice of seismic hazard and risk assessment. Their computation results from the integration of many different disciplines and should be conducted in coordinated fashion by seismologists, geologists, engineers and land-use planners (Fig. 2); the coordination among these

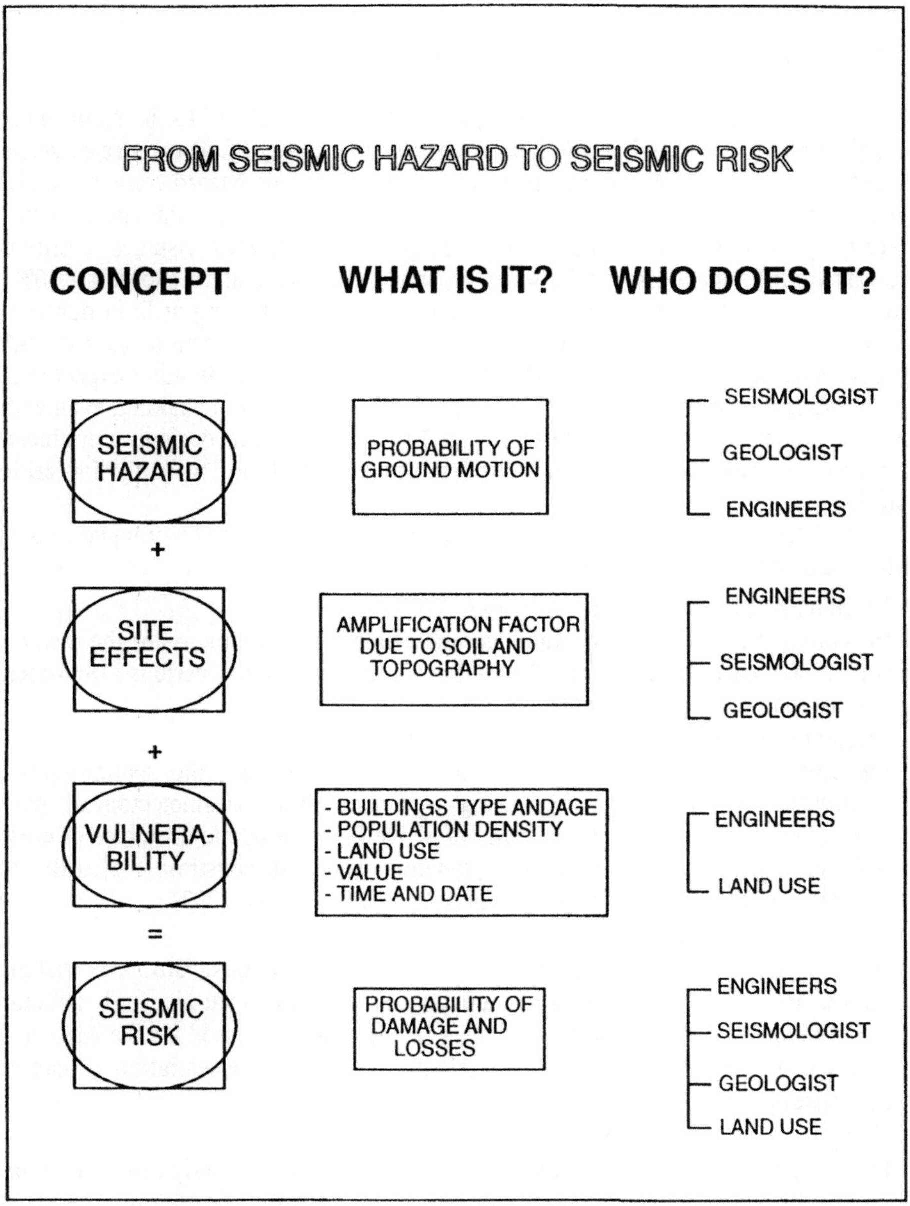

Figure 2. Basic concepts, definitions and required expertise behind the assessment of seismic hazard and risk.

groups has been so far not optimal in many countries. Secondly, seismic hazard is traditionally assessed at the national scale, while seismogenic structures and earthquake effects are not influenced by human boundaries. Several programs now exist to coordinate at regional and continental scale the multidisciplinary efforts in seismic hazard assessment. Here we review problems and limitations in the practice of seismic hazard assessment and the new global programs.

3. The practice of seismic hazard assessment

The seismic hazard is a measure of the ground shaking associated to the recurrence of earthquakes and is defined at any location as the probability of occurrence of seismic ground shaking in a specified time interval. A map of seismic hazard contours areas of equal seismic hazard expressed in terms of ground shaking parameters such as intensity, peak ground acceleration or velocity. Seismic hazard assumes a different meaning to different users; the 30-years probability of acceleration exceeding 10% of gravity characterizes the geographical extent of moderate shaking at high degrees of likelihood (e.g. 50-50), of high interest for the design and protection against highly probable shaking damage; on the other hand, the maximum acceleration expected at a 6% probability in 30 years measures the expected highest levels of shaking, even with a very low probability of occurrence (e.g. 1 in 20), and is used in designing engineered structures to withstand a once-in-a lifetime level of shaking, typically for critical facilities.

The basic elements of modern seismic hazard assessment can be grouped into four main categories (Cornell, 1968):

- Earthquake Catalogues and Data Bases:
  the compilation of a uniform data-base and catalogue of seismicity for the historical (pre-1900), early-instrumental (1900-1964) and instrumental periods (1964-today) (Johnston and Halchuck, 1993).
- Seismotectonics and Earthquake Source Zones:
  the creation of a master seismic source model to explain the spatial-temporal distribution of the seismicity, using evidences from seismotectonics, paleo-seismology, geomorphology, mapping of active faults, geodetic estimates of crustal deformation, remote sensing and geodynamic models to constrain the earthquake cyclicity in different tectonic provinces (Fig. 1; Muir-Wood, 1993).
- Strong Seismic Ground Motion:
  the evaluation of ground shaking as function of earthquake size and distance, taking into account propagation effects in different tectonic and structural environments and using direct measures of the damage caused by the earthquake (the seismic intensity) and instrumental values of ground acceleration (Boore and Ambraseys, 1993).
- Computation of Seismic Hazard:
  the computation of the probability of occurrence of ground shaking in a given time period, to produce maps of seismic hazard and related uncertainties at appropriate scales (McGuire, 1993a).

The quality and availability of the basic data needed for seismic hazard assessment varies greatly around the world and influences the quality of the hazard.

Instrumental seismic catalogues covering the last 30 years exist for all areas of the world; the accuracy of hypocentral locations and especially the determination of earthquake size (more than one hundred magnitude scales and regressions exist) strongly depend on the availability of monitoring networks. Fig. 3 shows the distribution of stations operating in the Middle East in the last 30 years; uneven station coverage results in uneven completeness of regional catalogues for moderate-size events.

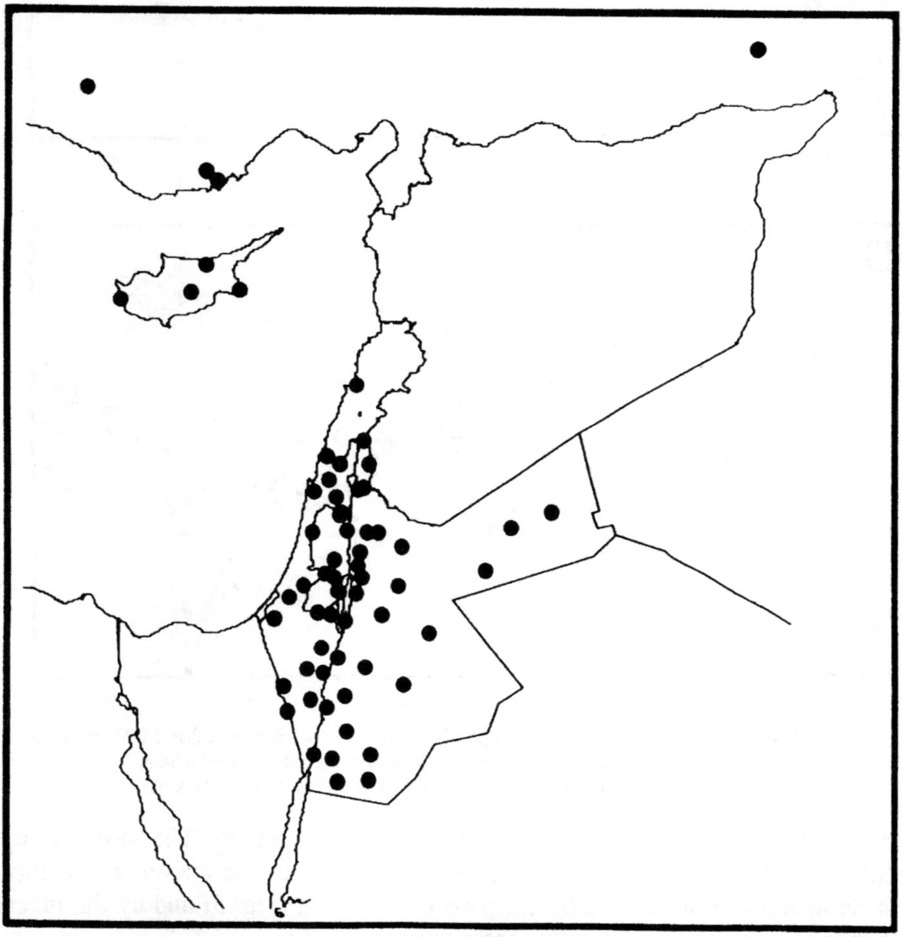

Figure 3. Distribution of seismic stations in the Middle East reporting to ISC in the last 30 years.

Even when the instrumental catalogue is accurate, it will span only a few decades, while the recurrence of large earthquakes in active areas may need characteristic period of hundreds or thousands of years (Fig. 1). A comparison of the instrumental seismicity of the last 30 years with the location of the historical earthquake disasters in the Mediterranean basin (Fig. 4) shows how some areas have experienced devastating

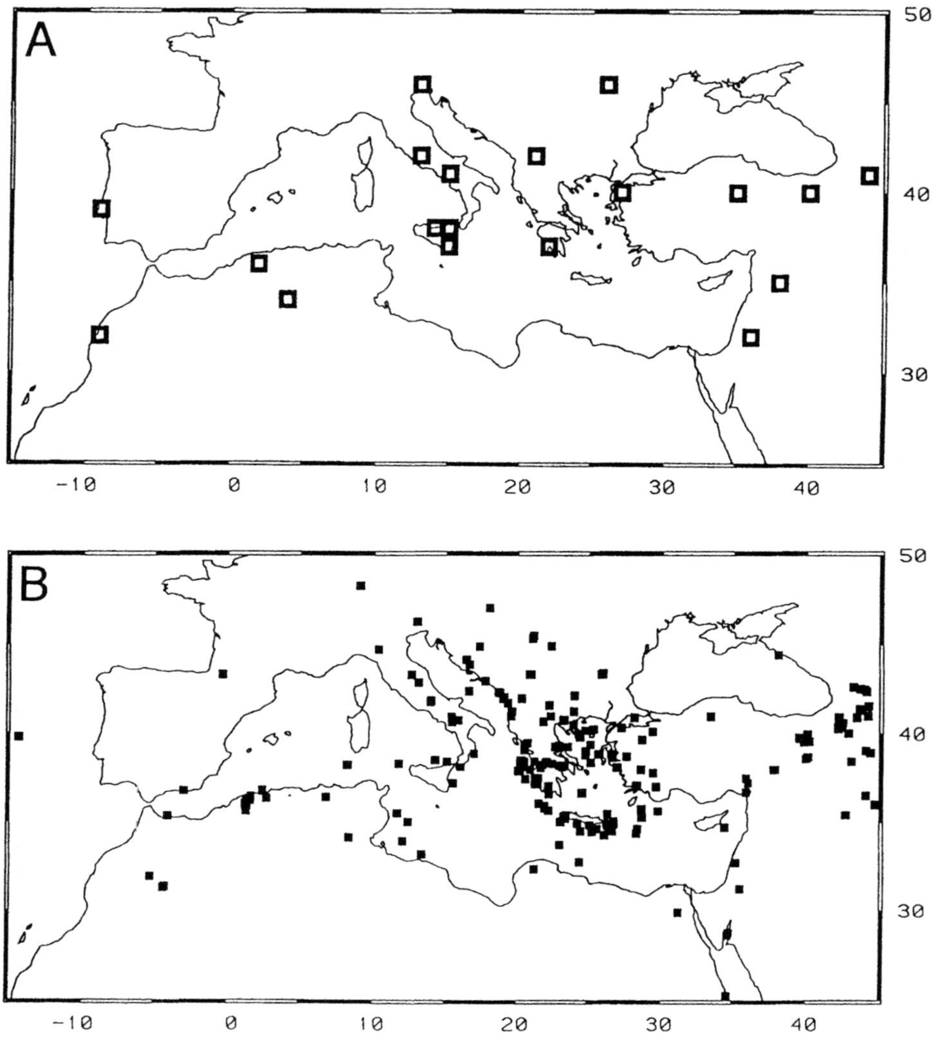

Figure 4. Comparison of the distributions of (A) the earthquakes causing more than 1.000 casualties in historical times and (B) the recent seismicity of the Mediterranean basin (219 CMT hypocenters from the Harvard dataset 1977-1994; h<50 km).

earthquake in the past, while displaying little or no recent activity. The need to compile accurate catalogues of historical earthquakes is hindered by the uneven availability of reliable historical sources in different parts of the world (Fig. 5) and by the inherent difficulty of searching and analysing these sources (Fig. 6).

The most complete catalogue will not provide all the information needed to characterize the seismogenic process; this is especially true for areas of active plate interiors and diffuse continental deformation and when only an insufficient historical record is available (Fig. 1). Seismic zoning attempted without sufficient background information from geology proves to be very uncertain (Fig. 7). Today geology is

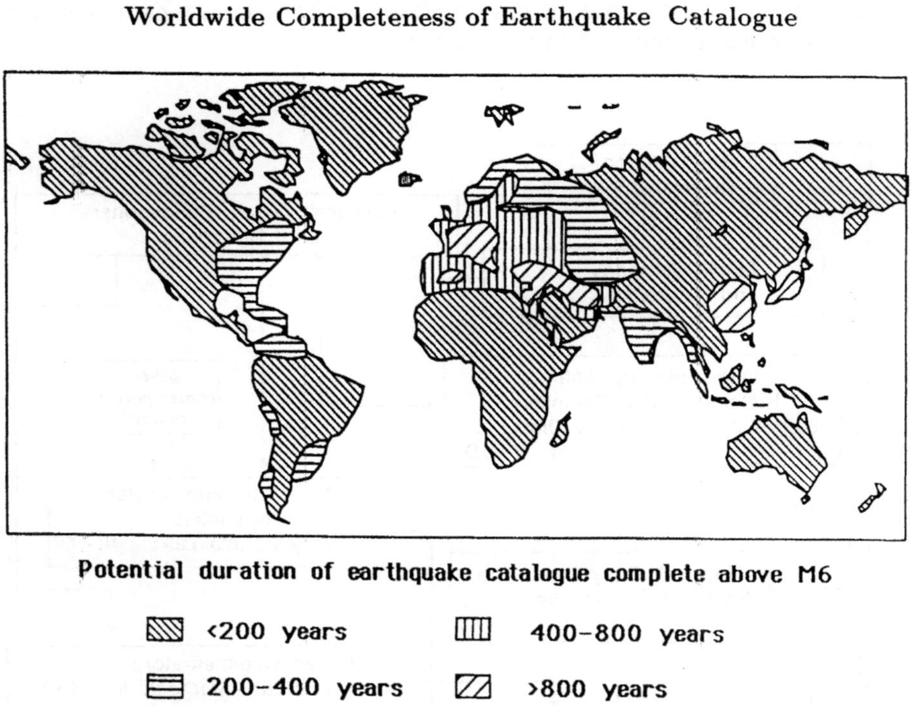

Figure 5. Geographical distribution of completeness level for historical earthquake catalogues (modified after Muir-Wood, 1993).

providing key input, allowing to associate the historical earthquakes to specific seismogenic features (through evidences from seismotectonics, paleo-seismology, geomorphology, mapping of active faults, geodetic estimates of crustal deformation, remote sensing and geodynamic models) and to build alternative models of seismic zonation (Fig. 8).

The quality of the data available from the catalogues and seismotectonics conditions the hazard assessment and the choice of methodology to be used for the hazard assessment. Four different approaches are available and commonly used (see McGuire, 1993b, for a global compilation):

1. Historical Determinism maps the maximum intensity of earthquake effects recorded in known historical times, to represent, with appropriate corrections, the highest intensity to be expected in the future.
2. Historical Probabilism builds a statistical model of seismogenic sources to reproduce the historical record of seismicity (location in space and time, frequency-size distribution).
3. Seismotectonic Probabilism incorporates geological evidence (prehistoric record of paleoseismic activity, geomorphology, rates of crustal deformation from land and space geodesy, neotectonic and geodynamic modelling) to supplement the historical record of seismicity in building a seismic source model covering earthquake cycles up to a few thousands years.
4. Time-Dependent Seismotectonic Probabilism: the use of non-Poissonian statistics allows to take into account not only the periodicity of earthquake recurrence but

also the time elapsed since the last significant earthquake, as a most significant parameter in assessing the future seismic activity.

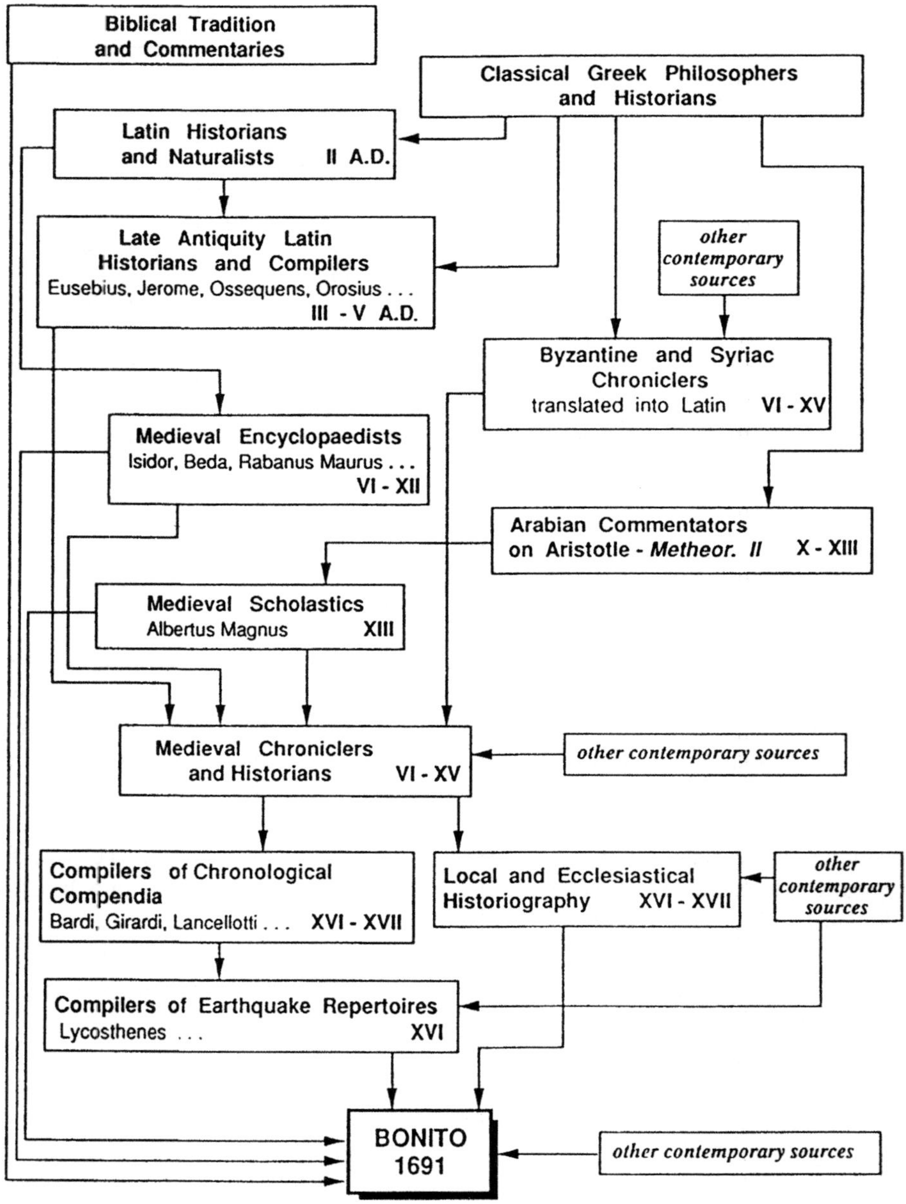

Figure 6. Relation tree for one source (Bonito, 1691) reporting historical earthquakes in Italy (modified after Guidoboni and Stucchi, 1993).

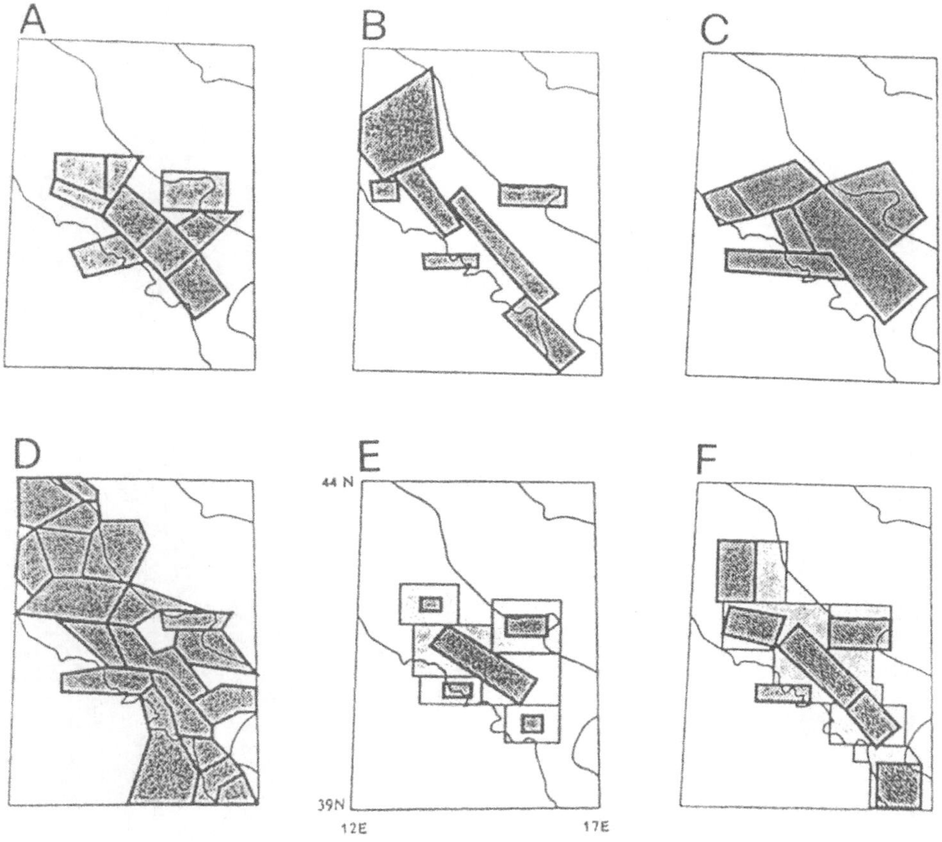

Figure 7. Comparison of seismic zonations by different members of the CEC-TERESA working group for the Sannio-Matese area of southern Italy: (A) Barbano *et al.*, (B) Mayer-Rosa, (C) Garcia and Egozcue, (D) Schenk *et al.*, (E) Lapajne *et al.*, (F) Siro and Slejko (modified after Barbano *et al.*, 1989).

The adoption of these different approaches leads to largely different expectations of short-term seismic hazard for locations characterized by similar seismo-tectonic setting and different seismic histories. For example (Fig. 9) Valdivia is affected by large earthquakes about once every 80 years, the last one occurring recently, while

Portland has not experienced a large earthquake in the last two centuries; the use of time-dependent analysis and seismotectonic constraints will lead to low short-term hazard for Valdivia, in open contrast with historical probabilism. Seismic hazard is assessed in some nations still using historical probabilism, relying essentially on the catalogue and often only on the instrumental catalogue, while others are already experimenting with time-dependent hazard assessment (see McGuire, 1993b).

Finally, an important element of division across boundaries lies in the geographical size of the country. Seismic hazard is assessed by most nations as the preliminary step toward the adoption of national zonation for building construction

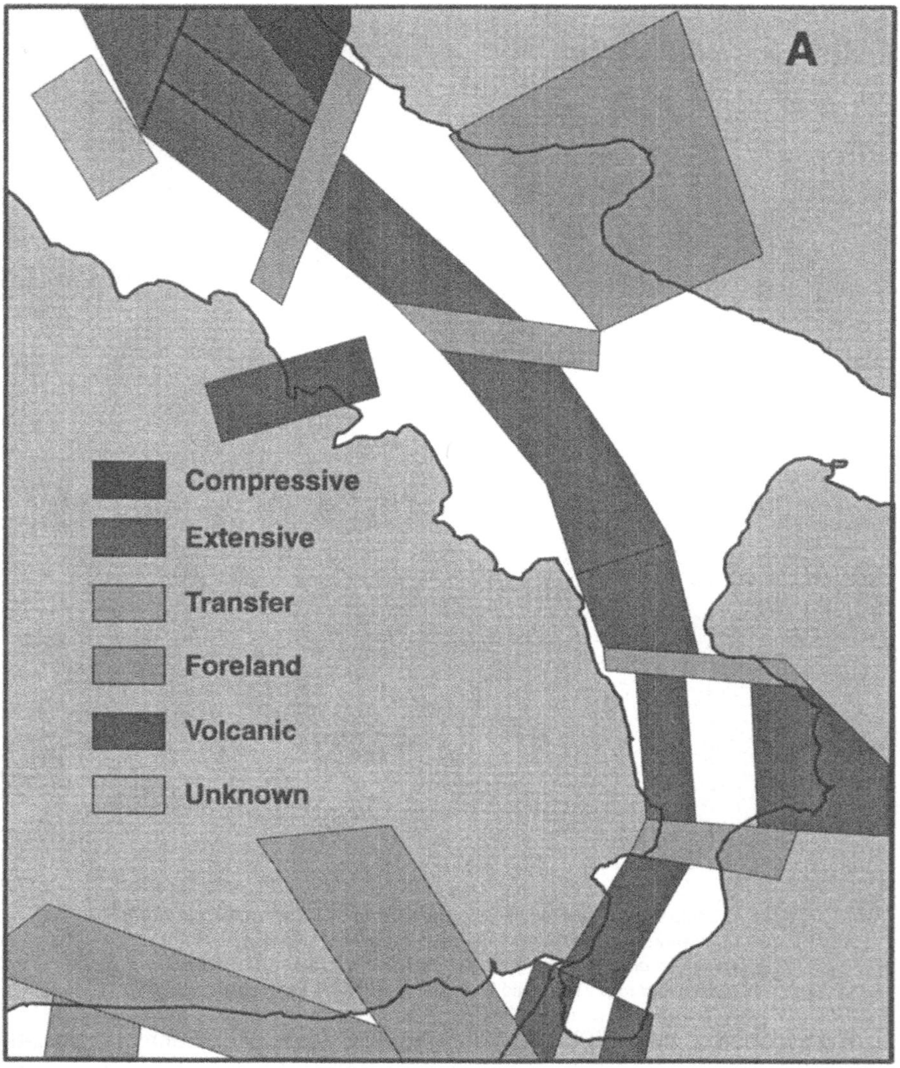

Figure 8. Comparison of (A) the preliminary source zoning proposed by the Italian CNR Group of Defense from Earthquakes for southern Italy (courtesy of Slejko, 1994) and (B) a map of seismogenic structures for the same area (courtesy of Valensise and Pantosti, 1994).

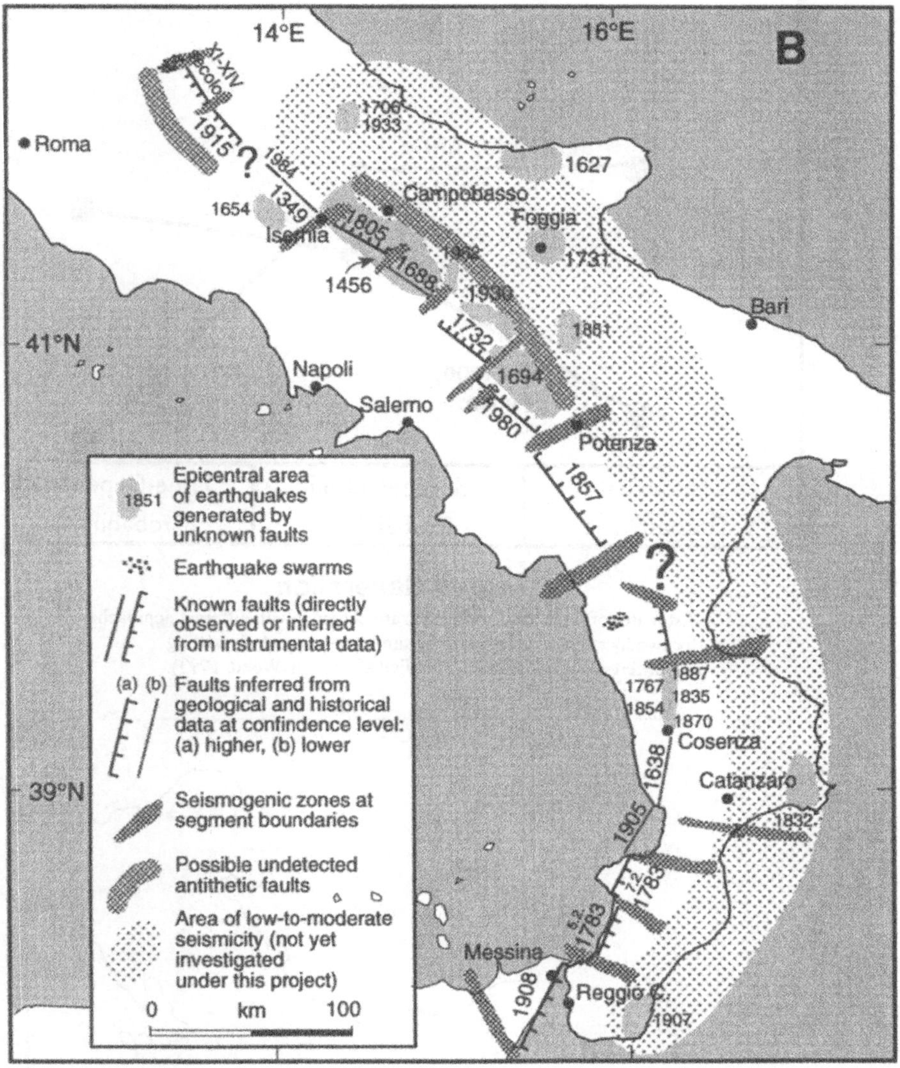

codes; depending on the size of the country, the national map product will be on a scale ranging between 1:500.000 and 1:6.000.000 (see for example the different sizes of China and Nepal, Fig. 10). Maps produced at so different scales will be incompatible, as the relevant information governing the seismic hazard varies at different scales.

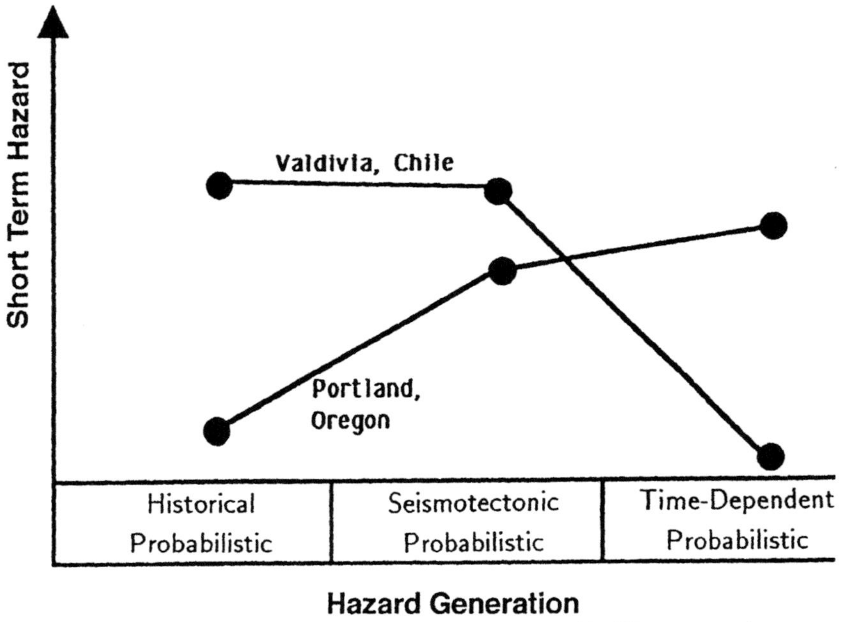

Figure 9. Estimates of short-term seismic hazard obtained with different approaches for two locations characterized by similar seismotectonic setting but different seismic history (modified after Muir-Wood, 1993).

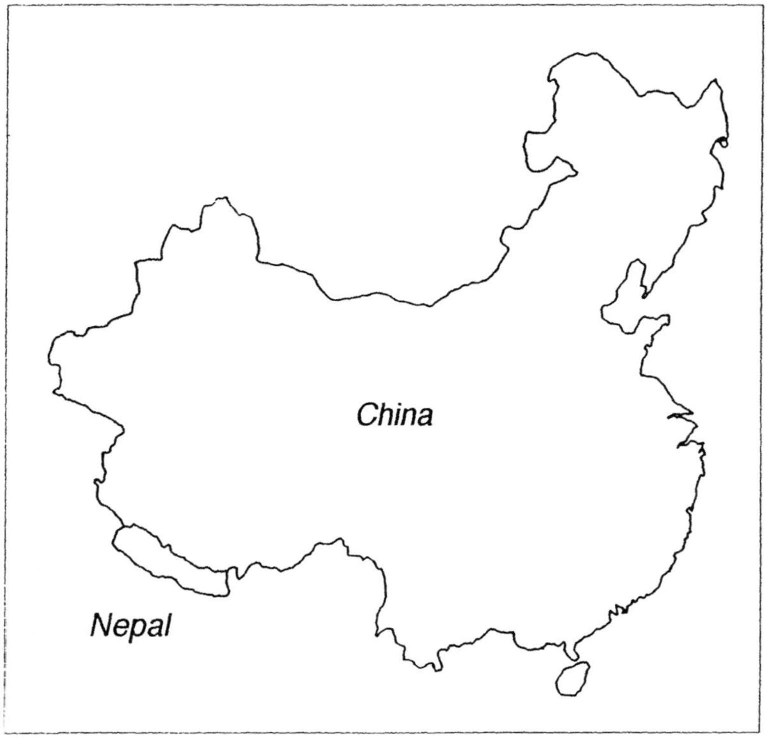

Figure 10. Geographical boundaries of China and Nepal.

4. A regional approach to global seismic hazard assessment

The need for improved coordination across boundaries and for the implementation of a multidisciplinary approach is resulting in many cooperative initiatives in seismic hazard assessment around the world.

The United Nations, recognizing natural disasters as a major threat to human life and development, designated the 1990-2000 period as the International Decade for Natural Disaster Reduction (UN/IDNDR; UN Res. 42/169/1987); the Decade goals are to increase worldwide awareness, foster the prevention and reduce the risks of natural disasters, through the widespread application of modern science and technology. As the first, necessary measure toward the implementation of risk reduction strategies, the Scientific and Technical Committee of the UN/IDNDR has endorsed international demonstration projects designed to improve the assessment of natural hazards (earthquakes, volcanoes, tropical hurricanes, floods, etc.). Among the spearhead programs endorsed in the UN/IDNDR context is the proposal of the International Lithosphere Program (ILP) for a Global Seismic Hazard Assessment Program (GSHAP), with the sponsorship of the International Council of Scientific Unions and the support of international scientific agencies (International Union of Geodesy and Geophysics, International Union of Geological Sciences, International Association of Seismology and Physics of the Earth's Interior) and of UNESCO.

The primary goal of GSHAP is to ensure that national agencies be able to assess seismic hazard in a regionally coordinated fashion and with the most advanced methods. The program is implemented at regional and local scale, with a regionalized approach based on the establishment of Regional Centres to assist national efforts, compile homogeneous regional data bases, ensure the needed coordination in across-boundaries hazard assessment, and provide a framework for data exchange and the implementation of unified assessment procedures.

GSHAP is coordinated by the Istituto Nazionale di Geofisica in Rome; the Regional Centres are hosted by main geophysical institutions in all continents (Fig. 11): North and Central America (UNAM, Mexico City), South America (CERESIS, Santiago), Northern Europe (GFZ, Potsdam), the Mediterranean basin (CNCPRST, Rabat), Continental Africa (University of Nairobi), the Middle East (IIEES, Teheran), Northern Eurasia (JIPE, Moscow), Central-Southern Asia (SSB, Beijing), East Asia (PHIVOLCS, Manila) and Oceania (AGSO, Canberra).

Activities in multinational test areas and programs in seismic hazard assessment are rapidly expanding, covering large continental areas, promoted by GSHAP and by other programs. Fig. 11 summarizes the status of cooperative programs around the world, as described in the following.

In South America CERESIS will produce a reference hazard map for the whole continent in 1995 and a European Community program will finance the GSHAP implementation in the Andean countries, starting in 1995.

In Central-North America a network of regional programs covers Canada, the US, Mexico, Central America (south of Mexico) and the Caribbeans, the last three with the support of Norway and Canada. Continental inter-program coordination is improving, with large-scale hazard tests planned for 1995. A workshop is planned for late 1995 in Mexico, where seismologists and engineers will test and compare methodologies for

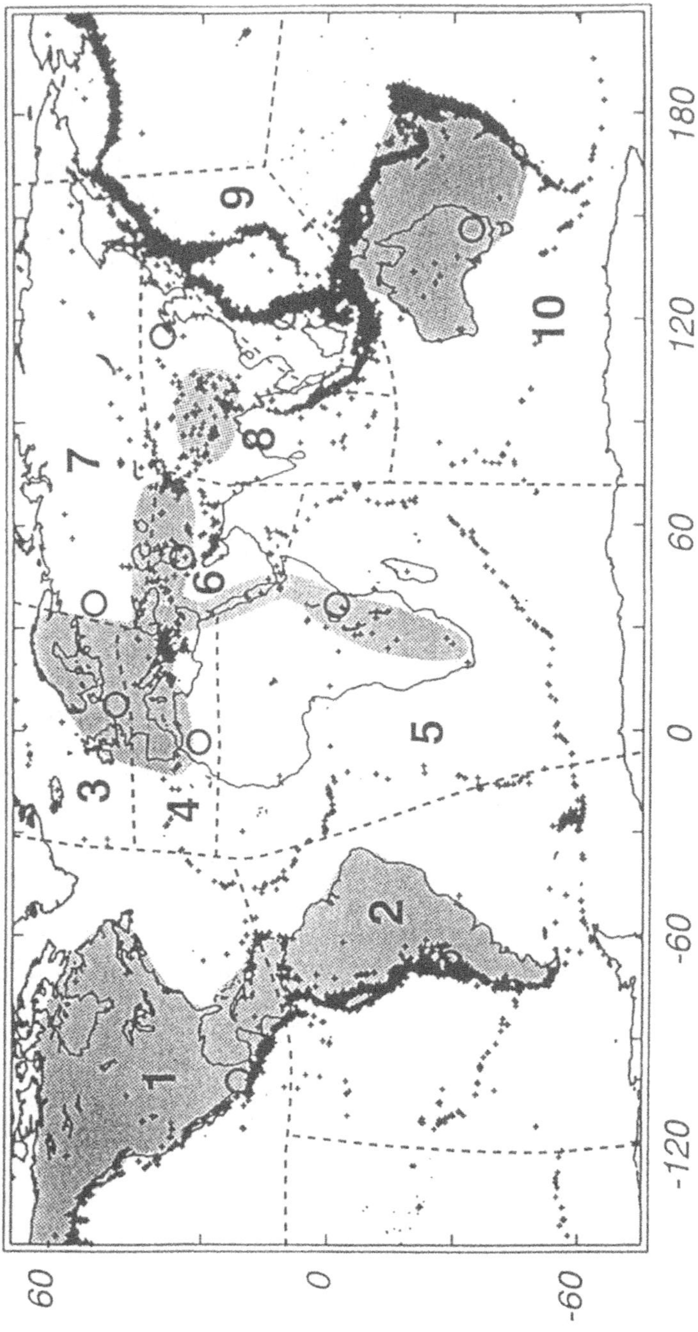

Figure 11. Survey of global seismicity and cooperative programs in seismic hazard assessment. Shaded areas identify areas where programs in multinational and multidisciplinary assessment are active; circles and dashed lines illustrate the distribution and regional operation of the GSHAP Centres. The seismicity is from the ISC catalogue (1964-1992; m>5).

seismic hazard assessment using the same datasets for a selected US/Mexico border area.

In the European-Mediterranean region a network of independent test-areas for seismic hazard assessment has been set up: the Ibero-Maghreb program, the Adria plate, the D-A-CH countries (Germany, Switzerland, Austria), Fennoscandia, the program for Reduction of Earthquake Losses in the Eastern Mediterranean Region (sponsored by UNESCO and the US Geological Survey), the Caucasus-Kopet Dag area, a European Community program for the compilation of a European-Mediterranean historical and instrumental seismic catalogue and database. The regional activities are lead by the GSHAP centres in Potsdam, Moscow and Rabat, under the overall coordination of the European Seismological Commission; the center of Rabat is becoming the center for seismic hazard assessment for the Open Partial Agreement on Natural Disasters of the European Council.

In Sub-Saharian Africa GSHAP has joined activities with the effort supported by Norway and Sweden aimed at coordinating the activities of individual seismological observatories and networks. A preliminary regional earthquake catalogue has been produced and will be finalized in a forthcoming workshop (Ethiopia, 1/95), leading to tests in regional seismic hazard computation planned for the summer of 1995 in South Africa. The interest of several European agencies and of South Africa has being formalized in a action framework for the next few years.

In Central Asia, leading national agencies from Iran, Turkey and the CSI have joined efforts in 1995 to launch a regional test area for seismic hazard assessment in the Caucasus-Kopet Dag. Work is in progress toward the compilation of regional seismic catalogues, seismotectonic maps and seismic hazard maps.

In Eastern Asia activities have started on the designed test area in the Assam-Bangla Desh-Burma-SW China region. A two-years program is under way, leading to the coordinated assessment of regional seismic hazard for 1996. In addition, the Natural Hazards Mapping program led by the Geological Survey of Japan is in progress; the GSHAP Regional Centres of Beijing and Manila are coordinating the seismic hazard assessment in that framework. The newly forming Asian Seismological Commission under IASPEI will promote the implementation of these regional efforts.

In Oceania, seismic hazard assessment has already been carried out through the whole region in coordinated manner and a workshop (Melbourne, November 1995) will review ongoing activities.

5. Concluding Remarks

While short- and mid-term earthquake prediction may one day be able to reduce significantly the death toll of earthquakes, the environmental effects (collapse of buildings and infrastructures, disruption of the productive chain, human resettlement) can be reduced only through a long-term prevention policy in earthquake-prone areas based on the assessment of seismic hazard and risk, the implementation of safe building construction codes, the increased public awareness on natural disasters, a strategy of land-use planning taking into account the seismic hazard and the occurrence of other natural disasters.

The UN/IDNDR provides an important chance to improve the global seismic hazard assessment by coordinating national efforts in multi-national, regional projects, reaching a consensus on the scientific methodologies for the seismic hazard evaluation and ensuring that the most advanced methodologies be available worldwide through technology transfer and educational programs.

The ILP's Global Seismic Hazard Assessment Program embodies many of the strategies and priorities of the IDNDR, filling a critical gap cited by many countries in attempting to assess properly the seismic hazard of their territory. The program promotes a regionally coordinated, multidisciplinary approach to seismic hazard evaluation. Cooperative initiatives in seismic hazard assessment are under way in many regions of the world, through the efforts of GSHAP and other regional programs.

The implementation of sound seismic hazard estimations into policies for seismic risk reduction will allow a focus on the prevention of earthquake effects rather than intervention following the disasters.

6. References

Barbano, M. S., J. J. Egozcue, M. Garcia Fernandez, A. Kijko, J. Lapajne, D. Mayer-Rosa, V. Schenk, Z. Schenková, D. Slejko, and G. Zonno (1989) Assessment of seismic hazard for the Sannio-Matese area of Southern Italy: a summary, *Natural Hazards* **2**, 217-228.

Boore, D. M., and N. N. Ambraseys (1993) Some notes concerning prediction of ground motions for GSHAP, *Annali di Geofisica* **36**, n.3-4, 169-180.

Cornell, C. A. (1968) Engineering seismic risk analysis, *Bull. Seismol. Soc. Am.* **58**, 1583-1606.

Giardini, D., and P.W. Basham (1993) The ILP's global seismic hazard assessment program for the UN/IDNDR, *Annali di Geofisica* **36**, n.3-4, 257 pg.

Guidoboni, E., and M. Stucchi (1993)The contribution of historical records of earthquakes to the evaluation of seismic hazard, *Annali di Geofisica* **36**, n.3-4, 201-216.

Johnston, A. C., and S. Halchuck (1993) The seismicity data base for the GSHAP, *Annali di Geofisica* **36**, n.3-4, 133-152.

McGuire, R. K. (1993a) Computations of seismic hazard, *Annali di Geofisica* **36**, n.3-4, 181-200.

McGuire, R. K. (1993b) *The Practice of Earthquake Hazard Assessment*, International Association of Seismology and Physics of the Earth's Interior, 284 pg.

Muir-Wood, R. (1993) From global seismotectonics to global seismic hazard, *Annali di Geofisica* **36**, n.3-4, 153-168.

IMPORTANCE OF GEOLOGICAL DATA IN PROBABILISTIC SEISMIC HAZARD ASSESSMENTS: A CASE STUDY FROM ETNE, WESTERN NORWAY

K.ATAKAN

Institute of Solid Earth Physics, University of Bergen

Allégt.41, N-5007 Bergen, Norway

M.R.KARPUZ and S.O.DAHL

Geological Institute, University of Bergen

Allégt.41, N-5007 Bergen, Norway

Abstract. In intraplate areas of low seismicity, any possible information about the existence of a large ($M_L > 6.0$) earthquake based on the local geology is quite valuable in estimating the true seismic hazard potential of the area of interest. Recent examples of destructive earthquakes from the seismically quiet areas such as India and Egypt, have demonstrated the often underestimated hazard potential of these areas. In conventional probabilistic hazard analyses, the seismic source zones are usually defined based on the instrumental catalogues which cover only the last 40-50 years, with a possible extension of a few hundreds of years from the historical data. The only source of information about the earlier events, is therefore dependent upon the detailed knowledge of the local geology of the area. In this study we demonstrate the importance of geological data in probabilistic seismic hazard assessments through an example from Etne in western Norway.

Detailed lineament studies conducted on satellite images, have revealed possible active fault zones in the Sunnhordland District, including the Etne Fault Zone (EFZ). The parallelism between a recently discovered NW-SE trending regional aeromagnetic lineament and the EFZ, implies that the fault zone may be related to a regional deep-seated zone of weakness which can be followed through the Precambrian basement as well as the allochthonous units. Various geological and geomorphological observations on mezoscopic and microscopic scales, as well as the positive correlation with the recent seismicity help to establish the EFZ as an active source zone. Seismic hazard map of the Sunnhordland area, where the hazard is expressed in terms of peak ground acceleration (PGA) for 10^2, 10^{-3} and 10^{-4}/years recurrence periods, indicate relatively high values that distinguish the seismic hazard potential of the EFZ. The contribution of EFZ to the seismic hazard estimates is clearly illustrated by the reduced PGA values obtained, when EFZ as an active seismic source zone is not included.

Key words. Peak ground accelerations, Etne Fault Zone, seismic hazard, geological data, western Norway.

V. Schenk (ed.), Earthquake Hazard and Risk, 169–198.

1. Introduction

In recent years there has been an evolution in the approach toward evaluation of seismic hazards. Deterministic estimates of maximum earthquake size and the associated ground motion that are based on a restricted data base are gradually being replaced by probabilistic assessments of future earthquake potential that incorporate information on the earthquake recurrence intervals, displacement per event, fault slip-rate, fault segmentation, and uncertainties in these parameters. This is occurring in large part, because of the progress that has been made in obtaining and using geologic data to quantify fault behaviour and earthquake process.

Neotectonic studies in intraplate areas have lately focused on tectonic interpretations of seismicity data obtained from regional seismic networks which provides details of the seismicity patterns (e.g. Johnston *et al.*, 1985; Adams and Basham, 1987; Dawers and Seeber, 1991; Schweig *et al.*, 1991; Kulhánek, 1989; Henkel, 1989; Bungum *et al.*, 1991; Karpuz *et al.*, 1991). An improved understanding of fundamental aspects of intraplate seismogenesis is critical for better assessments of seismic hazard.

The present study emphasizes the importance of geological data in seismic hazard analysis in areas of low seismic activity in a global scale (i.e. intraplate areas). In such regions, where direct correlation of surface rupture data with earthquakes are lacking, delineating the active fault zones is critical in defining the potential seismic source zones. In the following, we demonstrate the effective use of different geological and geophysical data in the assessment of reactivated earlier zones of weakness. In doing this, we first explain the general geological setting which provides the basis for the major structural trends in the area. These structural trends, when investigated in detail, reveal the important structures that are repeatedly reactivated throughout different orogenic episodes. Selected structures are further investigated by the latest geological and geomorphological development in Quaternary and Holocene. In this study, we chose one of the specific fault zones (EFZ), and checked with the field observations which we finally correlated with the recent seismicity. As a result, an active seismic source zone is assigned along this fault zone, and its contribution to earthquake hazard estimates is illustrated by the resulting peak ground acceleration (PGA) contour maps, obtained from standard probabilistic seismic hazard analysis.

2. Geological setting

Sunnhordland area is situated within one of the most geologically complex parts of Norway, which has undergone different orogenies since Precambrian. The geology of the area is dominated by an autochthonous Precambrian basement overlain by the Caledonian nappes of the "lower allochthon" (Figure 1). Metasupracrustals and granitic

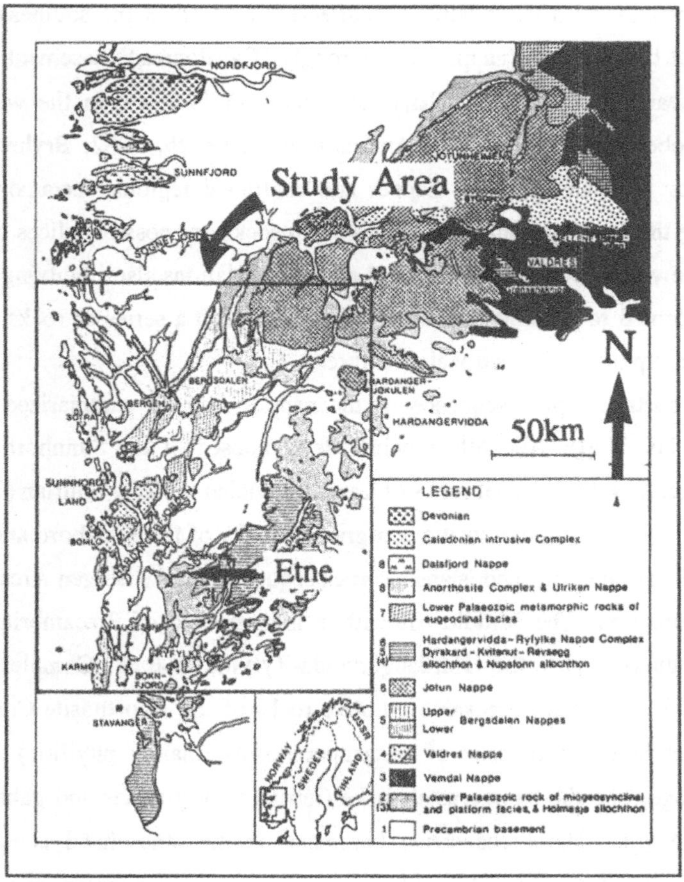

Figure 1. Geological map showing the main tectonostratigraphic units in the Sunnhordland District (modified from Andresen and Færseth, 1982).

gneisses of plutonic origin comprise the basement rocks. The Caledonian nappes comprise rocks of migmatitic Precambrian gneisses and phyllitic metasupracrustals of assumed Cambro-Ordovician age (e.g.Sigmond, 1978; Bryhni and Sturt, 1985). Several sets of dykes are found, ranging from Precambrian acid and mafic (Naterstad and Gabrielsen, 1981; Sigmond *et al.*, 1984) and Palaeozoic to Mesozoic mafic/ultramafic dykes (Færseth *et al.*, 1976; Færseth, 1978).

The Caledonian nappes in southwestern Norway comprise a complete range of orogenic elements from the relatively unaffected foreland in the southeast through a succession of thrust sheets (nappes) into strongly "Caledonized" basement overlain by obducted oceanic floor (Middle Cambrian-Middle Ordovician) in the west (Solli *et al.*,1978; Roberts and Sturt, 1980; Andresen and Færseth, 1982; Bryhni and Sturt, 1985). Major translations of the nappes and continued regional metamorphism took place during the Late Silurian. The Precambrian rocks are possibly slices of reworked basement, now seen as windows. Several nappe translations also involving the already existing deformed rocks of Precambrian origin, have left a series of rocks that exhibit a complex polyphase deformational structures.

The tectonostratigraphic sequence of the nappes may be summarized as follows (Figure 1). The Upper Allochthon, which is composed of the Sunnhordland Nappe Complex, includes supracrustal rocks of Late Ordovician to Early Silurian (Dyvikvågen Group), and gabbroic, granodioritic and granitic rocks of the Sunnhordland Batholith, together with the ophiolitic and island arc rock sequences of the Bergen Arcs (Gullfjellet Ophiolite Complex). The Middle Allochthon is composed of Precambrian basement migmatitic gneisses of the Hardangervidda-Ryfylke Nappe Complex, and the Precambrian basement metamorphic eruptive rocks of the Anorthosite Complex. The Lower Allochthon is composed of metasupracrustals (mainly phyllitic) of Cambro-Ordovician age, and the supracrustal rocks (intruded by granites and gabbros) of the Bergsdalen Nappe. The Precambrian basement on the other hand is composed of Øygarden Gneiss Complex (Precambrian gneisses and Caledonian migmatites and intrusions), Haugesund Basement Unit (weakly foliated granites, metamorphosed meta-volcanics and sediments, metagabbros) and the Sunnhordland Basement Unit (granitic gneisses, porphyric granites, migmatitic gneisses, gabbros).

The geology of the Etne Region is dominated by the nappe sequences (Sunnhordland Nappe Complex) of the Upper Allochthon and the reworked Precambrian basement rocks that outcrop mainly along the Etne river valley.

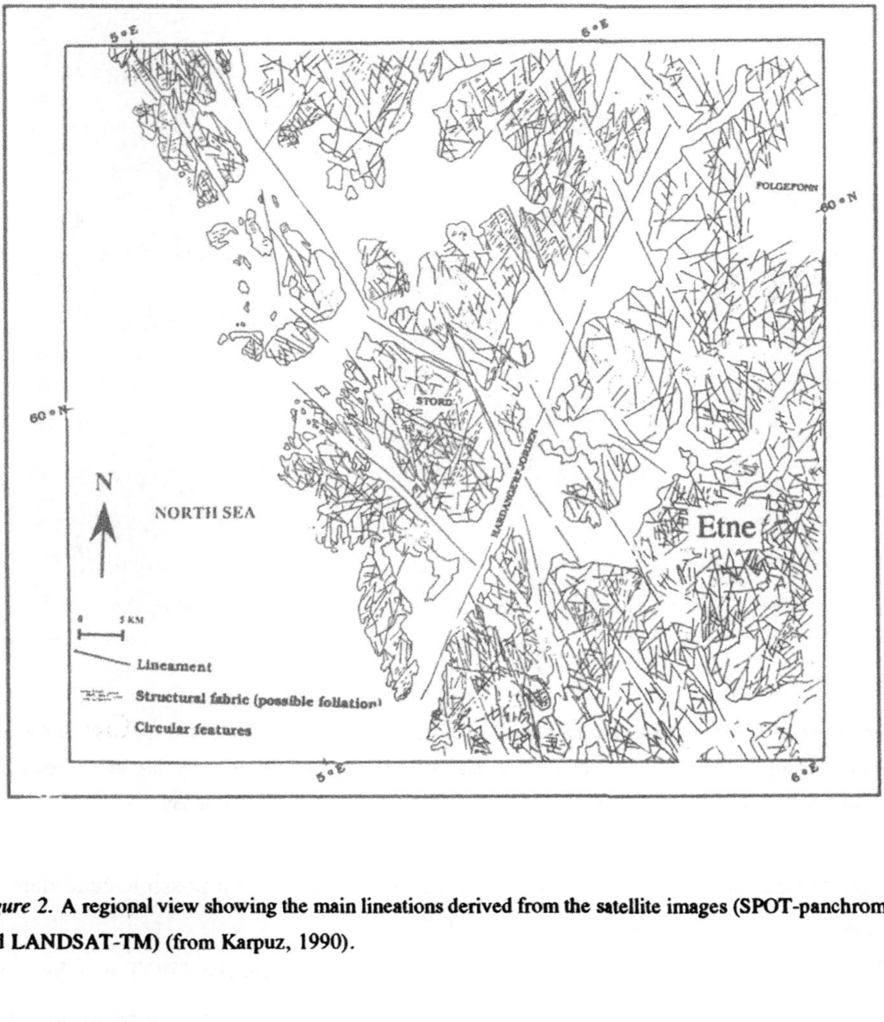

Figure 2. A regional view showing the main lineations derived from the satellite images (SPOT-panchromatic and LANDSAT-TM) (from Karpuz, 1990).

3. Structural trends

The regional lineament and fracture patterns of Southern Norway and the adjacent offshore areas have previously been studied (e.g. Ramberg *et al.*, 1977; Gabrielsen and Ramberg, 1979; Færseth, 1984; Rathore and Hospers, 1984). Recent detailed lineament studies conducted on the satellite images (Karpuz, 1990; Karpuz *et al.*, 1990) have revealed the identification of the possible fault zones in the Sunnhordland District and in

Figure 3. Residual aeromagnetic anomaly map displayed as an image (Karpuz *et al.*, 1991). Note the major anomaly running correlated with the EFZ in the NW-SE direction, marked by the two arrows. The aeromagnetic image was produced and enhanced by the Geological Survey of Norway.

particular one fault zone, Etne Fault Zone (EFZ), seems to be a possible candidate for the neotectonic activity (Karpuz *et al.*, 1991).

The regional interpretation of the lineaments mapped from the SPOT-panchromatic and LANDSAT-TM images, shows a complex pattern of cross-cutting relationships of several trends (Figure 2). The major lineament trends are in the NW-SE, NNW-SSE, NNE-SSW and NE-SW directions. Some of the aeromagnetic anomalies are correlated with the lineament trends, which tend to suggest that these lineament zones are deep -seated structures (Karpuz, 1990) (Figure 3). Detailed lineament mapping based on the SPOT-panchromatic sub-scenes of the Etne area (Figure 4), and on the aerial-photographs, have revealed a specific zone that trends NW-SE. This zone includes a previously known structure, a NW-SE trending fault (Sigmond, 1978), named Etne

Fault Zone (EFZ) by Karpuz *et al.* (1991). The field studies of this zone, which extends approximately 7km laterally, indicate that the lineaments within the zone are mostly parallel or sub-parallel to the NW-SE trend (Figure 5). The parallelism between a NW-SE trending regional aeromagnetic lineament and the EFZ, implies that the fault zone may be related to a regional deep-seated zone of weakness which can be followed

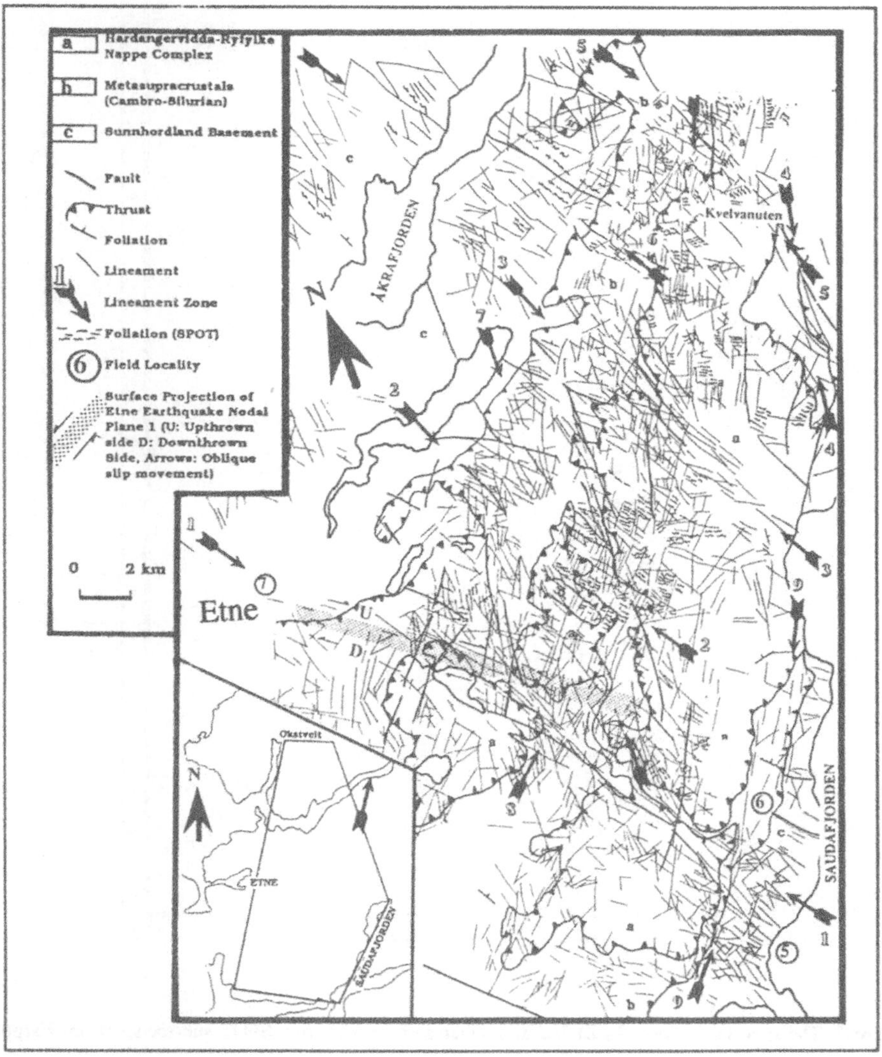

Figure 4. **Detailed lineament map based on the SPOT-panchromatic sub-scenes of the Etne area (reduced from 1:50 000) (from Karpuz, 1990).**

through the Precambrian basement as well as the allochthonous units. This trend may tentatively be associated with subparallel zones of weakness (Naterstad *et al.*, 1973; Gorbatchev, 1985), that are related to the Precambrian units. The zone is intersected

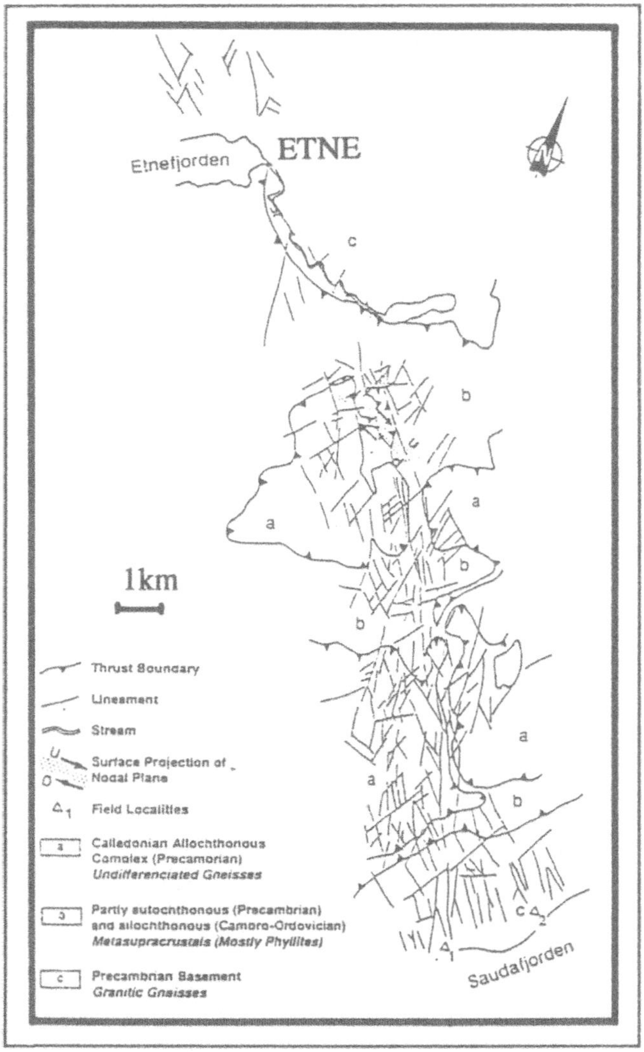

Figure 5. The Etne Fault Zone (EFZ) and its segments as mapped from SPOT sub-scenes (from Karpuz *et al.*, 1991).

by the lineaments in the NE-SW, NNW-SSE and NNE-SSW directions that are probably associated with the major Caledonian or Mesozoic structural trends (Karpuz *et al.*, 1991).

4. Quaternary geology

The present topography is the result of Tertiary uplift of an older mature landscape deeply dissected by fjords and valleys, combined with renewal by erosion during later Tertiary and Quaternary.

The Quaternary geology of the area has previously been investigated in detail by Dahl *et al.* (1989). A geological map showing the Quaternary deposits summarizes the major units in the Etne Region (Figure 6). The area southeast of Etne is dominated by the glaciofluvial deposits that are overlain by the fluvial sediments of the meandering Etne elva and its tributaries. In addition, there are also different tills and glaciolacustrine sequences with some marine deposits that outcrop occasionally. At several locations the basement rocks of metagabbros are seen to outcrop along the river bed.

Geomorphologically, most significant features are the moraine ridges and terraces. The latter is found in three different levels, the highest one being at about 75m, and interpreted to be the remnants after a glaciofluvial delta/sandurdelta. The terraces are suggested to represent the postglacial marine limit in the area, and the deposits are later eroded and redeposited as fluvial sediments by the meandering Etne elva and its tributaries Nordelva and Sørelva (Figure 6). On top of the glaciofluvial terraces, two terminal moraines are mapped in front of both Stordalen and Litledalen, interpreted to be the result of two mainly seperated valley glaciers. In the upstream direction for the terminal moraines, the glaciofluvial terraces are covered by till overlain by ablational moraine, and with glaciolacustrine sediments on top close to the old proximal slope of the delta. In stratigraphic positions below the glaciofluvial deposits, a blue marine clay outcrop occasionally along the rivers. At a locality along the northern side of Nordelva (Figure 6), shells (Mya truncata) found in the blue clay were ^{14}C dated to about 12,000yr B.P. (Anundsen, 1985). This locality is now destroyed. The thickness of the blue marine clay is uncertain, apart from a core from a well close to where the tributaries Sørelva and Nordelva join together, which was registered to be about 20 m.

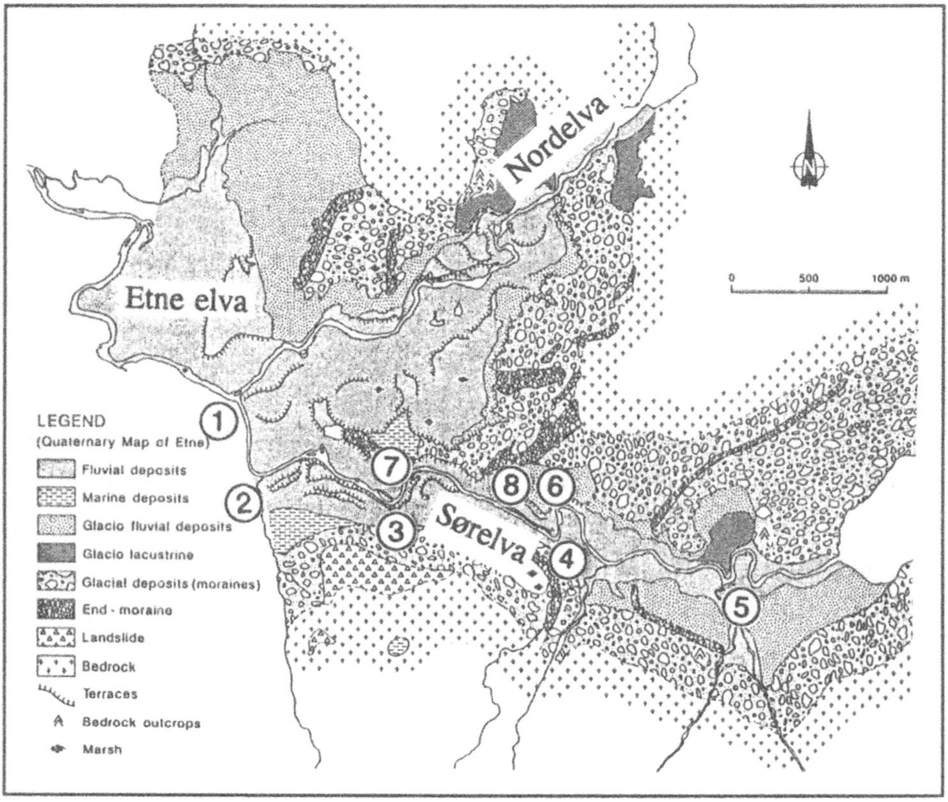

Figure 6. Quaternary geology map (M-1:10 000) of the area southeast of Etne (Dahl *et al.*, 1989). Numbers refer to the sampling localities. Note the drainage pattern anomalies (i.e. straight line segments) on Sørelva and the terraces.

Based on the ^{14}C dating from the blue marine clay, the area was first deglaciated close to 12,000 yr B.P., while the glaciofluvial delta was deposited during the Younger Dryas event (11,000-10,000 yr B.P.). Just before the final deglaciation of the area, the delta was partly overlapped by the valley glaciers.

Except for some few gullies, the glaciofluvial deposits north of the Etne elva are primarily in situ. South of the Etne elva, and especially along the Sørelva, the marine and glaciofluvial deposits are much more irregular due to more landslides. However, the thickness of marine clays is probably not greater in this area than to the north of the Etne elva.

The drainage pattern of the Etne elva and especially of the Sørelva (a southeastern tributary), follows anomalously straight lines that strike in the NW-SE direction, and forms a series of parallel to subparallel segments (Figure 6). These directions are correlated with a lineament zone that runs parallel to the general trend of the valley. These linear drainage anomalies are partly controlled by structural characteristics of metagabbros of the Precambrian basement to the east of the town of Etne. The relative age of these linear drainage anomalies, assuming that they have occurred due to some tectonic activities along the lineament zone parallel to Sørelva, must be less than 12,000 yr B.P., as these glaciofluvial rocks were deposited after the glacier have reached to its marine limit. Further, it may also be speculated, that these drainage anomalies in fact indicate an intense faulting within the same period, which may then be used as an evidence for palaeoseismicity, with several large events ($M>6.0$). Existence of older landslides and the number of terraces along the southern part of Sørelva, assuming that they were tectonically induced, also support this (Figure 6).

5. Field observations

The structural trends along the NW-part of the Etne Fault Zone (EFZ), when studied on a mezoscopic scale, show a close similarity to the general trends obtained from the large scale lineament studies of the aerial-photographs and the satellite images. The lineaments that fall into the area of study are investigated in detail. The eastern lineament is subdivided into three segments that strike parallel to the general trend of the Etne Fault Zone (NW-SE). The same trend is also coincident with the Sørelvavalley, in which the three step-like segments of the river may be correlated with the segments of this lineament. At several locations, especially at the sharp bends along the erosive side of the meandering river, the bedrock (metagabbros) is outcropped (Locations 4,5,6,7,and 8; Figure 6). There are two main fracture sets on the gabbroic rocks. The dominant orientation is the one which is more or less parallel to the NNW-SSE trend (168^0-170^0), whereas the conjugate direction is between 050^0-070^0. This dominant set corresponds well with the direction of the river at Locality 4 (Figure 6). At Locality 5 (Figure 6), the fracture pattern is more complex, with well developed lens structures. The general trend of the orientation of the x-axes of the lenses is approximately 050^0.

The gabbroic rocks are cut by pegmatitic dykes at several locations (e.g.Locality 8; Figure 6), which mainly lie parallel to the river segment (118^0-122^0). This may correspond to an earlier fracture pattern in which the pegmatitic dykes have been developed. The later fractures cut both through the gabbroic rocks and the dykes. The main direction is 164^0, and the conjugate direction is 056^0. The thicknesses of these dykes vary from max.50cm to min.1cm. It is noted that the orientation of the dykes are nearly perpendicular relative to each other.

The western, a NNW-SSE trending lineament (Figure 6; Locations 1 and 2), is in the quartz-mica schists. There is no systematic fracturing observed along the lineament, apart from a marked elevation difference (ca. 50m) across the small stream that runs along the lineament (Locality 2; Figure 6). The same trend (NNW-SSE) is also followed along the last segment of Sørelva, before it joins the Etne elva. Fracturing is observed within the quartz veins that develop along the earlier cracks that follow the foliation (Locality 1; Figure 6; 154^0). There are other fractures that are filled with quartz veins which cut across the foliation (Locality 2; Figure 6; 178^0). This is probably due to the fact that on more ductile areas the quartz veins follow the foliation, whereas on the brittle areas the fractures may cut across the foliation and hence the quartz veins developed are observed accordingly.

Fracturing on a microscopic scale is also observed on the thin-sections of the gabbroic rocks and the pegmatitic dykes mentioned above. Fracturing is less frequent on the gabbro compared to the pegmatites, but instead there are zones of microcrystalline grains that may define the earlier fractures. On the thin-sections from the granitic dykes, fracturing is more obvious and tends to follow the latest trend (160^0-170^0), which was observed in the field. This latest fracturing is concentrated along the zones parallel to the orientation of the dykes and therefore may indicate the reactivation of earlier fractures. Cross cutting relationships of the microfractures in the thin-sections from the mica-schists at Locality 1 (Figure 6), indicate displacements in the order of a milimeter. Microfractures developed in the pegmatitic dykes show signs of lense structures similar to those observed in the field (Locality 5; Figure 6).

6. Seismicity

The seismic activity in Scandinavia and the adjacent areas is related to intraplate seismicity which is relatively low compared to that of the active plate margins (Husebye *et al.*, 1978; Zoback *et al.*, 1989; Gregersen *et al.*, 1991; Bungum *et al.*, 1991). In Scandinavia the largest seismic activity is found along Western Norway (Figure 7) (Havskov *et al.*,1989; Kvamme and Hansen, 1989; Engell-Sørensen and Havskov, 1987; Lindholm and Marrow, 1990). Several mechanisms for this activity have been

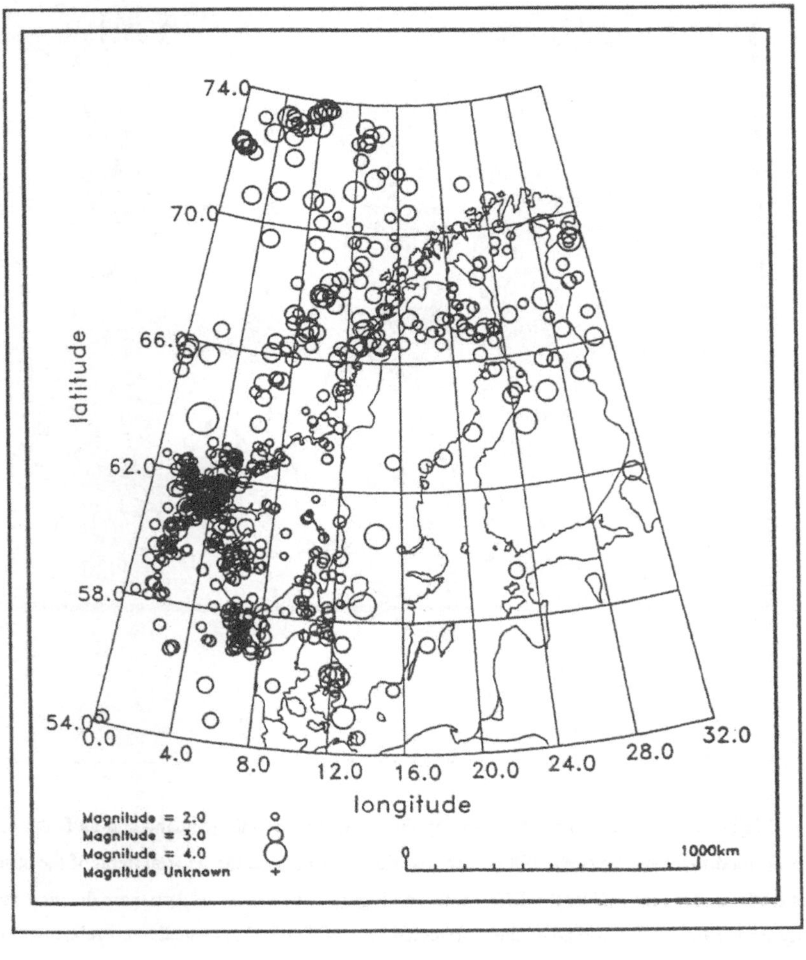

Figure 7. Epicenter distribution in Scandinavia and the adjacent areas within the period 1980-1990. Note that the highest concentration of earthquake epicenters are found in the western part of Norway.

proposed. Among these, some large-scale tectonic forces, such as the "ridge-push" effects from spreading along the Mid-Atlantic margin (Stephansson *et al.*, 1986; Slunga, 1989; Clauss *et al.*, 1989), local tectonic stresses such as the postglacial uplift (Lundquist and Lagerbäck, 1976; Muir Wood, 1989; Anundsen, 1985; 1989), and

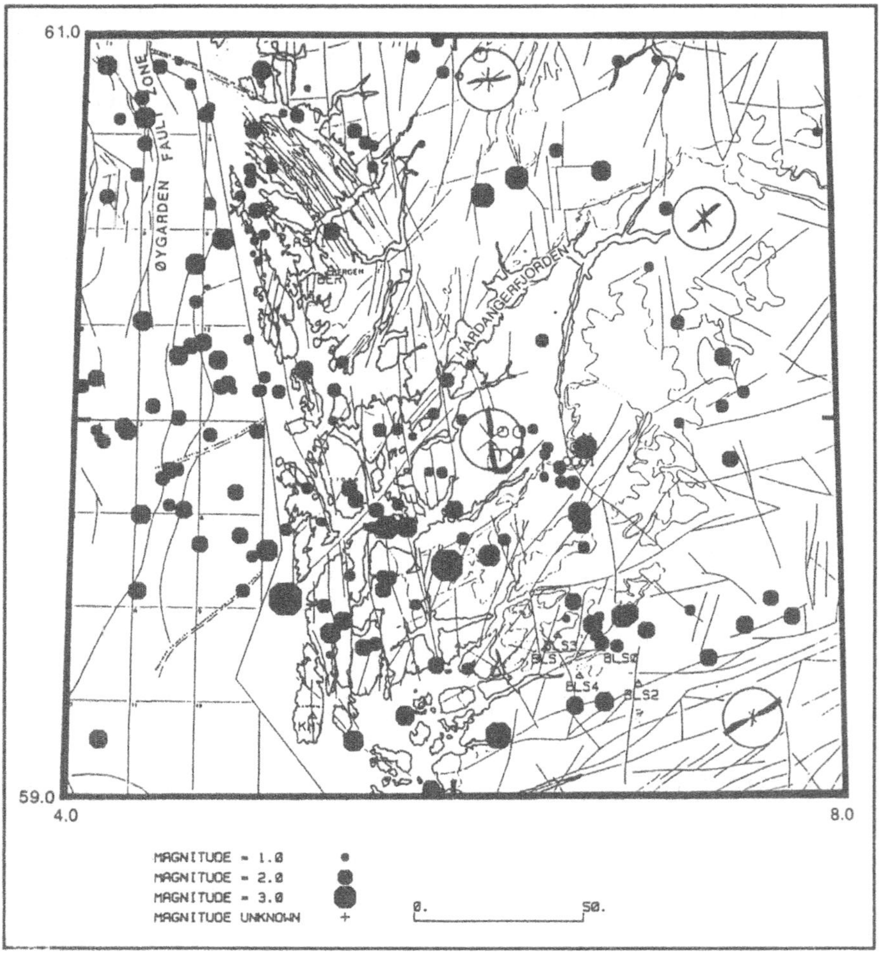

Figure 8. **Major tectonic features as expressed by the lineaments on the mainland and as faults on the off-shore areas (modified from Færseth, 1984). The rose diagrams indicate the general trend of the structures in different sections of the map (Note the differences with Figure 2, which is based on more detailed study). The seismicity from 1980-1990 recorded by the Norwegian Seismic Network, is added to correlate with the major structures.**

lithospheric loading and unloading due to sedimentation, especially since Tertiary (Hamar and Hjelle, 1984; Stein *et al.*, 1989), are regarded as important. Other possible forces are the membrane stresses resulting from the changes in the principal radii of curvature when the assumed rigid plates move over the surface of the imperfectly spherical earth, and remnant stresses in the lithosphere due to previous orogenic cycles (Husebye *et al.*,1978; Bungum and Fyen,1979).

Sunnhordland district includes the area with one of the highest earthquake occurrence rates in the Norwegian mainland (Figure 7). Recently there has been few relatively large or moderate earthquakes in the area (e.g. March 8, 1983 M_L=4.7 Stord earthquake; March 17, 1988 M_L=3.3 Hardangerfjord earthquake; Oct.20, 1988 M_L=3.5 Folgefonn earthquake; Jan.29, 1989 M_L=4.4 Etne earthquake; Feb.25, 1990 M_L=3.0). None of the larger earthquakes had any recorded aftershock sequence, apart from the Stord event, which was followed by six aftershocks within the next 24 hours after the main shock at 18.43 (GMT) on March 8, 1983.

There are six focal mechanisms for the major events that have previously been determined (Bungum *et al.*, 1991). The majority of these solutions indicate normal faulting with the principal axes of compression lying approximately in ENE-WSW direction. Normal faulting and this ENE-WSW trend of the principal stress axes has previously been interpreted as being related to the "ridge-push" from the plate movements and to the post-glacial uplift and lithospheric loading effects (Bungum *et al.*, 1991). Results from the Scandinavian Rock Stress Data Base (Stephansson *et al.*, 1986; 1991), also support this trend. Apart from the Stord event (March 8, 1983; M_L=4.7), and the Hardangerfjord event (March 17, 1988; M_L=3.3) which show reverse oblique faulting, all four events indicate normal oblique faulting. This anomaly in the Stord and Hardangerfjord events may be related to their geographical location, being closer to the off-shore structures. The focal mechanism of the off-shore events mostly indicate reverse faulting which may imply that the geological structures in the off-shore Western Norway are controlled by the plate movements (i.e. ridge-push effect). The boundary between these two different stress regimes may be speculated to correspond to a major fault zone running through the Hardangerfjord (Figure 1). This fault zone (Faltuns-graben) (Goldschmidt, 1912), has previously been investigated by using the deep-seismic reflection data, which provided evidence for mainly northwest dipping reflectors that indicate several episodes of reactivation during different geological times (Hurich *et al.*, 1989).

In the following interpretations, the digital database available from 1980 until 1991 of the Norwegian Seismic Network, operated by the Institute of Solid Earth Physics,

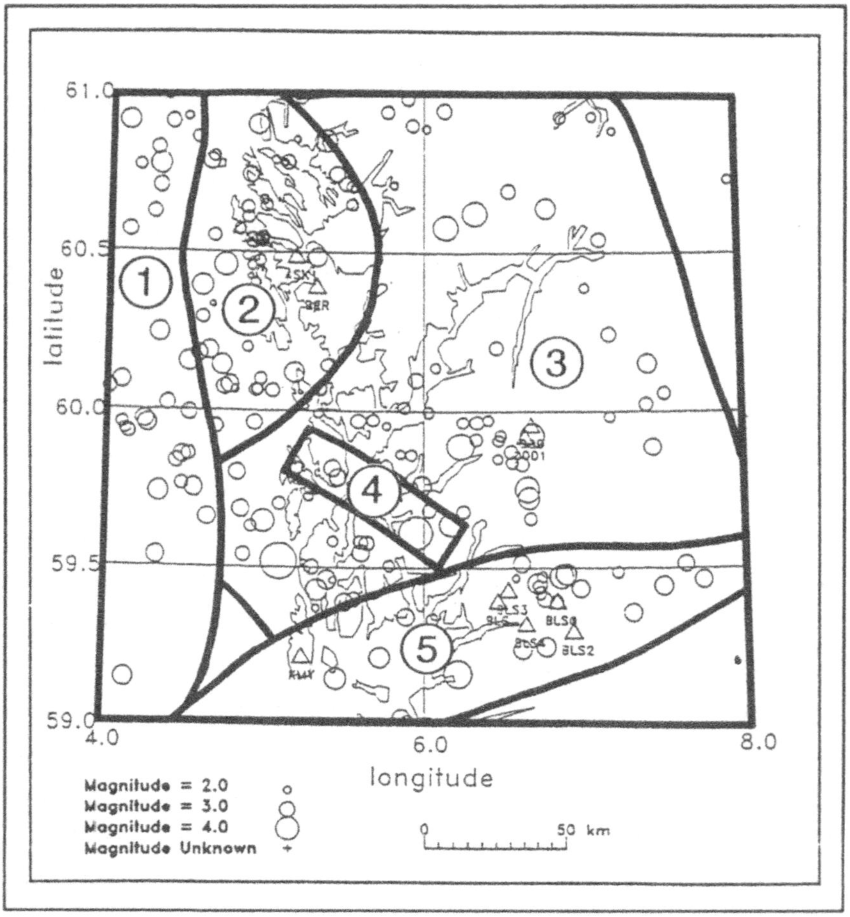

Figure 9. Five earthquake source zones defined, based on the epicenter clusters and the major geological structures that are related to these. Zone 1: Øygarden Fault Zone (ØFZ)/off-shore; Zone 2: Bergen Arcs(BA); Zone 3: Caledonian Terranes (CT); Zone 4: Etne Fault Zone (EFZ); Zone 5: Caledonian-Precambrian Transition Zone (C-PTZ). Seismicity from 1980-1990 recorded by the Norwegian Seismic Network, is added to correlate with the source zones. Triangles indicate the permanent seismograph stations.

University of Bergen, is used. Additional data from the neighbouring countries are also used. Magnitudes for the 1980-1984 are the original M_L as given by NORSAR, and the magnitudes from 1985-1990 are Bergen M_C. In addition, the historical events are also included (from the period 1883-1980) in the seismic hazard calculations. These are compiled basically from the Muir Wood catalogue from the ELOCS report (Muir Wood and Woo, 1987), which is considered reasonably complete for the $M>4$ events from 1883.

7. Seismic hazard analysis

The seismic hazard analysis performed for the Sunnhordland District is based on the software developed by Ordaz (1990). The whole area is divided into five source zones according to the earthquake occurence, local gelogy, and the major tectonic features (Figure 9). The westernmost of the epicenters in the off-shore areas are probably related to the N-S running Øygarden Fault Zone (Province 1). The second zone which is adjacent to the first one, seems to be defined by a concentration of earthquake epicenters which is related to the Bergen Arcs (Province 2), a geologically dominant arcuate structure around the Bergen area. The third zone, although not directly related to any clustering, may be defined roughly by taking the area restricted mostly to the Caledonian structures (Province 3). Within this zone there seems to be a cluster of epicenters around the Etne Region, which lies in a NW-SE direction that is regarded as the fourth zone (Province 4). As discussed in the earlier sections, further support for this source zone comes from the remote sensing and field geological data interpretations. The last (fifth) zone is an elongated area, lying in the ENE-WSW direction, which may be related to the transition zone between the Caledonian structural trends in the north and the Precambrian structural trends in the south (Province 5), (Figure 9). The resonable match between major lineaments, as compiled from Færseth (1984) (original lineament study is by Gabrielsen and Ramberg, 1981), and the present-day seismicity (1980-1990), seem to support this zonation (Figure 8). Here, it should be noted that the lineament map in Figure 2, shows some differences especially around the Etne Region, when compared to the one shown in Figure 8. This is due to the differences in the resolution of the satellite images used for the two maps. However, the

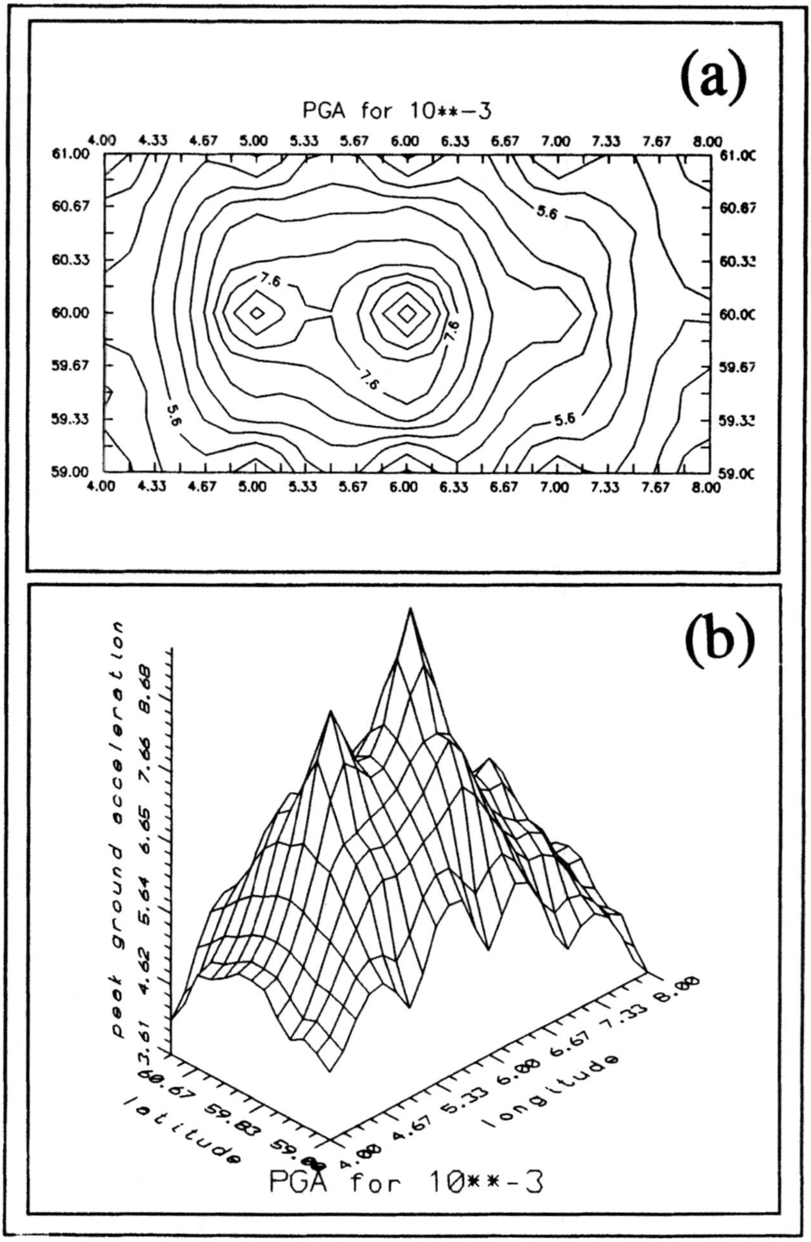

Figure 10. Contour maps showing peak ground acceleration (PGA) *(a)* for a recurrence period of 1000 years in the Sunnhordland District. Note the high values that correspond to the Etne Region, *(b)* 3-D presentation of the peak ground acceleration for a recurrence period of 1000 years in the Sunnhordland District. Note the high values that correspond to the Etne Region.

regional lineament zones in Figure 8, that are correlated with the present-day seismicity, are still useful in defining the tectonic provinces.

The magnitude - frequency-of-occurrence relationship for the Sunnhordland District for the period 1980-1989, was previously determined by Engell-Sørensen *et al.* (1989), where the best fitting curve for the interval M_L=3.5 to 4.2 gives the equation $\log N = 3.4 - 0.55M_L$, (logN is the common logarithm of the number of events with magnitudes larger than or equal to M_L, the local magnitude). The magnitude - frequency-of-occurrence relationships are also determined separately for each source zone and the following b-values are found for the period 1980-1990; 0.43, 0.64, 0.59, 0.43 and 0.47, for Provinces 1,2,3,4 and 5, respectively.

The details of the attenuation of the ground motion in Norway are discussed by Singh *et al.* (1990), in which they use an earthquake source model and results from Random Vibrations Theory (RVT) to estimate attenuation of ground motion parameters as a function of moment magnitude, M_W, and hypocentral distance, R (see also Hanks and McGuire, 1981; Boore, 1983; Boore and Atkinson, 1987; Toro and McGuire, 1987; Ordaz *et al.*, 1988; Singh *et al.*, 1989). Although a more recent attenuation relation is developed by Dahle *et al.*(1991) (also H.Bungum, pers.comm.), where the anelastic attenuation was modelled by the equation $Q(f) = 468 * f^{0.63}$ (Kvamme *et al.*, in press), for our purposes of illustrating the relative contribution of Province 4 (Zone 4), we use the attenuation relation given in Singh *et al.*(1990). For each zone the attenuation relations and the crustal structure are assumed to be uniform.

The maximum observed magnitudes are 4.1,4.6,4.9,4.9 and 4.1 for the Provinces 1-5, respectively. The maximum expected magnitude is set to 6.0 for all provinces, and the number of events above the treshold magnitude of 3.5, are determined from a catalogue time span of 107 years (1883-1990). The specific b-values are assigned from the instrumental catalogue for the period 1980-1990 (see also the previous section). Seismic hazard calculations are performed for 100, 1000 and 10,000 years recurrence periods, at a grid interval of 0.5^0 latitude and 1^0 longitude, with 200km maximum distance for integration at 0.1 magnitude increments.

The maximum peak ground accelerations (PGA) for the Sunnhordland District, are found to be roughly 1.8, 9.5 and 50.5 gals, for 100, 1000 and 10,000 years recurrence periods respectively. The distribution of the peak ground accelerations throughout the area is presented by contour diagrams. Obviously these absolute PGA values are

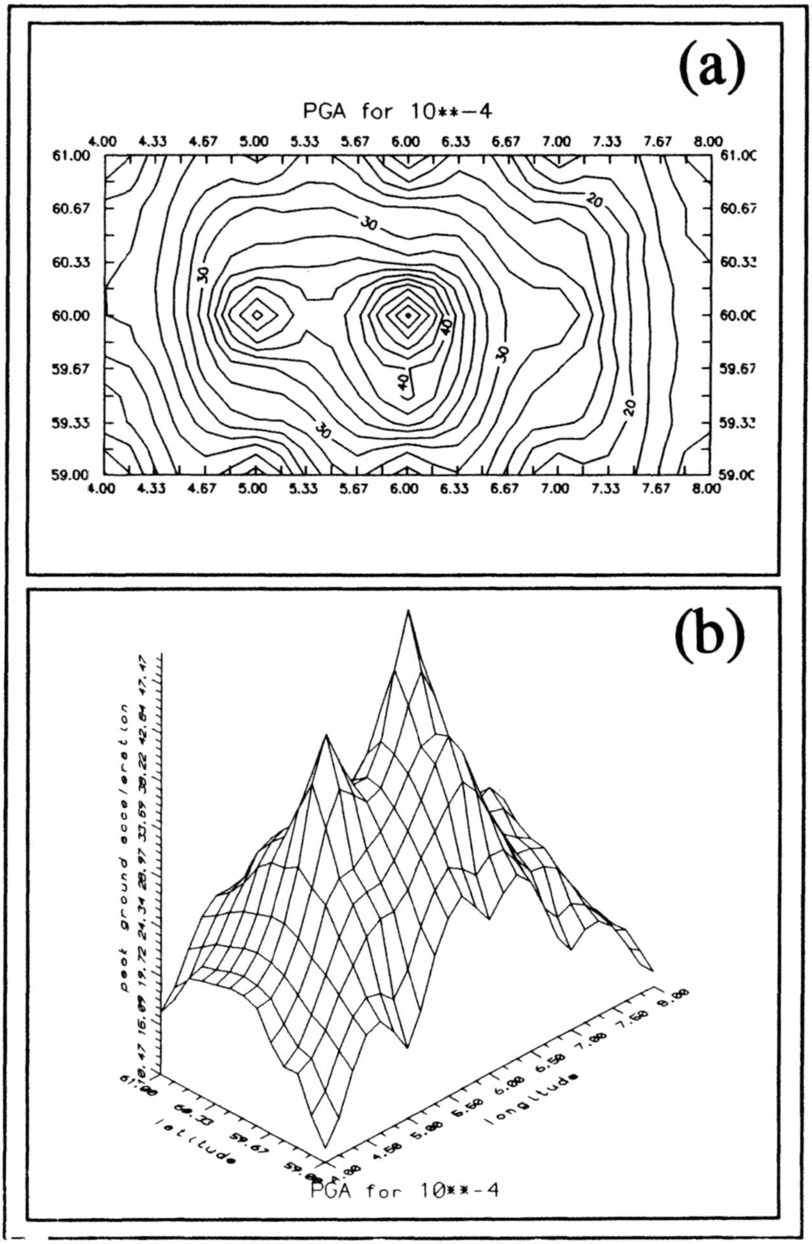

Figure 11. ***(a)*** **Contours showing peak ground acceleration for a recurrence period of 10,000 years in the Sunnhordland District. Note the high values that correspond the Etne Region.** ***(b)*** **3-D presentation of the peak ground acceleration for a recurrence period of 10,000 years in the Sunnhordland District. Note the high values that correspond the Etne Region (see text for discussion).**

dependent on several factors such as the attenuation relations, magnitude ocurrence relations etc. (Atakan *et al.*,1991). In this exercise, we are not concerned with the absolute values, as our purpose is mainly to illustrate the contribution of an additional seismic source zone (Zone 4), based on the detailed geological information. In this context, the relative values of PGA (i.e. the distribution of PGA within the study area at different recurrence intervals), are considered far more important.

In order to illustrate the contribution of the EFZ in the seismic hazard analysis program, Province 4 (EFZ), which was assigned based on the geological observations, is removed from the data and the events found within this zone are used in Province 3 instead. The resulting PGA contour maps and the 3-D presentations for 100 years recurrence period, show slightly higher peaks around the Etne Region compared to the results with Province 4, included. On the other hand, for the 1000 and 10,000 years recurrence periods (Figures 10 a,b and 11 a,b respectively), the high values of PGA around Etne Region seem to be reduced markedly (see Figures 12 a,b and 13 a,b, for comparison). This observation tend to suggest that the influence of Province 4 (and hence the gelogical observations related to the EFZ) is higher, at higher recurrence periods compared to the 100 years recurrence period.

8. Discussion

In intraplate areas of low seismicity, where one to one correlation of surface ruptures to earthquake records are lacking or not observed very frequently, identification and analysis of master fracture systems to identify possible active faults, become essential parts of the seismic hazard assessments. Usually, the instrumental catalogues cover only the last 40-50 years, whereas the earlier macroseismic information extends a few hundreds of years at the most. The only possible source of information about earlier events, is therefore dependent upon the detailed knowledge of the local geology of the area. In this context, any possible information about the existence of a large ($M_L > 6.0$) earthquake based on the local geology, is quite valuable in estimating the true seismic hazard potential of the area of interest.

The seismic hazard potential of the Sunnhordland District is evaluated by using the different data sets that help to determine the geological characteristics of the area. Major

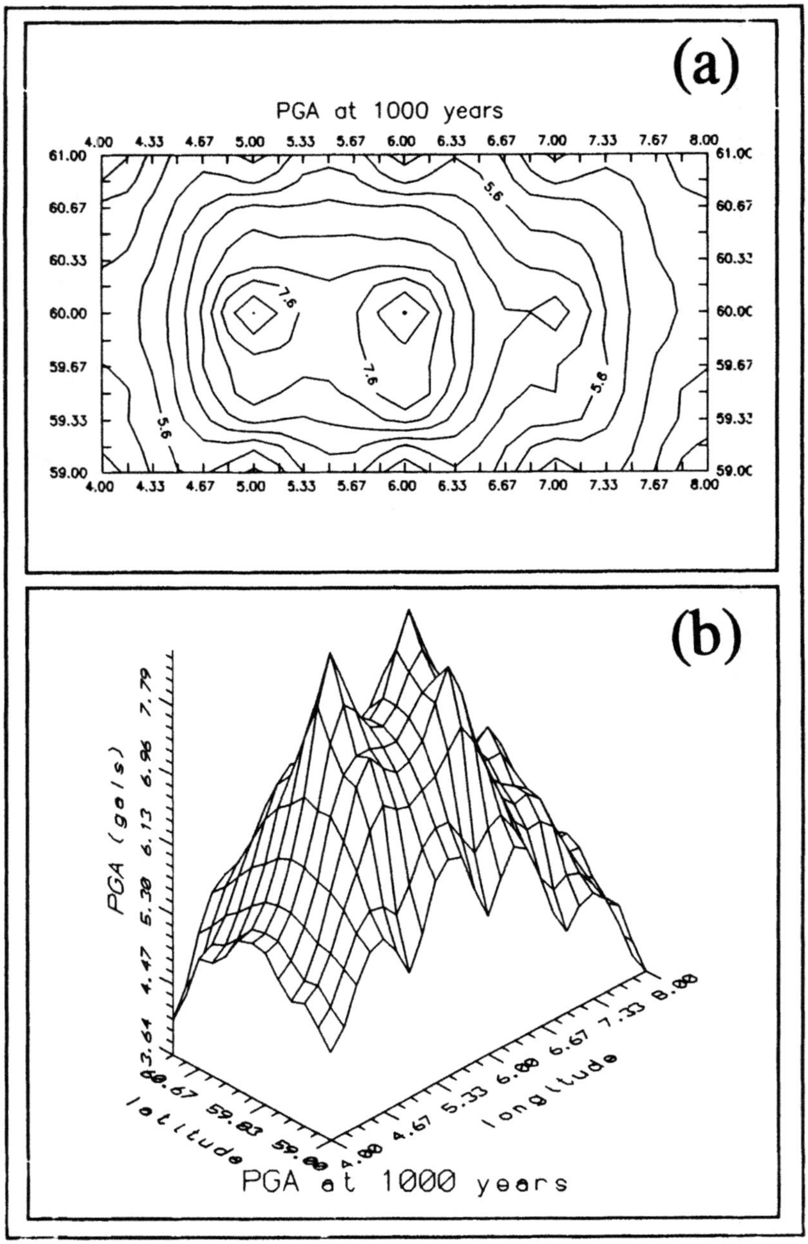

Figure 12. Contours showing peak ground acceleration (PGA) (Province 4, is removed from the data set in order to illustrate the effect of geology (see text for discussion)), (see Figure 10 for comparison), *(a)* PGA for a recurrence period of 1000 years, *(b)* 3-D presentation of PGA for a recurrence period of 1000 years. Note that the high peak that was concentrated around the Etne Region is somehow reduced (see text for discussion).

tectonic lineaments were located by using remote sensing data, and field controll was established. Correlating these zones to the present-day seismicity, gave rise to the idendification of potentially active fault zones in the area and one fault zone, the Etne Fault Zone, seems to be the possible candidate for the neotectonic activities. Geomorphological and geological observations that are interpreted as being the potential indicators of neotectonic activities in the Etne region, may be summarized as follows;

1. The major lineament zones are identified from the satellite images and aerial-photographs, and a lineament zone passing through the Etne Region in a NW-SE direction (Etne Fault Zone), is found to be the posibble candidate for the neotectonic activities.
2. A pervasive aeromagnetic anomaly is found to be correlating well with the general trend of the lineament zone along the Etne Region (Karpuz *et al.*, 1991).
3. These potential zones of weakness, when correlated with the present-day seismicty, have revealed an epicenter cluster of some moderate events ($M_L > 3.5$) along the same trend.
4. The investigation of these lineament zones have indicated that the linear drainage anomalies that were observed along the Sørelva, correlate well with the general trend of the lineament zone.
5. Terraces and landslides occurring frequently along the southern part of the Sørelva, further support the existence of neotectonic activities.
6. The lower age limit of these activities, is constrained by the ^{14}C dating of the blue marine clay, which is overlained by the glacio-fluvial deposits where the inferred neotectonic activities have occurred.
7. Field observations of the orientation of joints and fractures, as well as the microstructures in thin sections, showed similarities to the general trend of the lineament zone.
8. Additional information from marine shallow seismic profiles (Bøe and Hovland, 1990; Bøe *et al.*, 1992), geodetic measurements (Anundsen, 1985; 1989) and fault plane solutions (Bungum *et al.*, 1991), available from other studies, support the existence of neotectonic activities in the area.

The use of the above information empirically in the seismic hazard calculations poses

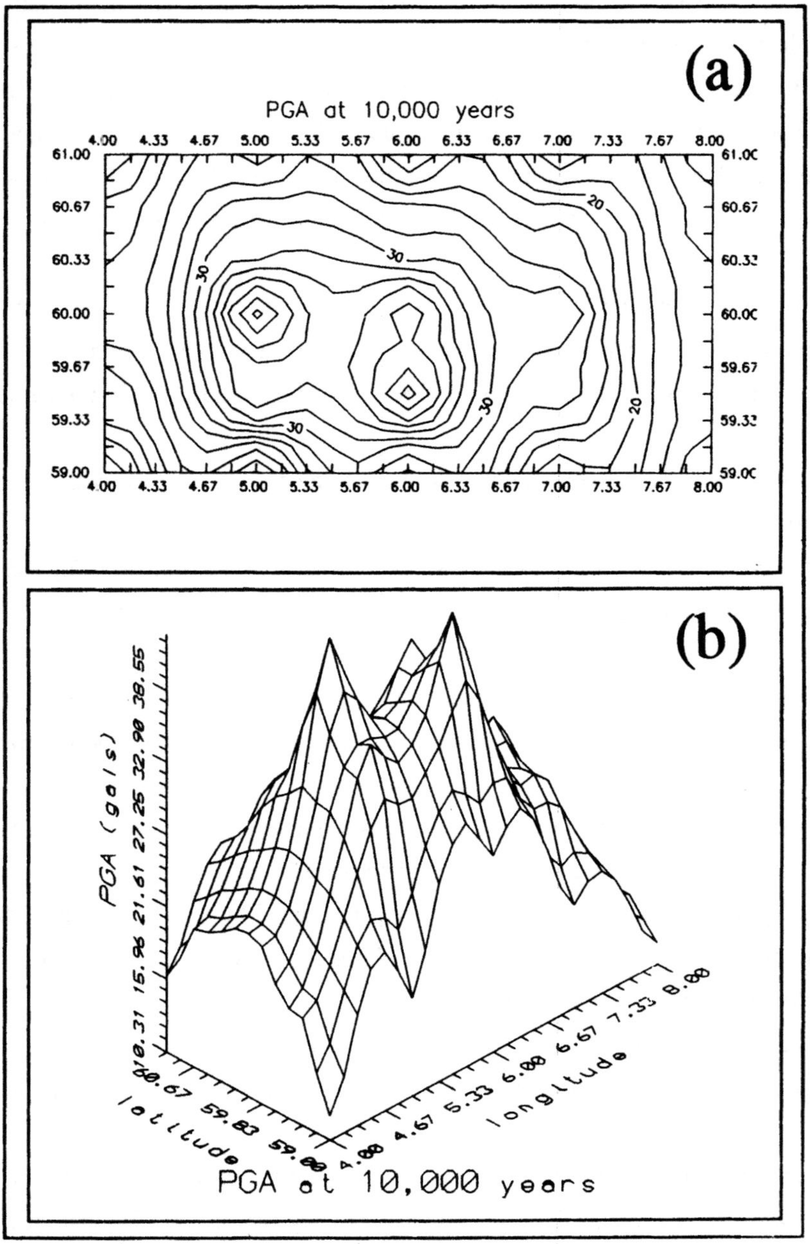

Figure 13. Contours showing peak ground acceleration (PGA) (Province 4, is removed from the data set in order to illustrate the effect of geology (see text for discussion)), (see Figure 11 for comparison), *(a)* PGA for a recurrence period of 10,000 years in the Sunnhordland District, *(b)* 3-D presentation of the PGA for a recurrence period of 10,000 years.

problems, as it is not possible to quantify the number and the size of the earthquakes that may be related to these local geological observations. Although, it is not directly used in the seismic hazard calculations, the existence of large earthquakes ($M_L > 6.0$) within the last 10,000 years along the Etne Fault Zone in the Sunnhordland District, may be inferred as suggested previously in section 4. Unfortunately, neither on the outcrops, nor on the open-pit which shows a whole section (ca. 10m), of glaciofluvial deposits, was it possible to observe any direct evidence (e.g. syn-sedimentary tectonic deformation, faulting or liquifaction) of palaeoseismic activity, which could have been measured and quantified. Therefore, assigning a higher level of activity rate to the EFZ, does not seem to be possible, and this information can only be used indirectly as an indication of relatively higher palaeoseismic activity in the Etne Region compared to other parts of the Sunnhordland District, without actually quantifying it. Nevertheless, this information is important in identifying possible zones that are tectonically active, and later, in designating the tectonic provinces described earlier (Provinces 1-5). The comparison of the contour maps (and the relevant 3-D presentations) for 1000 and 10,000 years recurrence periods, with and without the Province 4 (EFZ), clearly indicates this (Figures 10,11 and 12,13).

Although these PGA values do not directly indicate a very high seismic hazard for the Sunnhordland District, they certainly make it possible to discriminate the areas with higher seismic hazard potential. The Etne Region seems to be a potential seismic hazard zone within the Sunnhordland District. The reasonable match between the areas of maximum peak ground acceleration and the proposed active fault zone (Etne Fault Zone) in the Etne Region, suggests that it is possible to evaluate the seismic hazard potential of a specific area, even if the seismicity seems to be fairly low. Furthermore, Sunnhordland represents seismically one of the most active areas in Norway, with several moderate events ($M_L > 4.0$), occurring within the last 10 years. One of these events (Etne earthquake of Jan 29, 1989; $M_L = 4.2$) have triggered a small landslide, that resulted in surface ruptures in the Quaternary sediments in the northern part of Etnefjorden, (Engell-Sørensen *et al.*, 1989). Slope instabilities, that may be triggered by intermediate size earthquakes, especially when associated with heavy rainfall, make the area vulnerable to future events. In this respect, future studies concentrating on the systematic investigation of the slope instabilities related to earthquakes, as well as the local site response, would help estimating the true seismic hazard potential of this area.

Acknowledgements

The seismicity data used in this study was provided by the Western Norway Network which is operated by the Institute of Solid Earth Physics, University of Bergen and sponsored by the oil companies Norsk Hydro A/S, A/S Norske Shell and Statoil. Seismic hazard analysis program (CRISIS) was kindly provided by M.Ordaz. Authors are grateful to J.Havskov for his valuable comments on the manuscript. We also thank K.Anundsen, H.Bungum, and R.H.Gabrielsen for having read the earlier version of the manuscript and for their critical reviews. J.K. Ellingsen, from the University of Bergen, is thanked for her help with the illustrations.

References

Adams,J., and Basham,P. (1987) Seismicity, crustal stresses and seismo-tectonics of eastern Canada, in *Proceedings from the Symposium on Seismic Hazards, Ground Motions, Soil Liquifaction and Engineering Practice in North America.* K.Jacob (ed.), Natl. Cent. Earthq. Eng. Res. Tech. Rep., **NCEER-87-0025**, pp.300-317.

Andresen,A. and Færseth,R. (1982) An evolutionary model for the southwest Norwegian Caledonides. *American Journal of Science*, **282**, 756-782.

Anundsen,K. (1985) Changes in the shore-level and ice-front position in Late Weichsel and Holocene, southern Norway. *Norges Geogr, Tidsskr.*, **39**, 205-225.

Anundsen,K. (1989) Late Weichselian relative sea levels in southeast Norway: Observed strandline tilts and neotectonic activity. *Norges Geolog. Tidsskr.*, **111**, 288-292.

Atakan,K., Karpuz,M.R., Dahl,S.O. and Gabrielsen,R.H. (1991) Seismic hazard potential of the Etne region, western Norway. Seismol.Obsv., University of Bergen, *Seismo-Series* **55**, 69p.

Boore,D.M. (1983) Stochastic simulation of high-frequency ground motions based on seismological models of the radiated spectra. *Bull. Seismol. Soc. Am.*, **73**, 1865-1884.

Boore,D.M. and Atkinson,G.M. (1987) Stochastic prediction of ground motion and spectral response parameters at hard-rock sites in Eastern North America. *Bull. Seismol. Soc. Am.*, **77**, 440-467.

Bryhni,I. and Sturt,B.A. (1985) Caledonides of Southwestern Norway, in *The Caledonide Orogen-Scandinavia and Related Areas.* D.G.Gee and B.A.Sturt (eds.), John Willey, pp.89-107.

Bungum, H. and Fyen, J. (1979) Hypocentral distribution, focal mechanisms and the tectonic implications of Fennoscandian earthquakes, 1954-1978. *Geologiska Foreningens i Stockholm Forhandlinger*, **101**, 261-271.

Bungum,H., Alsaker,A.,Kvamme,L.B., and Hansen,R.A. (1991) Seismicity and Seismotectonics of Norway and Nearby Continental Shelf Areas. *Journal of Geophysical Research*, **96**, 2249-2265.

Bøe,R. and Hovland,M. (1990) Neotectonism and sediment instability along the Norwegian coast south of Karmøy. (Abstract), in *Post-Cretaceous Uplift and Sedimentation along the Western Fennoscandian Shield.* NGF's Tectonic and Structural Geology Study Group, TSGS 7th Annual Meeting, Stavanger.

Bøe,R., Sørensen,S. and Hovland,M. (1992). The Karmsundet Basin, SW-Norway: Stratigraphy, structure and neotectonic activity. *Norsk Geologisk Tidsskrift*, **72**, 281-283.

Clauss,B., Fuchs,K., and Marquart,G. (1989) Stress orientations in the North Sea and Fennoscandia, a comparison to the Central Europian stress field, in *Earthquakes at North Atlantic Passive Margins: Neo tectonics and Postglacial Rebound.* S.Gregersen and P.W.Basam (eds.), Kluwer Academic Publ., Dordrecht, pp. 277-287.

Dahl,S.O. (with the participants in the G-106 field course in Quaternary geology) (1989) Quaternary geology map of Etne, M-1:10 000 (unpbl.) Geological Institute, University of Bergen, Norway.

Dahle,A., Bungum,H., and Kvamme,L.B. (1991) Empirically derived PSV spectral attenuation models for intraplate conditions. *European Earthq. Eng.*, **3**, 42-52.

Dawers,N.H., and Seeber,L. (1991) Intraplate faults revealed in crystalline bedrock in the 1983 Goodnow and 1985 Ardsley epicentral areas, New York, in *Intraplate Deformation, Neotectonics, Seismicity, and State of Stress in North America.* J.-C.Mareschal (ed.), Tectonophysics, 115-131.

Engell-Sørensen,L., and Havskov,J. (1987) Recent North Sea seismicity studies. *Phys. Earth Planet. Inter.*, **45**, 37-44.

Engell-Sørensen,L., Atakan,K. and Aranda,C. (1989) Fracturing in the initial stages of a possible landslide triggered by the earthquake of January 29, 1989, at a locality in the Etne Region, Western Norway. Seismological Observatory, Institute of Solid Earth Physics, University of Bergen, Seismo-Series No.**29**, 45p.

Færseth,R.B. 1978. Mantle derived lherzolite xenoliths and megachryst from Permo-Triassic dykes, Sunnhordland, Western Norway. *Lithos*, **11**, 23-25.

Færseth,R.B. (1984) Tectonic map of the northeast Atlantic. Norsk Hydro a.s., 1984.

Færseth,R.B., McIntyre,R.M. and Naterstad,J. (1976) Mesozoic alkaline dykes in the Sunnhordland region, Western Norway: Ages, geochemistry and regional significance. *Lithos*, **9**, pp.332-345.

Gabrielsen,R.H., and Ramberg,I.B. (1979) Fracture patterns in Norway from LANDSAT imagery: Results and potential use, in *Proceedings of Norwegian Sea Symposium,* Tromsø, 1979, Norwegian Petroleum Society, NSS/**23**, pp.1-28.

Gabrielsen,R.H. and Ramberg,I.B. (1981) Regional geological, tectonic and geophysical features of Nordland, Norway. *Earth Evolution Sciences* 1/1981, 14-26.

Gabrielsen,R.H. (1984) Long lived fault zones and their influence on the tectonic development of the southwestern Barents Sea. *J. Geol. Soc. London*, **141**, Part 4, 651-662.

Gabrielsen,R.H. (1989) Reactivation of faults on the Norwegian Continental Shelf and its implications for earthquake occurrence, in *Earthquakes at North Atlantic Passive Margins: Neotectonics and Postglacial Rebound.* S.Gregersen and P.W.Basham (eds.), Kluwer Academic Publ., Dordrecht, pp.67-90.

Goldschmidt,V.M. (1912) Die kaledoniske deformation der sudnorwegischen urgebirgtafel. *Skr. Vidensk. Selsk.*, Christiania, **19**, 1-11.

Gorbatchev,R. (1985) Precambrian basement of the Scandinavian Caledonides, in *The Caledonide Orogen-Scandinavia and Related Areas.* D.G.Gee and B.A.Sturt (eds.), John Willey, pp.197-212.

Gregersen,S., Korhonen,H. and Husebye,E. (1991) Fennoscandiandynamics: Present-day earthquake activity, in *Imaging and Understanding the Lithosphere of Scandinavia and Iceland.* S.Björnsson, S.Gregersen, E.S.Husebye, H.Korhonen and C.-E.Lund (Eds.), Tectonophysics, **189**, pp.333-344.

Hamar,G.P., and Hjelle,K. (1984) Tectonic framework of the Møre Basin and the Northern North Sea, in *Petroleum Geology of the North European Margin.* E.M.Spencer *et al.* (eds.). Norwegian Petroleum Society, Graham and Trotman, London, pp.349-358.

Hanks,T. and McGuire,R.K. (1981) The character of high-frequency strong ground motion. *Bull. Seismol. Soc. Am.*, **71**, 2071-2095.

Havskov,J., Lindholm,C.D., and Hansen,R.A. (1989) Temporal variations in North Sea seismicity, in *Earthquakes at North Atlantic Passive Margins: Neotectonics and Postglacial Rebound.* S.Gregersen and P.W.Basham (eds.), Kluwer Academic Publ., Dordrecht, pp.413-427.

Henkel,H. (1989) Tectonic studies in the Lansjärv Region, in *Interdiciplinary study of Post-Glacial Faulting in the Lansjärv Area, Northern Sweden, 1986-1988.* G.Bäckblom and R.Stanfors (eds.), Swedish Nuclear Fuel and Waste Management Co.(SKB) Technical Report 89-31, pp.2:1-2:23.

Hurich,C.A., Palm,H.,Dyrelius,D. and Kristoffersen,Y. (1989) Deformation of the Baltic continental crust during Caledonide intracontinental subduction: Views from seismic reflection data. *Geology*, **17**, pp.423-425.

Husebye,E.S., Bungum,H., Fyen,J., and Gjøystad,H. (1978) Earthquake activity in Fennoscandia between 1497 and 1975 and intraplate tectonics. *Norsk Geol. Tidskr.*, **58**, 51-68.

Johnston,A.C., Reinbold,D.C., and Brewer,S.I. (1985) Seismotectonics of the Southern Appalachians. *Bull. Seismol. Soc. Am.*, **75**, 291-312.

Karpuz,M.R. (1990) Seismotectonic study of the Sunnhordland District, Southwestern Norway. Unpbl. Cand. Scient. Thesis., University of Bergen, 175p.

Karpuz,M.R., Anundsen,K., Gabrielsen,R.H., and Barton,R.H. (1990) Utilizing multi-sensor and geodata for analysis of neotectonics in Sunnhordland, Southwest Norway, in *Proceedings of the 7th Thematic Conference on Remote Sensing for Exploration Geology, Calgary, Canada. Oct. 2-6, 1989,* pp.713-733.

Karpuz,M.R., Gabrielsen,R.H., Engell-Sørensen,L., and Anundsen,K. (1991) Seismotectonic significance of the January 29, 1989 Etne earthquake, Southwest Norway. *Terra Nova*, **3**, 540-549.

Kulhánek,O. (1989) Earthquakes near the Lansjärv fault, in *Interdiciplinary study of Post-Glacial Faulting in the Lansjärv Area, Northern Sweden, 1986-1988.* G.Bäckblom and R.Stanfors (eds.), Swedish Nuclear Fuel and Waste Management Co.(SKB) Technical Report 89-31, pp.6:1-6:13.

Kvamme,L., and Hansen,R.A. (1989) The seismicity in the continental margin areas of Northern Norway, in *Earthquakes at North Atlantic Passive Margins: Neotectonics and Postglacial Rebound.* S.Gregersen and P.W.Basham (eds.), Kluwer Academic Publ., Dordrecht, pp. 429-440.

Kvamme,L., Hansen,R.A., and Bungum,H. (in press) Seismic source and wave propagation effects of L_g waves recorded in Scandinavia. *Geophys. J. Int.*

Lindholm,C.D., and Marrow,P.C. (1990) Ocean bottom seismometers in the North Sea: Experience and preliminary results with the Statfjord OBS. *Bull. Seismol. Soc. Am.*, **80**, 1014-1025.

Lundquist,J., and Lagerbäck,R. (1976) The Pärve Fault: A late glacial fault in the Precambrian of Swedish Lappland. *Geol. Fören. Stockholm Förh.*, **98**, 45-51.

Muir Wood,R., and Woo,G. (1987) The historical seismicity of the Norwegian Continental Shelf, in *Earthquake Loading on the Norwegian Continental Shelf (ELOCS), Earthquake Data Base, Technical Report*, Bungum,H. and Selnes P.B.(eds.), Norwegian Geotechnical Institute, Oslo, NTNF/NORSAR, Kjeller and Principia Mechanica Ltd., London.

Muir Wood,R. (1989) Extraordinary deglaciation reverse faulting in Northern Fennoscandia, in *Earthquakes at North Atlantic Passive Margins: Neotectonics and Postglacial Rebound*. S.Gregersen and P.W.Basham (eds.), Kluwer Academic Publ., Dordrecht, pp.141-173.

Naterstad,J., Andresen,A., and Jorde,K. (1973) Tectonic succession of the Caledonian nappe front in the Haukelisæter-Røldal area. *Norsk Geol. Tidskr.*, **243**, 99-121.

Naterstad,J. and Gabrielsen,R.H. (1981) Geology of the Bergen Haugesund District, Southwestern Norway, in *Excursion Guide to the 4th International Conference on Basement Tectonics*. B.T.Larsen (ed.), Oslo, August 1981. Nytt fra Oslofeltgruppen.

Ordaz,M., Singh,S.K., Reinoso,E., Lermo,J., Espinosa,J.M., and Dominguez,T. (1988) The Mexican earthquake of September 19, 1985 - Estimation of response spectra in the lake-bed zone of the Valley of Mexico. *Earthquake Spectra*, **4**, 815-834.

Ordaz,M. (1990) CRISIS40: A computer program for the seismic risk analysis. Universitaria, Coyoacan, Mexico.

Ramberg,I.B., Gabrielsen,R.H., Larsen,B.T., and Solli,A. (1977) Analysis of fracture pattern in southern Norway. *Geol. en Mijnbouw*, **56**, 295-310.

Rathore,J.S., and Hospers,J. (1984) A lineament study of southern Norway and adjacent off-shore areas. *Tectonophysics*, **131**, 257-285.

Roberts,D. and Sturt, B.A. (1980) Caledonian deformation in Norway. *J. Geol. Soc. London*, **137**, 241-250.

Schweig,E.S., Van Arsdale,R.B., and Burroughs,R.K. (1991) Subsurface structure in the vicinity of an intraplate earthquake swarm, central Arkansas, in J.-C.Mareschal (ed.), *Intraplate Deformation, Neotectonics, Seismicity, and State of Stress in North America*. Tectonophysics, 107-115.

Sigmond,E.M.O. (1978) Beskrivelse til berggrunnsgeologisk kartbladet Sauda 1:250 000. *Norges Geol. Unders.*, **341**, 1-94.

Sigmond,E.M.O., Gustavson,O. and Roberts,D. (1984) Berggrunnskart over norge M. 1:1 million. *Norges. Geol. Unders.*

Singh,S.K., Ordaz,M., Anderson,J.G., Rodriguez,M., Quaas,R., Mena,E., Ottaviani,M., and Almora,D. (1989) Analysis of near source strong motion recordings along the Mexican subduction zone. *Bull. Seismol. Soc. Am.*, **79**, 1697-1717.

Singh,S.K.,Ordaz,M.,Lindholm,C.D. and Havskov,J. (1990) Seismic Hazard in Southern Norway. Seismol. Observ. Univ.of Bergen Seismo-Series No.**46**, 33p.

Slunga,R. (1989) Focal mechanisms and crustal stresses in the Baltic Shield, in *Earthquakes at North Atlantic Passive Margins: Neotectonics and Postglacial Rebound.* S.Gregersen and P.W.Basham (eds.), Kluwer Academic Publ., Dordrecht, pp.261-276.

Solli,A., Naterstad,J. and Andresen,A. (1978) Structural succession in a part of the outer Hardangerfjord area, West Norway. *Norges Geol. Unders.*, **343**, 39-51.

Stein,S., Cloetingh,S., Sleep,N.H., and Wortel,R. (1989) Passive margin earthquakes, stresses and rheology, in *Earthquakes at North Atlantic Passive Margins: Neotectonics and Postglacial Rebound.* S.Gregersen and P.W.Basham (eds.), Kluwer Academic Publ., Dordrecht, pp.231-259.

Stephansson,O.,Särkkä,P., and Myrvang,A. (1986) State of Stress in Fennoscandia, in *Proceedings of the International Symposium on Rock Stress and Rock Stress Measurements.* O.Stephansson (ed.). Centek, Luleå, Sweden, pp.21-32.

Stephansson,O., Ljunggren,C. and Jing,L. (1991) Stress measurements and tectonic implications for Fennoscandia, in S.Björnsson, S.Gregersen, E.S.Husebye, H.Korhonen and C.-E.Lund (eds.), *Imaging and Understanding the Lithosphere of Scandinavia and Iceland,* Tectonophysics, **189**, 317-322.

Toro, G.R. and McGuire, R.K. (1987) An investigation into earthquake ground motion characteristics in Eastern North America. *Bull. Seismol. Soc. Am.*, **77**, 468-489.

Zoback,M.L., Zoback,M.D., Adams,J., Assumpçao,M., Bell,S., Bergman,E.A., Blümling,P., Brereton, N.R., Denham,D., Ding,J., Fuchs,K., Gay,N., Gregersen,S., Gupta,H.K., Gvishiani,A., Jacob,K., Knoll,P., Magee,M., Mercier,J.L., Müller,B.C., Paquin,C., Rajendran,K., Stephansson,O., Suarez,G., Suter,M., Udias,A., Xu,Z.H., and Zhizin,M. (1989) Global Patterns of Tectonic Stress. *Nature,* **Vol.341,** No.6240, 291-298.

EXPERT SYSTEM FOR EARTHQUAKE HAZARD ASSESSMENT: ESEHA

Zhu Yueqing① **Andreas Vogel**② **Hao Ping**①

ABSTRACT

Earthquake Prognostics is a concept and framework for previewing future earthquakes both with repect to the natural hazard and the associated risk on man and man–made structures. It has been put into practice by implementation of artificial intelligence techniques and expert systems. In this paper, first the significance and methodology for developing for the Expert System for Earthquake Hazard Assessment abbreviated ESEHA is described; secondly, its content composition, logical structure, knowledge– base, major algorithms, case study and its results are introduced; finally some conclusion and discussion are given.

Keywords:expert system, earthquake hazard assessment, reasoning

INTRODUCTION

Earthquake is one of the most serious natural disasters in the world. It kills many people and causes a lot of losses to mankind every year. With the increasing population density and industrialization, earthquake poses an increasing threat to humanity. In USA, in former USSR and in Armenia, in China and Mexico, in Japan and many other countries or regions, large earthquakes had just occurred or will occur. According to some prognostics there are several hundreds of potential disaster earthquake sources hiding underground in southern Europe, Asia, America, and Africa. Humanity and many man–made structures are facing a series of new serious impacts of earthquakes. So many governments are deeply concerned with earthquake disaster mitigation and earthquake loss reduction. But where will be the location? What is the magnitude? And what scale of the imending disaster will occur? People could not know what is the exact answer of all of these problems. So there are many governments and people who have no any preparedness for protecting the coming impacts. Many buildings are almost non–engineered and haven't the capability against the impacts of impending strong shocks.

①Centre for Analysis and Prediction, State Seismological Bureau, Fuxing Ave. No,63, Beijing, China.
②Mathematical Geophysics Group, Free University of Berlin, Podbielskiallee 62, D–1000 Berlin 33, Germany.

V. Schenk (ed.), Earthquake Hazard and Risk, 199–209.

Earthquake Prognostics is a very important concept and framework to preview the future earthquakes both with repect to the natural hazard and the associated risk on man and man–made structures. It is being developed into practice in the sense of a logistics by implementation of expert systems. One of the impotant efforts is the development of the Expert System for Earthquake PROgnostics abbreviated as ESEPRO. The global desin of it was completed in 1992 (A. Vogel and Zhu, 1992a). ESEPRO includes a series of sub–expert systems. In this paper, the authors presented one of the sub–expert systems in ESEPRO, which we called ESEHA (the Expert System for Earthquake Hazard Assessment).

METHODOLOGY FOR DEVELOPPING ESEHA

The Expert System for Earthquake Hazard Assessment ESEHA is based on the practises of the earthquake countermeasures in China, Europe and some other countries and regions (A. Vogel and Klaus Brandes, 1988). The purpose of it is to obtain the site seismic intensity distribution and its Certainty Factors (CF).

The methodology for designing the expert system ESEHA is as below:

First of all, based on the model simulation of source process, the parameters of the focal mechanisms, such as the magnitude, fault–angle can be obtained; secondly, we calculate the basic intensity distribution based on a certain seismic attenuation law. Here we used the elliptical attenuation law. If it is needed, it can be replaced with some other one by only changing some subroutines. Because the seismic attenuation law depends on the analysing regions, the multiple regression method is used to determine the coefficients of the formula by using the historical seismic intensity records of strong events. The medium in the analyzing region here is considered as homogeneous. With this step of analysis, the basic intensity distribution has been obtained.

But the real earth medium is not homogeneous, so it is needed to consider the different effects of different factors in different sites. Owing to that the whole analysis region is divided into grids, the intervals of each grid can be determined freely by the user experts before the system running.

Based on many researches, the site intensity will be mainly affected by geological, geographical, soil conditions and soil dynamical factors, et al. For all of these factors, we have summarized and refined many items of expert knowledge based on many books, papers, documents, experimental reports, and seismic codes.

After the summarization and refinement works, we checked and tested all the items of the knowledge obtained. Based on the final results of those works, we completed software designing. At last, all the main modules and the whole software system had been tested by using a set of data of some cases, then the modules and whole system had been improved step by step.

THE LOGICAL STRUCTURE OF ESEHA

The logical structure and flowchart of the expert system ESEHA is shown in Fig. 1. From this figure, we can see, there are 10 available function modules. The Master Control Module (MCM) manages the whole system by a set of menus. The Knowledge–Base Module (KBM) consists of knowledge–base management and knowledge–base substance. The Method Library Module (MLM) stored a set of methods for analysing some logical problems or completing digital calculations for evidence acquisition and reasoning. The Data–Base Module (DBM) stored data and information related to earthquake hazard

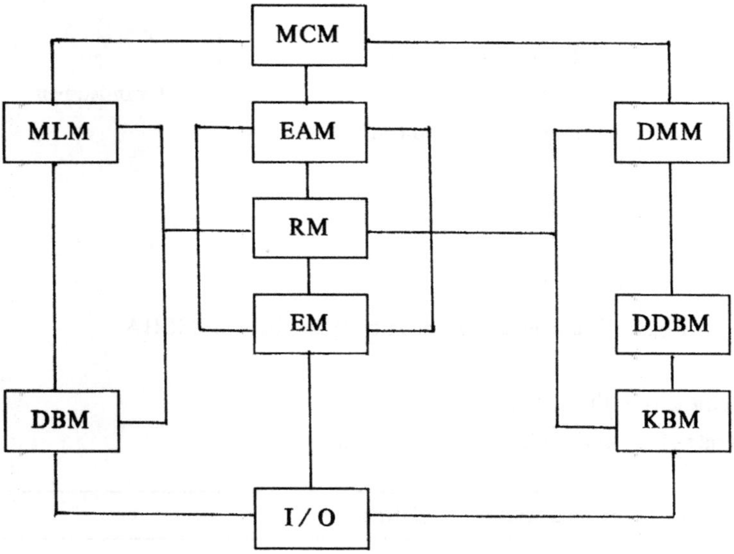

Fig. 1 Structure and flowchart of ESEHA

assessment. The Dictionary Management Module (DMM) is an intelligence bank for the whole system. The Dynamical Data–Base Module (DDBM) can store all the temporary results and the running message. The Evidence Acquisition Module (EAM) can complete the evidence acquisition work automatically by calling the programs from the module MLM and inquiring about the data from the DBM. The Reasoning Module (RM) will use all the items of the knowledge to do reasoning. During the reasoning time the system will search all the evidence trying to use them matching the items of the knowledge, then storing the results into the dynamical data–base. The Explanation Module (EM) can give the answers for the questions on " what is the conclusion? ", " how to get the conclusion? " and " why this conclusion is obtained? ". The I / O Module will input or output the needed information, so that users can have a friendly man– machine interface.

KNOWLEDGE–BASE AND KNOWLEDGE REPRESENTATION

(A). The structure of the knowledge– base

The structure of the knowledge– base is shown in Fig. 2.

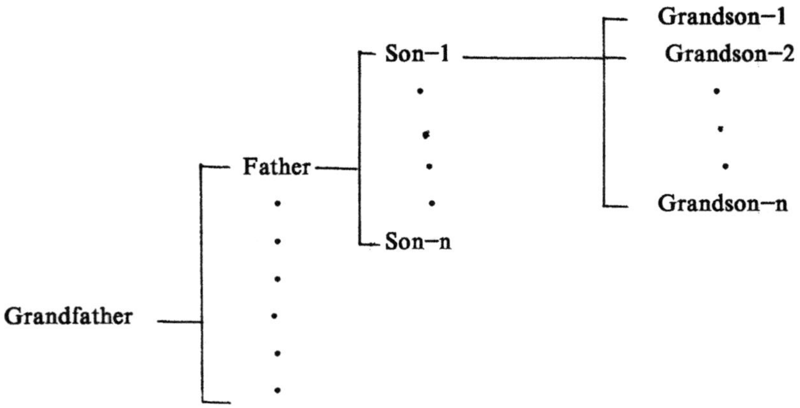

Fig. 2 The structure of the knowledge–base in ESEHA.

(B). The contents of the knowledge–base

The contents of the knowledge–base is shown in Fig. 3.

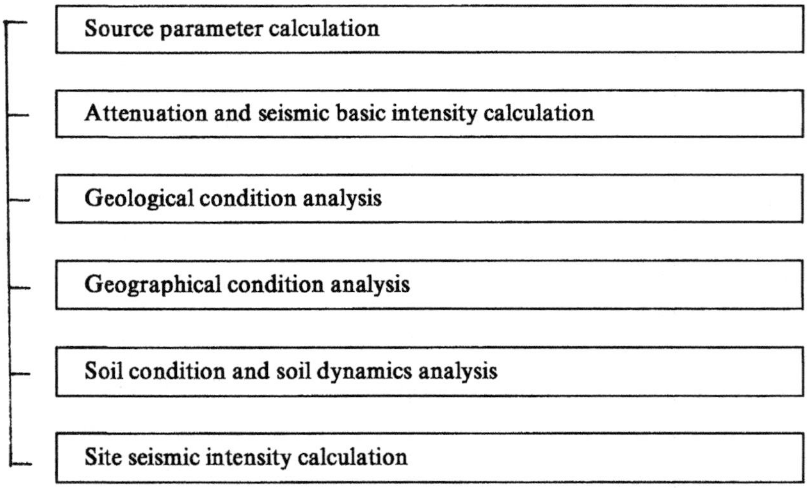

Fig. 3 The composition of the knowledge–base in ESEHA

(C). Knowledge representation

In order to meet the different needs of the different format and different contents of the knowledge items, we have used the following different modes of knowledge representation in ESEHA:

(1) Production rule

Most of the knowledge items are represented by production rules. The general format is as bellow:

$IF(C_1 \wedge C_2 \wedge ... C_k \vee C_{k+1} \wedge C_{k+2} \wedge ... \wedge C_n)$

$THEN((H_1,CF_1),(H_2,CF_2),...(H_m,CF_m))$.

Here the Ck is the condition, the Hm is the hypothesis or conclusion, and the CFm is the Certainty Factor of that Hm would be ture.

Here is a rule:

If (the effective intensity Ih > = Icr1 historically then sand liquefaction occurred)

Then (when Is> =Icr1 the sand liquefaction would occur), (0.75).

Where the contents in the former parentheses are the conditions, the contents in the later parentheses are the conclusion, and the value 0.75 is the Certainty Factor of that the conclusion might be ture.

With the production rules the predicates usually be used and its general formula follows the following definitions:

<rule> ::= <conclusion> :– <promise>

<promise> ::= <promise> :– <assertion> | <promise> ∧ <promise> |
<promise> ∨ <promise> | <promise>

<assertion> ::= <assertion> :– <expression> <type 1 relation>
<expression> | <type 1 relation> <expression>

<expression> ::= <expression> :– <value> | <term> | <expression>
<type 2 relation> <expression>

<conclusion> ::= <conclusion> :– <assertion> (<Certainty Factor>)

(2)Procedure mode

The procedure mode is usually used for doing some calculations and is usually represented as the following format:

<Procedure Name> (<Parameter List>):–
<Procedure Definition>,
<Parameter Definition>,
<Parameter Transfer>,
<Procedure Call>,
<Parameter Transfer>.

(3)Selection mode

This mode is usually used for doing some selection, such as answering "yes" or "no".

(4)Fact mode

The fact mode of the knowledge representation mode is used for describing some facts or evidences. The general fomula can be shown as bellow:

<Fact Name>(<Item Name 1> <Value 1> <CF_1>,
<Item Name 2> <Value 2> <CF_2>,
•
•
<Item Name n> <Value n> <CF_n>.

MAJOR ALGORITHMS

According to the purpose of ESEHA, the relevant algorithms as bellow:

(A) Seismic attenuation law and basic intensity ditribution

Seismic attenuation law describes the relationship between the variation of intensity with the magnitude and distance from the site. The magnitude can be assumed, or obtained from potential earthquake source recognition or from model simulation. Here the distance can be taken as the distance from the site to epicenter, source, surface fault, macro–epicenter, or the mean distance from the site to the both ends of the fault line. In this system we adopted the last one and marked it as R.

Generally there are the following two formulae of seismic attenuation law:

$$I = I_0 + A - BR - C\ln(R + R_0) \quad (1)$$

$$I = A + BM - C\ln(R + R_0) \quad (2)$$

Here A, B, C are regressive constants, the R_0 is an assumed constant, and I_0 is the intensity of epicenter. In our expert system the former law is adopted, but if it is needed we can easily change it.

The epicenter intensity I_0 is the function of magnitude M and depth h. We adopted the following formula:

$$M = 0.68 I_0 + 1.39 \ln h - 1.4 \quad (3)$$

Here the coefficients are obtained by regression with the historical records in a region of China. Users can easily change it based on the features of the analysing regions.

The calculation of R is as below:

$$R = a' \sqrt{1 + \left(\frac{h}{b'}\right)^2} - \frac{s}{2} \quad (4)$$

Where the s is the length of the real fault,

$$s^2 = 4(a^2 - b^2)(1 + h^2 / b^2) \quad (5)$$

and the a and b are the long axis and short axis of the alternative circle which has the equivalent area with that encircled by the isoseismic line of intensity I. For eastern and western China the average a and b have been listed in a table stored in the Data–Base of ESEHA.

(B) Site intensity analysis

Based on the seismic attenuation law we can only get the basic intensity distribution. But for any site there are many affected factors mentioned above and the intensity in these points will be changed. So ESEHA possesses a set of knowledge for analysing them.

(1) Site soil conditions

We divided the soil into three classes. The first class is competent rock, for which the intensity will not be different or only be a little bit change from the basic intensity. The 2nd class is the general hard or moderate stiff soil, for which the basic intensity will has about 0.5 to 1.2 degree changes. The 3rd class is the soft, loose, muck soil with high level of underground water or the silt soil, with which soil dynamical factorssand liquefaction index will also considered. For this class, the basic intensity will be changed with 0.8 to 2.0 degree. Many items of knowledge have been stored in the Knowledge–base, and a detailed table with all the relevant data has been stored in the Data–base for inquiring.

(2) Site geostructure analysis

The faults in ESEHA have been divided into two types. One is seismogenic fault, another one is non–seismogenic fault. The first one will release energy when event occur so it is located in meizoseisal region and has the highest intensity, the later one will change the basic intensity with about 0.3 to 1.3 degree.

(3) Topographical facotor and other factors analysis

These factors include landslide, ground fissure, old channel, kop or isolated hill, et al. For that the basic intensity will be change with about 0.3 to 1.5 degree.

(C) Uncertainty analysis and imprecise reasoning

It is obviously that all the items of knowledge (such as rules) and evidences (such as geostructure factors) could not be absolutely certainty. It can only be one with a certain Certainty Factor(CF). So all the items of knowledge and evidences must possess their own CF. In the Knowledge–base in ESEHA every item of the knowledge has its CF value. When a user uses ESEHA and prepare his evidences he has to give the CF value for any item of it.

It is just because of that, in ESEHA imprecise reasoning strategy has been used. The definition and algorithms for that are as bellow:

* Certainty Factor (CF) definition:

$$CF(H,E) = \begin{cases} \dfrac{P(H/E) - P(H)}{1 - P(H)}, & When\ P(H/E) > P(H), \\ 0, & When\ P(H/E) = P(H), \\ \dfrac{P(H/E) - P(H)}{P(H)}, & When\ P(H/E) < P(H). \end{cases} \qquad (6)$$

* CFs of evidence and rule:

$$CF(E) = \prod_{i=1}^{m} CF(i) \qquad (7)$$

$$CF'(H,E) = CF(H,E) * CF(E) \quad (8)$$

* Imprecise reasoning and CF transmitting

$$CF_{ij}(H) = CF'(H,E_i)+CF'(H,E_j)-CF'(H,E_i) * CF'(H,E_j) \quad (9)$$

MENU DRIVING AND MENU TREE IN ESEHA

ESEHA is programmed by Prolog language combing with C and Fortran languges on a micro computer. It is driven by a menu set. Users can easily operate the whole system by using the menu set step by step. All the steps can be completed only by moving the cursor and pressing the "return" key on the keyboard. The structure of the menu resembles a tree as shown in Fig.4.

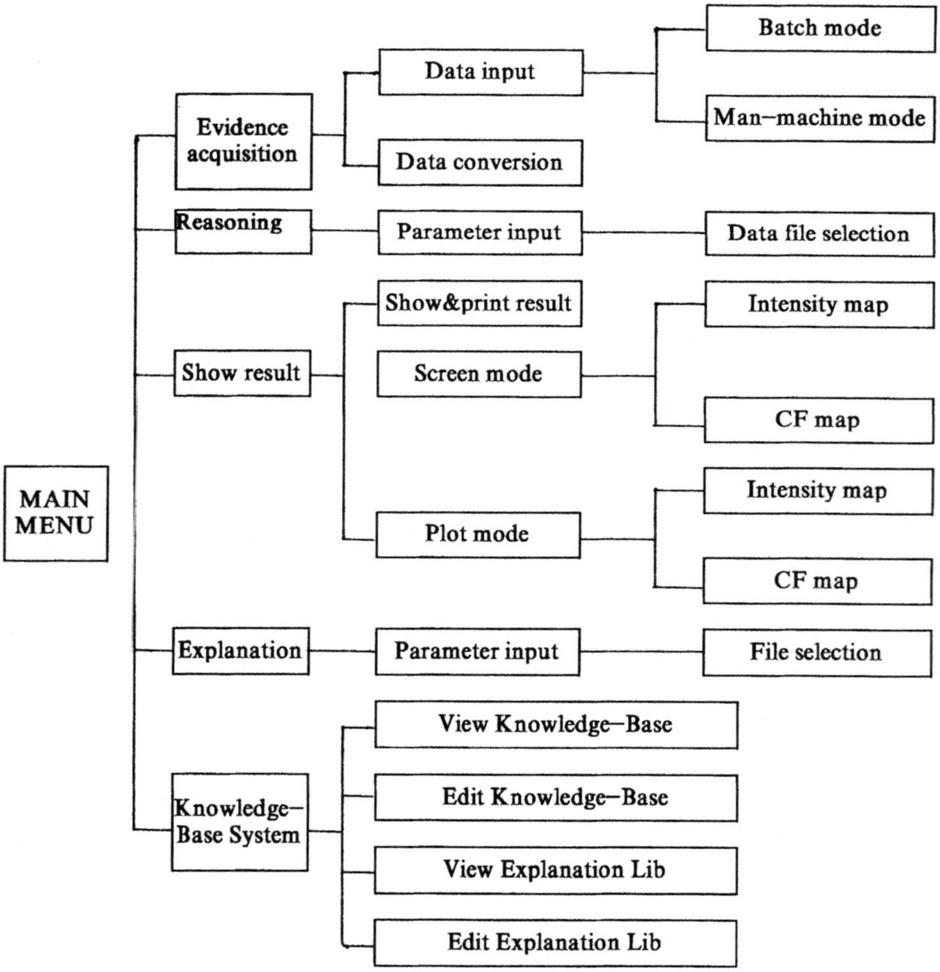

Fig.4 Menu structure and Menu tree

CASE STUDY

After completed all the research works and software implementation, some cases study had been completed. Here is one.

One region in China is adopted and a set of data files had been prepared. It includes a certain earthquake with magnitude 6.0 and many items of data for each grid of the analysing region, in which there are 6 different sites with special diferent kinds of site conditions, such as faults, special class of soil conditions and soil dynamical features, special local topographic form, for some points liquefaction occured historically, et al. After completing machine reasoning by using the ESEHA, the results have been obtained and shown in Fig. 5, in which the values in each grid are site intensities, the interval of the contours of site intensities is 0.2, the latitude and longitude have been ignored here.

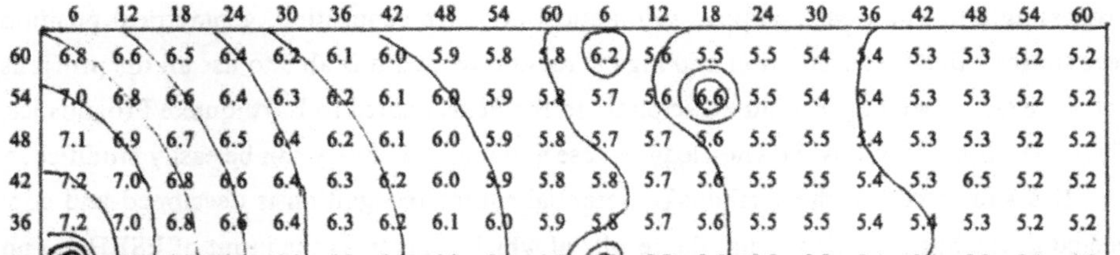

Fig. 5 The site intensity distribution

DISCUSSION AND CONCLUSION

The purpose of development of the expert system ESEHA is for both practical earthquake prognostics and disaster mitigation and methodology study.

The main functions and features of it are as below:

It is a high–level artificial intelligence software system. It has a good knowledge–base system which can edit, manage and view all the reasoning knowledge and explanation knowledge conveniently. It can do reasoning based on the current conditions and situations of the problem. The reasoning is performed on like the brain of the human experts, but not like a general software system which can only do what the programmer had arranged before it runs.

It is a multi–disciplinary and comprehensive practical software system for earthquake disaster mitigation. It is based on the concept and framework of Earthquake Prognostics, and combines the related disciplines and methods for doing what the human experts hope to do.

It possesses very strong software functions on data and knowledge management, data processing and knowledge editing, automatical evidence acquisition, geometrical position recognition, digital computation and logical reasoning, and it is able to use all the previous software achievements of multi– disciplines in the fields related to Earthquake Prognostics. The obvious merit of it is the knowledge– base and method library can be easily modified.

If a subsystem on the earthquake potential source recognition is developed and connected at the head of this system, the results of which is taken as the input of ESEHA; and another sub– expert system (or a segment of software module) on earthquake loss estimation is connected at the end of this system, the result of ESEHA is taken as the input of it, then a set of analyses can be done automatically and conveniently. It will be helpful for earthquake disaster mitigation.

Although the authors had done their own best for making the knowledge– base and method library to be complete, but owing to the regionality and complexity of the problem on earthquake hazard assessment, so it has to be improved step by step for meeting different regions.

Acknowledgement

The Authors would like to give many thanks to Professor Wu Shuming and Dr. Li Faxiang for their early stage works in our research group.

REFERENCES

Andreas Vogel and Klaus Brandes(Eds.),1988, Earthquake Prognostics,Vieweg, Berlin.

Andreas Vogel, Zhu Yueqing, Klaus Brandes, 1992a, Earthquake Prognostics Expert system and Artificial Intelligence Techniques as a part of a Logistical Approach to Earthquake Disaster Mitigation (Preprint).

Zhu Yueqing,Andreas Vogel, Klaus Brandes, Satyo P Bindra, Elka v.Hoyningen, Xing Ruying, 1992b, The Global Design of the Expert System for Earthquake PROgnostics: ESEPRO (Preprint).

Zhu Yueqing, 1991, Expert System for Comprehensive Earthquake Prediction ESCEP, Seismological Press, Beijing.

Zhu Yueqing, Mei Shirong and Yong Yurong, 1988, Design of the Expert System for seismic Disaster Relief, Proceedings of the International Conference on Earth–quake Countermeasures, Scientific & Technologic Press, Beijing.

Zhu Yueqing, 1991, Primary Study on the Imprecise Reasoning Method of the Expert System for the Potential Source Recognition and Seismic Damage Analysis, in: Expert System for Comprehensive Earthquake Prediction ESCEP, Seismological Press, Beijing.

ESTIMATING LOSSES FROM FUTURE EARTHQUAKES IN CHINA

CHEN YONG, CHEN XINGLIAN, FUZHENGXIANG
YIN ZHIQIAN, YAN MANDONG
State Seismological Bureau
Beijing, 100036, China

1. Introduction

The quantitative estimation of the hazard due to future earthquakes comprises the foundation on which earthquake hazard prevetion, countermeasures and relief work are based. The loss caused by earthquakes is determined by the following two factors: a) The temporal and spatial distribution of seismic activities which is dealt with by the seismic hazard analysis and b) The degree of damage done by earthquakes on the society, economy and people is the target for the seismic vulnerability analysis. The map of the expected losses caused by earthquakes is precisely the synthetic representation of the above-mentioned two factors, and such a map has manifested the ultimate goal of seismological work to serve the economic construction and social progress.

It is the common wish of scientists all the world as well as of government at respective levels in China to predict the loss due to future earthquakes, and such an estimation on the future earthquake hazard is also one of the important activities of the International Decade for Nature Disaster Reduction (Consultative Committee of IDNDR, 1988). Scientists both in and outside China have made energetic and effective efforts in this respect. For example, the four federal organizations in the U. S. , FEMA, USGS, NSF and NBS, have set as their long-term task to fund research work relevant to hazard prediction. The Group of Experts for Estimating Earthquake Losses got funding in 1989 and worked out its report " Estimating the Losses Due to Future Earthquakes " (1989). The Chinese scientists are making rapid progress in the field of estimation of future earthquake losses owing to that they have a fairly good foundation in this connection. The present paper shows the newest results of their studies.

V. Schenk (ed.), Earthquake Hazard and Risk, 211–220.

2. Seismic Hazard Analysis

Main tasks for seismic hazard analysis are resonable idetification of locations of future potential earthquakes (usually known as hypothetical earthquake or potential earthquake source) and predicting the magnitude and distribution of ground motion resulting from the earthquakes in a specified region.

In the late 1960's, Cornell (1968) first used the method of probability to analyze earthquake hazard and given probabilistic expression of the ground motion in a given site within time T in the future. Requisite data for such probabilistic analysis of seismic hazard are generally obtained by taking the following steps: a) locating potential earthquake focal regions (faults); b) determining probabilities of occurrence of different magnitude earthquakes in each of focal region; c) evaluating ground motion or intensity attenuation outward from each focal region; d) by summarizing the above-mentioned data, calculating hazard curves of site ground motion to give the probability that ground motion A exceeds the give value a within unit time (Panel on earthquake loss estimation methodology, 1989):

$$P(A \geq a) \cong \sum_i \nu_i \iint P[A \geq a | m, r] f_{MR}(m,r) dmdr \tag{1}$$

Within time T, $P_T(A \geq a) = 1 - [1 - P(A \geq a)]^T$

where $P[|\ ]$: condition probability;

$f(,)$: probability density;

ν_i : the rate of earthquake occurrence in potential focal region within unit time ;

m and r : magnitude of earthquake and distance from the site;

$\sum$: sum of all earthquake sources.

It is evident that within time T, the probability of occurrence of ground motion ranging (a, a+1) at a site is

$$P(a) = P(A \geq a) - P(A \geq a + 1) \tag{2}$$

The ground motion usually can be described in terms of ground acceleration, velocity, displacement and so on. Because historical earthquake in most areas have been measured in intensity, and because currently, instrumentally recorded data of strong motion are limited, therefore, intensity I is still used in most studies and the intensity is basic parameter for estimation of ground motion and seismic damage. Thus, (2) became

$$P(I) = P(A \geq I) - P(A \geq I + 1) \tag{3}$$

In 1992, a new seismic zoning map (1990) was issued (State Seismological Bureau, 1992), which gave the intensities with exceeding probability of 10% in the future 50 years for different parts of the country under an average soil condition. In preparing the seismic zoning map of 1990, the method of seismic probability analysis and the experiences in earthquake prediction reseach were used in attempt to embody the time-space inhomogeneity of seismic activities in China.

3. Vulnerability Analysis

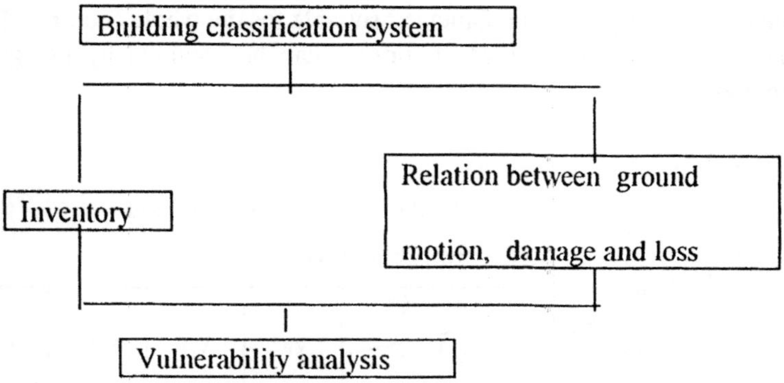

Figure. 1. Block diagram of building vulnerability analysis

As for building vulnerability analysis, firstly, a building classification system must be established based on resistance of building to earthquakes when exposed to seismic ground motion. Using the data of a nationwide general survey of houses and buildings in China in 1986, we classified the urban buildings as 4 types: a) civil brick houses, including all brick houses except for factory buildings; b) factory buildings, mainly including buildings used for production and storage; c) high rise reinforced concrete buildings; d) one-story old civil houses, mainly including one-story old brick houses, one story column and wooden frame adobe houses. Rural houses in China can be categorized into 2 types: a) brick houses newly built in recent years, of which the most are one-story and only the minority are two-story; b) adobe houses, including the ordinary adobe and earth caves in northeastern China.

Economic loss and casualty caused by earthquake disasters are directly related to seismic damages od buildings. Therefore, seismic vulnerability analysis for buildings is one of the major subjects of earthquake loss research. To describe the degree of damage and calculate economic losses of buildings from an earthquake, we usually grade to 5 degrees the damages of buildings that possibly occur when an earthquake hits: a) wholly intact; b) light damage; c) moderate damage; d) heavy damage; e) complete destroyed. Each of these 5 degrees has its explicit definition (Yin Zhiqian, 1990).

The probability that a certain type of buildings suffers seismic damage(Dj) in the comming T years is

$$P\left[D_j\right] = P[I]P\left[D_j|I\right] \tag{4}$$

where $P[I]$: probability of occurrence od seismic intensity I within T years;

$P\left[D_j|I\right]$: seimic damage matrix, i.e., probability the buildings suffer damage of degree j in case of intensity I.

Table 1-8 give 8 seismic damage matrixes of buildings in urban and rural areas in China. These tables are obtained on the basis of seismic damage data of a large amount of earthquakes occurred in China. These 8 matrixes are the statistical mean value matrixes of seismic damage, suitable for the whole country. The seismic hazard matrixes of other types of buildings can be obtained by using similar statistical methods.

TABLE 1. Seismic damage matrix of factory building P[Dj | I] (%)

Degree of damage(Dj) \ P[Dj | I] \ Intensity(I)	VI	VII	VIII	IX
Intact	73	57	24	10
Light damage	20	28	28	20
Moderate damage	7	12	31	40
Heavy damage	0	3	15	25
Destroyed	0	0	2	5

TABLE 2. Seismic damage matrix of multistory reinforced concrete structures P[Dj |] (%)

Intensity(I)	VI	VII		VIII		IX	
Degree of damage(Dj) \ P[Dj | I] \ Protection condition	No	Yes	No	Yes	No	Yes	No
Intact	75	45	34	34	2	15	0
Light damage	25	46	55	55	49	65	9
Moderate damage	0	9	11	11	42	14	50
Heavy damage	0	0	0	0	7	6	36
Destroyed	0	0	0	0	0	0	5

Note: Structures in Bejing are considered as seismically protected. In other cities, the mean values of the protected and unprotected are taken.

TABLE 3. Seismic damage matrix
of brick buildings P[Dj |] (%)

Degree of damage(Dj) \ Intensity(I) P[Dj | I]	VI	VII	VIII	IX
Intact	92.0	76.5	56.1	34.2
Light damdge	7.0	16.8	23.1	23.8
Moderate damage	0.9	4.3	12.9	22.7
Heavy damage	0.1	1.9	6.2	14.1
Destroyed	0	0.5	1.7	5.2

Note: Suitable for Heilongjiang, Jilin, Liaoning, Inner Mongolia, Qinghai and Xinjiang province.

TABLE 4. Seismic damage matrix
of brick buildings P[Dj |] (%)

Degree of damage(Dj) \ Intensiyt(I) P[Dj | I]	VI	VII	VIII	IX
Intact	68.9	43.3	24.3	11.9
Light damage	19.0	21.7	17.4	11.7
Moderate damage	7.2	15.2	20.6	19.5
Heavy damage	3.4	12.1	21.1	27.1
Destroyed	1.5	7.7	16.6	29.8

Note: Suitable for Yunnan, Hubei, Hunan, Zhejiang, Sichuan, Guizhou, Guangxi, Guangdong, Jiangxi, Hainan, Fujian and Shanghai provincs.

TABLE 5. Seismic damage matrix
of brick building P[Dj | I] (%)

Degree of damage(Dj) \ Intensity(I) P[Dj | I]	VI	VII	VIII	IX
Intact	83.7	62.0	40.1	21.5
Light damage	12.4	21.4	22.9	19.0
Moderate damage	2.8	9.0	18.5	24.1
Heavy damage	0.8	5.4	12.7	22.0
Destroyed	0.3	2.2	5.8	13.4

Note: Suitable for Beijing, Tianjin, Hebei, Shanxi, Shandong, Anhui, Henan, Ningxia, Gansu, Tibet and Shaanxi province.

TABLE 6. Seismic damage matrix of one-story old civil house P[Dj | I] (%)

Degree of damage(Dj) \ Intensity(I) P[Dj | I]	VI	VII	VIII	IX
Intact	50	20	15	5
Light damage	28	42	25	5
Moderate damage	15	25	30	35
Heavy damage	5	8	20	30
Destroyed	2	5	10	25

TABLE 7. Seismic damage matrix of brick buildings in rural area P[Dj | I] (%)

Degree of damage(Dj) \ Intensity(I) P[Dj | I]	VI	VII	VIII	IX
Intact	84	62	39	22
Light damage	12	21	23	19
Moderate damage	3	9	19	24
Heavy damage	1	6	13	22
Destroyed	0	2	6	13

TABLE 8. Seismic damage matrix of adobe houses in rural area P[Dj | I] (%)

Degree of damage(Dj) \ Intensity(I) P[Dj | I]	VI	VII	VIII	IX
Intact	30	11	8	2
Light damage	38	32	22	15
Moderate damage	25	29	30	28
Heavy damage	5	18	25	30
Destroyed	2	10	15	25

According to the cost of restoration of earthquake-damaged buildings in recent years in China, earthquake loss rate of buildings can be obtained, as shown in Table 9. Numbers in the table are rate of restoration cost to the current cost of building construction. Seismic damage matrix and total cost of construction of a building can be calculated by use of this table.

TABLE 9. Earthquake loss rate of buildings (%)

<table>
<tr><th>Degree of damage
Loss rate
Structural type</th><th>Intact</th><th>Light damage</th><th>Moderate damage</th><th>Heavy damage</th><th>Destroyed</th></tr>
<tr><td>Reinforced concrete structures</td><td rowspan="4">0</td><td rowspan="4">5-10</td><td rowspan="4">10-40</td><td rowspan="4">40-70</td><td rowspan="4">70-100</td></tr>
<tr><td>Brick structures</td></tr>
<tr><td>One-story old house</td></tr>
<tr><td>Adobe house</td></tr>
<tr><td>Industrial buildings</td><td>0</td><td>4-8</td><td>8-35</td><td>35-70</td><td>70-100</td></tr>
</table>

4. Losses from future earthquakes in China in the forthcoming 50 years

4.1. EXPECTED ECONOMIC LOSS OF SEISMIC HAZARD

In the light of earthquake hazard, probability analysis and building damage matrix, expected loss of seismic hazard may be obtained. In year T and at site K, the absolute value of expected economic loss of the type S buildings is E_k

$$E_{KS} = \sum_{D_j}\sum_{I} P_{KS}(D_j|I) \bullet P_K(I) \bullet b_S(D_j) \bullet B_S \tag{5}$$

where $P_{KS}(I)$: probability of intensity occurring in year T at site K;

$P_{KS}(D_j|I)$: probability of occurrence of degree D_j damage of type S buildings
at site K in case of intensity I;

$b_S(D_j)$: mid value of damage ratio in case that type S buildings experience degree D_j damage;

B_S : total value of reconstruction for type S buildings.

The total of expected economic loss of all types of buildings at site K is E_{KS} .

Figure. 2. Distribution of the total expected economic losses for 4 type buildings from earthquakes in China in the forthcoming 50 years

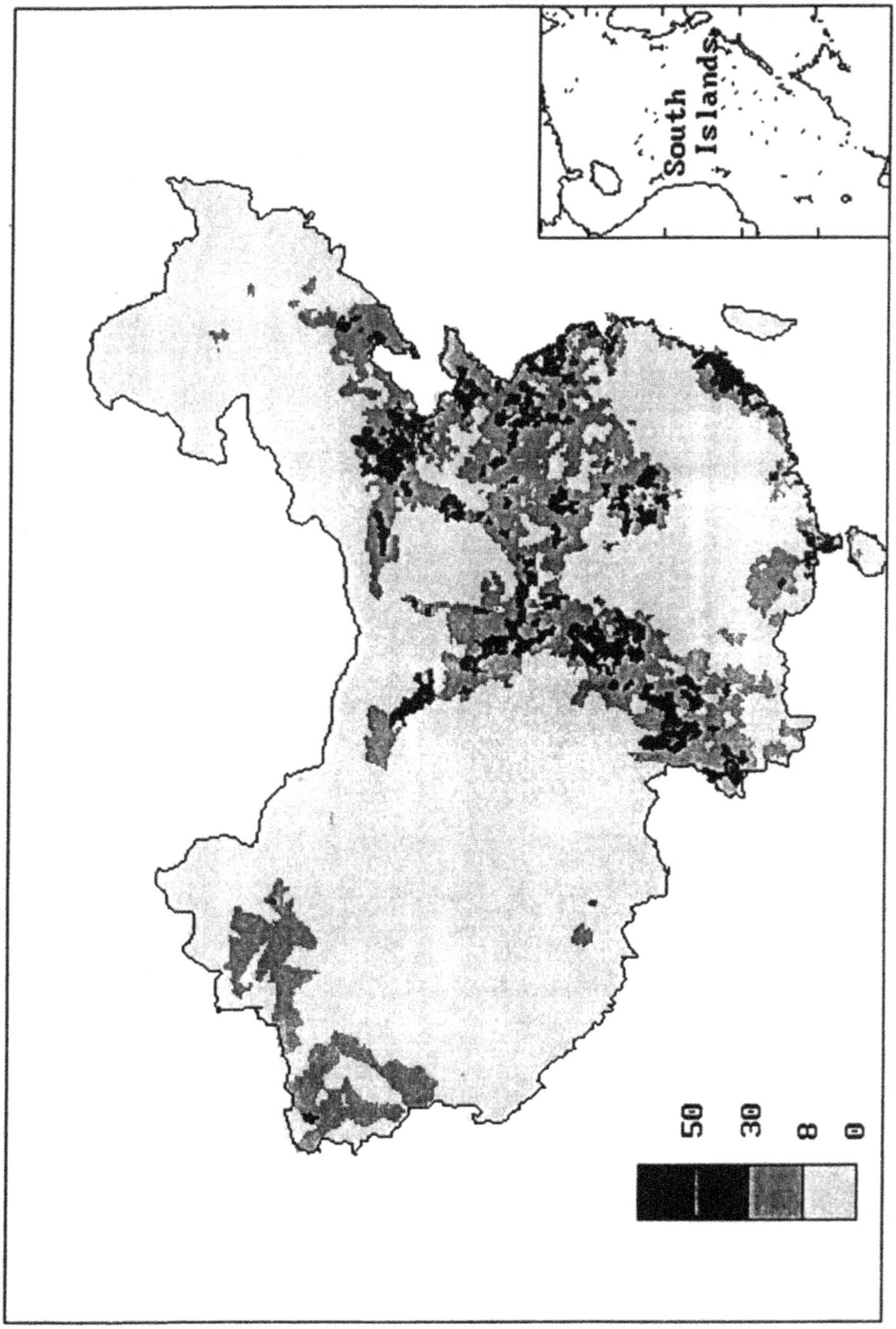

4.2. DISTRIBUTION OF THE EXPECTED ABSOLUTE ECONOMIC LOSS

The area of various types of buildings are taken from the hand-collected data of the first general investigation of cities and towns of China conducted in 1985. The construction costs for the 4 types of buildings are 120 yuan / m^2 for older hauses, 300 yuan / m^2 for RC high-rise buildings respectively.

figure 2 shows the distribution of the sums of the expected economic loss(in RMB yuan) in unit areas caused by earthquakes in the future 50 years for the 4 types of buildings(old houses, brick houses, industrial buildings and RC high-rise buildings). In the figure 2, different levels of losses are expressed by the gray scale.

If we divided mainland China into the eastern and western parts by 105° east longitude, the frequency ratio of strong earthquakes of magnitude≥7 between the eastern and western parts is 1 : 7°; the corresponding ratio of energy released in earthquakes is 1 : 25. Such a background of seismic activity will play a decisive role in a very long time period, that is, seismicity is low in the east whereas high in the west. the expected losses due to earthquake damages in the future shows that the economic losses are larger in the east part than that in the west. It goes without saying that this is because the population is denser and the economy is better developed in the eastern part of mainland China. Therefore, it provides the objective basis that the focus of the work of earthquake hazard prevention and hazard reduction in China should be laid in the east part. Among them, North China area should be put under intensified monitoring.

It should be pointed out that the above calculated results are only the economic losses of damage on buildings, they do not include the losses of damages on lifeline facilities, the losses of equipments and property, the losses of interruption of enterprises and production. The estimation of the losses of earthquake damage has not considered the factors such as the population and economic development in the future 50 years. That is to say, all the economic losses caused by earthquakes will exceed the figure 2. Therefore, the study on the prediction of the earthquake losses should be further deepened.

Acknowledgements

The authors wishes to acknowledge the fund from National Science Foundation of China, also acknowledge the valuable support of Gao Wenxue and Shi Zhenliang, who provide us the new seismic zoning map (1990) as the basis of seismic hazard analysis in our research. It also wishes to thank the following individuals who served as technical consultant to this research: Hu Lixian, Yan Jiaquan and Zhang Yumin. In particularly the Panel wishes to thank Ms. Li Geping (project officer).

References

1. Consultative Committee of International Decade for Natural Disaster Reduction (1988), Chinese version translated by Xie Lili and Zhou Yongnian, *Facing natural disaster,* Seismological Press.
2. Panel on Earthquake Loss Estimation Methodology (1989), *Estimating losses from future earthquakes,* National Academy Press, Washington D.C.
3. Cornell, C. A. (1968), Engineering seismic hazard analysis, *B.S.S.A*. 58, 1583-1606.
4. State Seismological Bureau (1992), *Seismic intensity zoning map of China,* Seismological Press.
5. Yin Zhiqian et al. (1990), Method for estimating seismic damages and losses, *Earthquake Engineering and Engineering Vibration,* 10, No. 1.
6. Yin Zhiqian et al. (1991), Prediction of earthquake damage and earthquake damage class identification, *Earthquake Research in China*, 7, No. 1.
7. Ministry of Urban and Rural Construction of the People's Republic of China, *Hand-Collected Data on the First Nationwide Investigation of the buildings in Urban and Rural Areas of China,* 1986, 12.
8. Ministry of Public Security of the People's Republic of China, *Population Statistics of Countries and Cities of the People's Republic of China (For the Year of 1985).*

EARTHQUAKE RISK ASSESSMENT IN GREECE

G.A. PAPADOPOULOS[(1), (2)] & A. ARVANITIDES[(2)]
[(1)] *Geodynamic Institute, National Observatory of Athens, 11 810 Athens, Greece.*
[(2)] *Hellenic Air Force Academy, Athens, Greece.*

Abstract

A quantitative approach of the earthquake risk, in a long-term and large-scale sense, has been made for 23 distinct regions of Greece. The risk, Ri, in each region, i, has been determined as a convolution of earthquake hazard, buildings oldness coefficient, average population density and regional Gross Domestic Product per capita. A relative risk scale was adopted by assinging each region with $R_{ri} = R_i/\min\{R_i\}$. The final product is a map of large-scale, relative earthquake risk distribution in Greece, which allows the elaboration of strategic plans for the risk mitigation.

1. Introduction

The term "natural hazard" refers to a potentially damaging or even destructive natural event, such as an earthquake. Any measure of the earthquake hazard describes only natural processes, that is the probabilistic expectancy for the earthquake occurrence in a 3-D sense: space, time, size. On the other hand, expected earthquake losses are approached by the assessment of earthquake risk which is a description of the measurable impact of the earthquakes on the human society (e.g. Caputo et al., 1974; State Seismological Bureau of China, 1992).

The area of Greece is characterized by high seismicity and seismotectonic complexity (Papadopoulos *et al.*, 1986; Papadopoulos, 1993). Methods for the earthquake hazard assessment in Greece and its adjacent regions have been elaborated by several investigators. However, only a very limited number of papers related to the earthquake risk description have appeared in Greece (Galanopoulos, 1977; Ambraseys and Jackson, 1981; Pomonis, 1987).

V. Schenk (ed.), Earthquake Hazard and Risk, 221–229.

In the present paper an effort has been made to develop a methodology appropriate for assessing the earthquake risk and to present, for a first time, a map of the relative earthquake risk in the several regions of Greece.

2. The Method

A quantitative expression of earthquake risk, R, is generally adopted as the convolution of earthquake hazard, K, vulnerability, V, and value at risk, VA:

$$R = K * V * VA \tag{1}$$

However, different measures of K, V, and VA may be used when R is considered in different time and space scales. Therefore, methods of risk assessment are scale-dependent. Figure 1 is an example of nine different time-space combinations that could be conventionally considered. In this paper the earthquake risk is determined in the long-term and large-scale sense. Moreover, our main purpose is to produce a relative earthquake risk scale in Greece. As a consequence, appropriate measures of K, V, and VA have been introduced.

3. Earthquake Hazard

One of the most recent approaches of the earthquake hazard in Greece is that presented by Papadopoulos and Kijko (1991). In this approach the maximum likelihood estimation of earthquake hazard parameters has been made on the basis of a procedure which utilizes mixed data containing incomplete files of large historical events and complete files of instrumental earthquakes. The entire earthquake catalog used covers the period from 479 B.C. to 1985 A.D.

The maximum regional magnitude, M_{max}, as well as the mean return period (and respective probabilities), T_m, of earthquakes with a certain lower magnitude $M \geq m$ are among the parameters determined for eighteen distinct seismogenic regions. Assuming that K is a function of the form $\Theta(M_{max}, T_{6.0})$, increasing with M_{max} and decreasing with $T_{6.0}$, Papadopoulos and Kijko (1991) were able to create a relative hazard scale and then to classify regions by hazard level based on their M_{max} and $T_{6.0}$; where $T_{6.0}$ is the mean return period of earthquakes of magnitude equal to or larger than 6.0. For $M_{max} < 7.2$, $7.2 \leq M_{max} < 7.6$ and $M_{max} \geq 7.6$ the function $\Theta(M_{max})$ is defined as equal to 2, 4, and 6, respectively. For $T_{6.0} \geq 30$, $30 > T_{6.0} \geq 17.5$ and $T_{6.0} < 17.5$, $\Theta(T_{6.0})$ is defined as equal to 2, 4, and 6, respectively. The arithmetic mean $K = [\ \Theta(M_{max}) + \Theta(T_{6.0})\ ]/2$ signifies the adopted relative seismic hazard of a particular region. K takes values 2, 3, 4, and 5 leading to the definition of four classes of relative seismic hazard: low (K = 2), intermediate (K = 3), high (K = 4), and very high (K = 5).

Space \ Time	Long-term	Mid-term	Short-term
Large-scale	X		
Mid-scale			
Small-scale			

Figure 1. Conventional scale assignement for earthquake risk assessment. Short-, Mid-, and Long-term may indicate time scales of 1, 10 and 100 years, respectively. Small-, Mid-, and Large-scale of space may be considered as describing areas of surface of 1, 100 and 10,000 km^2, respectively. The position of the present earthquake risk approach is indicated by X.

Figure 2 shows the distribution of the relative earthquake hazard in the eighteen seismogenic zones of Greece and its adjacent regions. As one may observe the highest earthquake hazard manifests itself in the Western Hellenic Arc (zones 2 and 3), Marmara Sea (zone 16) and Central Macedonia (zone 18). This pattern of earthquake hazard distribution has been adopted for the earthquake risk assessment in the several seismogenic regions. The "white" regions of Figure 2 are of low seismicity. Due to the very low number of instrumental and historical seismic events the earthquake hazard has not been determined in these regions. For these reasons an earthquake hazard level of K = 1 has been conventionally assigned when these regions were considered in the earthquake risk assessment.

4. Vulnerability

The seismic vulnerability, V, of a particular region may be considered as a convolution of the form

$$V = g * q * p \tag{2}$$

where g=local ground conditions, q=quality of structures, p=population features. In the large scales of space considered in our approach, a reasonable assumption is that in each particular region different local ground conditions are found in different sub-regions or sites. Therefore, the average vulnerability factor, due to the ground conditions, is of the same order of magnitude in all regions considered.

As for the quality of structures, q, a similar suggestion could be made. However, there is a general trend towards an increased seismic resistance of the structures in the last decades. Namely, it was the year of 1959 when a National Antiseismic Building Code was established in Greece. This Code was partially revised in 1984 while a New Code was presented in 1992. On

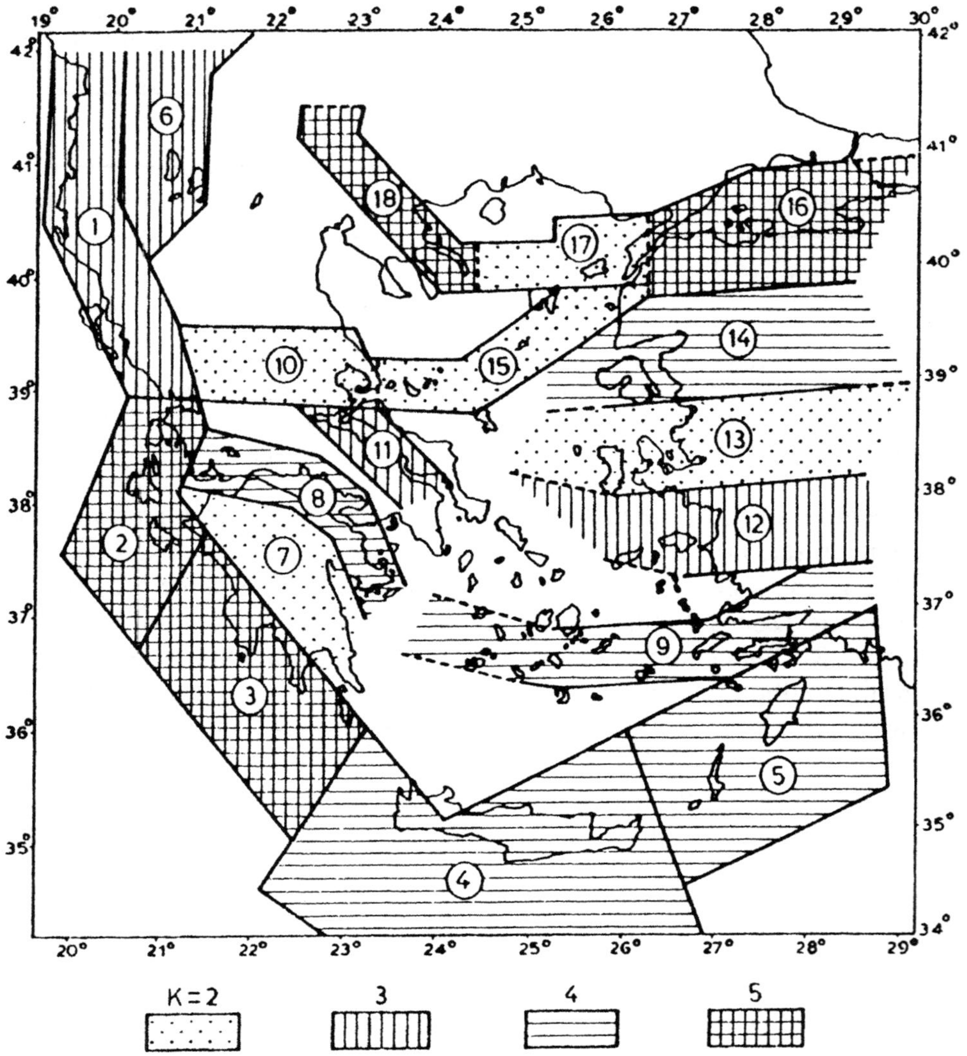

Figure 2. Distribution of relative earthquake hazard in Greece and adjacent region (after Papadopoulos and Kijko, 1991). Stipple = areas of low hazard (K = 2); vertical ruling = intermediate hazard (K = 3); horizontal ruling = high hazard (K = 4); cross - hatching = areas of very high hazard (K = 5).

this basis, and allowing for some time needed for an adjustment to the 1959 regulations, the year of 1965 has been considered as the turning point of the increased earthquake resistance or decreased vulnerability of structures in Greece. To express this in quantitative terms a coefficient, Σ, was introduced as

$$\Sigma = (A + 0.5B) / (A + B) \tag{3}$$

where A=number of buildings constructed before 1965, and B=number of buildings constructed during or after 1965. Σ, varying between 0.5 and 1.0, is an expression of the average oldness of a population of buildings and, therefore, a measure of the average vulnerability of structures. This consideration of vulnerability is justified by an independent study of Papadopoulos and Archontis (1995). They showed that the ratio of dead over injured people due to earthquakes in Greece has been gradually decreased from 0.75 in the 1950's to 0.61 in 1980's, while the ratio of dead people over damaged buildings has been gradually decreased from 0.09 in the 1950's to only 0.02 in the 1980's.

Population density, d (in population/10 km^2), has been considered as the most important measurable population feature affecting vulnerability. In summary, the seismic vulnerability in each region examined has been determined as the convolution

$$V = g * \Sigma * d \quad (4)$$

with $g = 1$.

5. Value at Risk

The earthquake impact on the human society is either direct, such as damages in buildings and their content, critical facilities and life lines, and/or indirect which includes social, economic, psychological and even political consequences. However, estimating quantitatively the expected earthquake losses is a hard task due to the many uncertainties involved in the problem. In our approach, aiming to produce a map of the relative earthquake risk in Greece, an assumption has been made that the relative value at risk, VA, is proportional to the regional Gross Domestic Product (GDP) per capita, G.

6. Risk Assessment and Data Used

The previous analysis implies that the final expression used for the earthquake risk assessment in each region, i, is

$$R_i = K * \Sigma * d * G \quad (5)$$

Expression (5) was applied in the several regions shown in Figure 2 with some modifications so that the whole Greek terittory, and only this, be covered. In this sense the risk assessment has been made for the seventeen out of eighteen regions as well as the "white" regions shown in Figure 2. The region 16 has not been considered as being located mostly outside the Greek terittory. Totally, twenty three regions were taken into account for the risk assessment. They are numbered from 1 to 24 in Figure 3 and Table 1 because the region 16 has been ommitted. That is, $i=1, 2, ..., 24$ with $i \neq 16$.

The relative earthquake risk, R_{ri}, in each region has been determined as

$$R_{ri} = R_i / R_m \quad (6)$$

where $R_m = \min \{Ri\}$.

As already explained, the values used for the earthquake hazard, K, are those determined by Papadopoulos and Kijko (1991) for the regions 1 to 18, while K = 1 was conventionally assumed for the "white" regions (see Table 1). Building, population, and economic statistics of the National Statistical Service of Greece were adopted for the rest parameters introduced in (5). Namely, the 1980 building inventory, the 1981 population census, and the 1988 determinations of regional GDP per capita were taken into account for the estimation of the parameters Σ, d, and G, respectively. Areas estimations of the regions examined are according to the 1963 national planimetry as revised in 1973 and 1983.

7. Results and Discussion

Table 1 summarizes the results of the earthquake risk assessment in Greece in terms of R_i and R_{ri}. R_{ri} ranges between 1.00 and 56.13. On the basis of the R_{ri}, distribution, five classes of relative earthquake risk were suggested: very low (R_{ri} = 1.00 – 1.79), low (R_{ri} = 2.55 – 3.77), intermediate (R_{ri} = 4.44 – 5.63), high (R_{ri} = 6.69 – 7.35) and very high (R_{ri} = 13.11 – 56.13). This classification is illustrated in Figure 3 where for purposes of (1) comparison with the hazard map of Figure 2, and (2) better visualization of the risk distribution, not only land areas exposed to risk, but also adjacent sea areas, not actually exposed to earthquake risk, are represented.

From Table 1 it results that the risk classification, is generally consistent with the hazard classification. Particularly one may observe that in 11 out of 23 regions the same hazard and risk classes have been assigned. In other six regions a deviation of only one class is observed, while a deviation of two classes has been obtained for four regions. However, completely incosistent results were obtained for the metropolitan regions of Athens (region 19, population ~ 3.2 millions) and Thessaloniki (region 21, population ~ 0.78 millions). Although the earthquake hazard appear as being very low in both regions, the earthquake risk turns to be the highest. This result is not far from being realistic as the recent seismic experience in both regions indicates. In fact, large earthquakes of M = 6.5 (20 June 1978) and M = 6.7 (24 February 1981) occurred some 25 km and 75 km to the east of Thessaloniki and to the west of Athens, respectively. Apart from the extensive damages and vistims created, both areas suffered for a long time as these earthquakes produced a series of indirect social and economic problems.

Table 1. Hazard, K, classification (after Papadopoulos and Kijko, 1991) and risk, R_{ri}, classification (present paper) of the several regions, i, of Greece (see Figures 2 and 3).

i	K	Hazard Class	R_i	R_{ri}	Risk Class
1	3	intermediate	690,337,362.4	3.77	low
2	5	very high	986,783,714.7	5.39	intermediate
3	5	very high	1,223,917,889.0	6.69	high
4	4	high	1,252,574,724.0	6.84	high
5	4	high	1,345,427,852.0	7.35	high
6	3	intermediate	273,683,831.9	1.50	very low
7	2	low	326,791,800.9	1.79	very low
8	4	high	1,327,936,960.0	7.26	high
9	4	high	1,029,491,494.0	5.63	intermediate
10	2	low	469,272,663.2	2.56	low
11	3	intermediate	821,606,030.7	4.49	intermediate
12	3	intermediate	813,341,783.0	4.44	intermediate
13	2	low	467,321,442.6	2.55	low
14	4	high	914,202,891.6	5.00	intermediate
15	2	low	254,699,788.6	1.39	very low
16	5	very high	–	–	–
17	2	low	238,825,844.3	1.31	very low
18	5	very high	935,666,073.0	5.11	intermediate
19	1	very low	20,543,924,048.0	56.13	very high
20	1	very low	221,917,766.4	1.21	very low
21	1	very low	4,799,281,602.0	13.11	very high
22	1	very low	192,174,999.8	1.05	very low
23	1	very low	198,178,947.5	1.08	very low
24	1	very low	183,023,809.3	1.00	very low

The position of Athens and Thessaloniki at the top of the relative risk scale is mainly due to their very high average population density, d (in people/10 km^2), being equal to 19,667.69 and 4,790.81, respectively. In the rest regions, d varies between 200.49 in region 6 and 629.49 in region 8. The building oldness coefficient, Σ, and the parameter G, expressing the regional Gross Domestic Product per capita, do not influence very much the results since Σ varies between 0.7560 and 0.9187, and G varies between 465,336 GDR and 801,897 GDR (GDR=Greek Drachma). Of considerable influence in the risk configuration is the earthquake hazard classification ranging between 1 and 5.

The results obtained are of great importance for practical applications. They allow to elaborate strategic long-term plans for the earthquake risk mitigation by designing priorities and allocating properly man power, funds and other means.

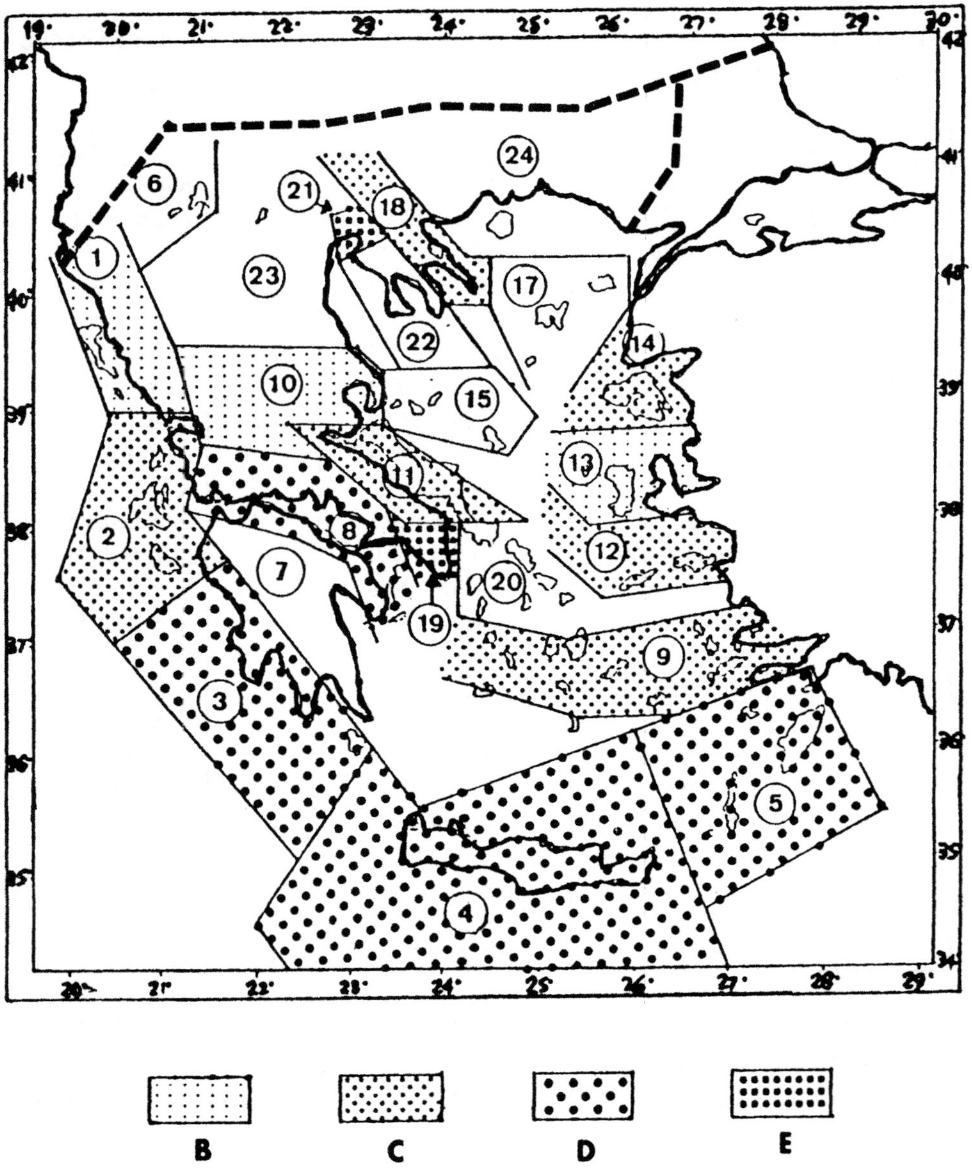

Figure 3. Distribution of relative earthquake risk in Greece. Risk classification: very low (white regions), low (B), intermediate (C), high (D), very high (E).

The procedure used to create a relative earthquake risk scale at a national level could be repeated to create a risk scale of regional or even local interest. This means that other time-space combinations should be selected from Figure 1 and some new parameters (e.g. intensity attenuation, local ground conditions) for the risk description may be introduced.

8. References

Ambraseys, N.N. and Jackson, J.A. (1981) Earthquake hazard and vulnerability in the northeastern Mediterranean: the Corinth earthquake sequence of February-March 1981, *Disasters* **5**, 355-368.

Caputo, M., Keilis-Borok, V.I., Kronrod, T.L., Molchan, G.M., Panza, G.F., Piva, A., Podgaetskaja, V.M. and Postpischl, D. (1974). The estimation of seismic risk for central Italy, *Annali di Geofisica* **27**, 349-365.

Galanopoulos, A.G. (1977) The earthquake expectancy of maximum intensity in the city of Athens, *Technika Chronika* **1**, 5-11 (in Greek with English abstr.).

Papadopoulos, G.A. (1993) Tectonic and seismic processes of various space and time scales in the Greek area, in E. Boschi, E. Mantovani and A. Morelli (eds.), *Recent Evolution and Seismicity of the Mediterranean Region*, Kluwer Academic Publishers, Dordrecht, pp. 239-249.

Papadopoulos, G.A. and Archontis, P. (1995) A study of the earthquake impact in Greece, to appear.

Papadopoulos, G.A. and Kijko, A. (1991) Maximum likelihood estimation of earthquake hazard parameters in the Aegean area from mixed data, *Tectonophysics* **185**, 277-294.

Papadopoulos, G.A., Kondopoulou, D.P., Leventakis, G.-A. and Pavlides, S.B. (1986) Seismotectonics of the Aegean region, *Tectonophysics* **124**, 67-84.

Pomonis, A.A. (1987) *A study on earthquake environment and seismic risk assessment in Greece*, Master Degree Thesis, Hokkaido University, Sapporo, Japan, 146 pp.

State Seismological Bureau, China (1992) Estimating losses from earthquakes in China in the forthcoming 50 years, *Seismological Press, Beijing*, 60p.

THE OPTICAL DISK "SPITAK EARTHQUAKE OF 1988".

G. SOBOLEV,
National Geophysical Committee, RAS,
Molodezhnaya, 3, 117296, Moscow, Russia
Yu. TYUPKIN, I. FROLOV, N. SERGEYEVA,
Geophysical Center, RAS,
Molodezhnaya, 3, 117296, Moscow, Russia
S. AREFEV, A. GOVOROV, E. ROGOZHIN, R. TATEVASYAN, M. ZHIZHIN,
Institute of the Physics of the Earth, RAS,
B. Gruzinskaya, 10, 123810, Moscow, Russia
Yu. SCHUKIN,
Research Institute for Geophysics, Moscow, Russia
P. LOCKRIDGE, L. WHITESIDE, R.E. HABERMANN, H. MEYERS,
National Geophysical Data Center, NOAA,
Boulder, CO, USA
A. MALITZKY,
Seismological Division, Institute for Petroleum Research and
Geophysics, P.O. Box 2286, Holon 58122, Israel.

Abstract

A vast amount of information is needed in order to develop strategies for mitigating impacts of catastrophic earthquakes. This includes aspects of the natural environment, the performance of man-made structures, and the earthquake itself. Recent advances in digital data storage have made it possible to collect these data onto a single compact disk, greatly simplifying the process of data access and distribution. The authors have endeavored to collect all available sets of observational data on the Spitak earthquake of 1988. This is the first attempt to build up an information collection for a comprehensive description of a disastrous earthquake. The database has the following subdivisions: Geology, Geophysics, Main shock, Aftershocks, Impact (epicentral area, towns, buildings), Elements of Prediction, and Seismic Waveforms. Information is represented by a set of computer maps with digital data and text used as a basis for the maps. The data management system makes the data easily available to the user. This version of the database is designed for IBM PC/AT compatible computers, and includes about 180 Mb of information and data.

V. Schenk (ed.), Earthquake Hazard and Risk, 231–238.

1. Introduction

The 1988 Armenian earthquake is unique both in catastrophic effects and in study coverage. Research groups from the former U.S.S.R., U.S.A., France, and Japan studied the earthquake area. The IDNDR Commission of the IASPEI has undertaken creation of the "Spitak Earthquake of 1988" database to make the enormous amount of data obtained by these research groups available to the global scientific community.

This project involves a number of diverse geophysical and computer science challenges. The resulting database contains practically all available material relevant to the Spitak earthquake and its aftereffects, and also the geological-geophysical information collected during the years prior to the event and which is important for the understanding of the event. So far as we know, it is the first database containing such a complete integrated description of a catastrophic earthquake.

The information included in the database comes in several forms, some familiar (local and teleseismic seismic waveforms of the main shock and of related events), and some unusual. A large part of the information is presented as images in the PCX format. These images were created at the World Data Center B in Moscow and include a variety of maps, cross-sections, and photographs, as well as x-y plots of data related to the earthquake and the region where it occurred. The data depicted in these images is, for the most part, not included in the database.

The primary benefit of this approach is that it significantly standardizes the information to be presented and, therefore, simplifies the process of presentation. Unfortunately, this benefit is gained at a substantial cost, much of the information is inaccessible as "data" in the typical usage of the word. However, some of the information on the disk is available as data, principally the waveforms, the seismicity catalogs, and some of the gridded parameters. We identify below these latter data sets with an asterisk.

2. The Database Contents

The database "Spitak Earthquake of 1988" has the following parts:
- General description.
- Geology and geophysics.
- Main shock.
- Aftershocks.
- Impact.
- Elements of prediction.
- Waveforms.

2.1. General description.

This part contains a brief historic, economic and geographic description of Armenia with a review of the seismic history of the region, including:
- Catalog of regional historic earthquakes (*);
- Isoseismal maps of the sixteen best studied earthquakes;
- Catalog of strong earthquakes (*);
- Instrumental catalog of earthquakes (*);
- The seismic zoning map of the U.S.S.R. showing the territory of the Caucasus.

2.2. Geology and geophysics.

This part is composed of information on the geologic, tectonic and geophysical fields of the region, including geologic data collected in the epicentral area after the earthquake.

The geologic-tectonic characteristics of the region includes:
- Map of major faults of the Caucasus;
- Map of lineaments and fractures of Transcaucasian Transversal Rise;
- Map of active faults;
- Map of largest active faults of the Near East and Transcaucasia.

The geologic-geophysical characteristics of the region include:
- Amplitudes of neotectonic movements (*);
- Gradient of velocity of post-Sarmatian vertical tectonic movements (*);
- Distribution of geological heterogeneity (*);
- Distribution of Mesozoic-Cenozoic faults (*);
- Density of faults of different ranks (*);
- Mohorovicic discontinuity relief (*);
- Surface of consolidated basement (*);
- Bouguer anomalies of gravity field (*);
- Isostatic gravity anomalies (*);
- Magnetic field anomalies (*);
- Heat flow anomalies (*);
- Temporal anomalies of first arrivals of crustal waves (*);
- Variations of upper mantle horizontal velocity (*);
- Surface relief (*).

These fields are included in a single digital database with the values of the indicated parameters averaged over 5'x5' grids. The user can display a map of a certain parameter and also obtain information on the value of any parameter at a given point or along a linear profile. These fields are also available as simple binary data files which can be imported into other visualization or mapping software systems.

The description of the epicentral zone includes:
- Structural-geologic map;

- Map of earthquake ruptures (on geologic basis);
- Structure of the zone of seismogenic ruptures in the man-made denudations in trenches;
- Morphology of seismic ruptures;
- Plicated deformation of Quaternary terraces of the Pambak River;

and also cross-sections of the following fields along the Section of the Epicentral Area:

- Seismic section;
- Anomalies of magnetic field;
- Bouguer anomalies;
- Topography;
- Heat flow.

2.3. The main shock.

This part is composed of:

- General information about the event;
- Fault plane solution;
- An example of near field records;
- An example of teleseismic records;
- Isoseismal map of the main shock;
- The model.

2.4. Aftershocks

This part includes the catalog of aftershocks and illustrations of the aftershock process, namely:

- Vertical cross-sections through aftershocks swarm;
- Fault plane solutions;
- Variations of seismic parameters;
- Radiation patterns.

2.5. Impact

This part includes four digital databases with information about geological and geophysical parameters which influence (or may influence) the effects of the earthquake, and about the effects of the Spitak earthquake themselves. It is arranged in such a way as to enable the user to have access to the information at different spatial scales ranging from the regional level to specific information on three cities: Spitak, Kirovakan, Leninakan.

This part also includes statistics of damage to multistory residential buildings for 23 cities in Armenia, photographs and brief descriptions of landslides, and, finally, data on certain engineering aspects of the earthquake. The latter information contains data on response spectra characteristics, data on angles of emergence of seismic rays, and information about the behavior during the earthquake of the major types of buildings (stone masonry bearing-wall buildings, precast-concrete-frame buildings, precast large-panel buildings, lift-slab buildings, and industrial facilities).

2.6. Elements of prediction

Before the Spitak earthquake, different methods were applied in the Caucasus to reveal precursors on a wide network of seismic stations and from many years of study of seismicity precursors. Other types of precursors were also studied, including observations of ground water level and geomagnetism, as well as experimental deformation studies.

Unfortunately, the social and political situation in Armenia in 1988 obstructed the systematic collection and processing of data on earthquake prediction. Even earthquake catalogs were published with a delay of one year. Because of this situation, no prompt observations of variations of prediction parameters were carried out. Therefore the occurring anomalies or precursors were not properly considered.

This part of the database contains a certain amount of information on seismological, hydrodynamic, electromagnetic and biological precursors, and on the deformation processes before the Spitak earthquake.

2.7. Wave Forms

Three independent databases of seismic wave forms are included:
- The database of seismograms of global seismic network (*);
- The database of seismograms of telemetric network of temporary stations (*);
- The database of digital records of strong motions (*).

3. Software

Two types of software, SDBS (Spitak Data Base Software) and GeoVu, provide access to the database. SDBS is the software specially developed to work with the database "Spitak Earthquake of 1988". As with any specialized software, SDBS provides quick and logically reasonable access to information. GeoVu, on the other hand, was created as a general tool for access to interdisciplinary data collections. The database "Spitak Earthquake of 1988" is one such collection. Naturally, universality is recompensed by longer time for reaching information, but the facilities offered by SDBS and GeoVu, as we believe, augment one another. A short description of the SDBS and GeoVu possibilities are presented below.

3.1. Spitak Data Base Software

The user has access to information through a menu system which allows:
- use of the next level menu for the same operations provided by the previous menu level;

- use the Ground Level Menu (GLM) to provide direct access to: textual information in ASCII files, information in ASCII tables with a special headline for displaying the titles of table columns, and graphic information presented in PCX format. In the latter case, the user can be supplied with an additional sub-menu giving access to several graphic and textual files;
- linkage of external executive programs (EEP) that provide access to databases through their own menu system.

Fig. 1 shows the logical scheme of user access to information. The "TEXT" and "DATA" buttons allow display of text, searches with a given set of symbols, and text output to a printer or a file. The access to digital databases is provided by EEPs. Some of these programs allow one to display not only the values of any of the fields in an elementary cell, but also to plot the variation of any of the fields along a linear profile. In the latter case, it is possible to draw plots of two different fields simultaneously.The selected information can be exported into a file indicated by the user.

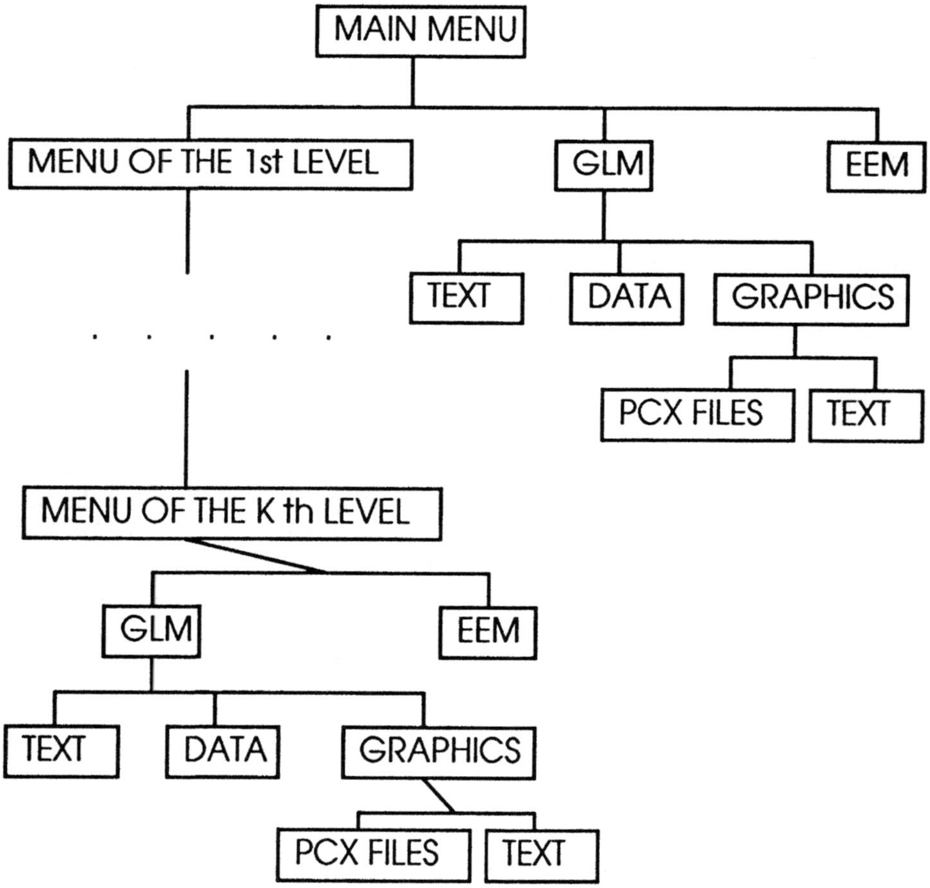

Fig. 1. Spitak Data Base Software: The logical scheme of user access to information.

EEM means External Executive Program; GLM means Ground Level Menu.

Two types of software are used in the mode of EEP for access to Waveform Databases. Access to the Global Digital Seismological Network Database is achieved by the WF software. This software has the ability to select a digital seismogram from the database, display it on the screen, mark wave arrivals, rotate the record, particle motion diagram, compute the complex transfer function of the digital recording system, execute polarization analysis, spectral analysis, filtration, and deconvolution of the seismic records. WF software can be used independently of SDBS. The seismic data used by WF software can be obtained also from a search of the NEIS - Event Data CD-ROM using the package SONIC.

Access to the database "Strong motions" and "Digital velocigram" is made through SMDB software, which permits:

- selection of the requisite record from the database;
- display of information about the event to which the selected record pertains;
- display of information about the recording instruments and about their locations;
- display of environmental setting;
- display of the graphic image of the record;
- cutting out part of the record;
- exporting the selected information into a user selected file.

In addition, because the number of the records in the databases can be large, the SMDB software has the capability of creating a virtual database from records satisfying the given conditions. These conditions are:

- the time interval,
- the geographical position of the recording stations, or their codes

As a result, during one session, the user can operate only with records of interest. This arrangement considerably shortens the time and simplifies the process of access to information. SMDB software also can be used independently of SDBS.

3.2 GeoVu

NOAA's National Geophysical Data Center has produced numerous diverse collections of data on CD-ROM's and has developed general tools for accessing such collections. This experience was brought to bear on this project after most of the data had been collected and converted to PCX images. This collection is similar to those on several CD-ROM's of natural hazards photographs produced at NGDC, so it was decided to experiment with access to this CD using GeoVu, a general data browser developed at NGDC.

Many of the differences between the GeoVu and SDBS implementations of this data collection reflect the fact that the data and the SDBS were developed hand-in-hand and that GeoVu was developed independently. The purpose of this experiment was to compare access to an existing data collection with specialized software, with access by a generic tool which can be used with numerous data collections. Such comparisons are rare, yet they yield important information on the trade-off between expensive development and maintenance efforts and data access.

GeoVu provides access to all of the data files included on the CD as well as to the PCX images. It also includes a Slide Show feature which allows quick browsing of sections of the CD. Guidance in data selection is provided by a menu system similar to that used by SDBS. The menu structures differ in several ways, reflecting differences in views of the data collection and flexibility which is built in to the GeoVu menu system. GeoVu is presently available for Microsoft Windows and will be available for some UNIX platforms and Macintosh's by the end of 1995. The most recent release of GeoVu and documentation are available via ftp in the Access_Tools directory at ftp.ngdc.noaa.gov, or via Mosaic at www.ngdc.noaa.gov.

Acknowledgments

The authors have benefited greatly from comments and support of Academician N.P. Laverov, vice-president of Russian Academy of Sciences.

In addition to the authors listed above, the information for the database was presented by V.V. Shteinberg, B.A. Borisov, B.M. Graizer, L.A. Delitsin, S.G. Molotkov, I.E. Parshin , K.G. Pletnev, P.A. Aleksin, A.A. Romanov, D.E. Loshtanov, I.S. Aptekman, T.V. Afim'ena, N.E. Shilova, and N.V. Shebalin, of the Institute of the Phisics of the Earth, RAS; V.I. Keilis-Borok, V. Kosobokov, A.V. Lander, O.D. Voevoda, of the International Institute of Earthquake Prediction and Mathematical Geophysics; G.L. Koff, Institute of Lithosphere, RAS; J.-C. Bousquet, H. Philip, Univercity of Montpellier, France; G. Ball, H. Blumentritt, A. Cisternas, C. Dorbath, L. Dorbath, H. Haessler, L. Rivera, J. Trampert, Institut de Physique du Globe, Strasbourg, France; M. Diament, A. Nercessian, Institut de Physique du Globe, Paris, France; J. S. Filson, National Earthquake Information Center, Denver, Co., USA.

The record of the mainshock in Gukasyan was obtained by the scientists of the Institute of Seismology of Armenian Academy of Sciences

The authors are grateful to F.M. Christie, Editorial Coordinator of magazine "Earthquake Spectra" for kind permission to scan a portion of the American Earthquake Reconnaissance Report into this database.

The authors would like to thanks G. Ivanova, E. Maksimova, A. Melkumyan, L. Zabarinskaya for assistance.

This work was supported by the grant of the Commission for IDNDR of IASPEI.

Part 3:

STRONG GROUND MOTION AND MICROZONATION

A STUDY OF SITE EFFECTS ON STRONG MOTION RECORDS OBTAINED FROM LOCAL NETWORK AROUND GULF OF CORINTH, (CENTRAL GREECE).

Makropoulos K., Diagourtas D., Voulgaris N., Drakopoulos J.
Department of Geophysics, University of Athens, Panepistimioupolis, Ilissia 15784, Greece

ABSTRACT

In July 1991, a strong motion network consisting of 8 SSA-1 accelerometric stations, was installed around the western part of gulf of Corinth, (C. Greece), in order to provide the basis for strong ground motion attenuation and site effect studies. Until now almost 550 3-component accelerograms from 130 events with epicentral distances ranging from 1 to 100 km having local Richter magnitudes ranging from 2.5 to 5.3 R, with maximum horizontal peak-to-peak acceleration values ranging from 2 to 360 milli-g, were recorded. Three of the stations were installed in the southern part of gulf of Corinth on Plio-quaternary sediments while the other five were installed on the northern part on alpine formations. The eighteen events which provided satisfactory record quality at most of the stations since 1991, have been analysed in order to define the importance of the site effects in a very geologically complex area like the gulf of Corinth.
The examination of the corrected accelerograms and corresponding response spectra at different stations shows relatively large variations in peak acceleration values as well as in frequency content at sites with different geological formations with comparable epicentral distances, underlining the importance of considering the role of site conditions in aseismic design studies.

INTRODUCTION

Within the framework of an EEC project aiming at studying the seismotectonic regime of Western Gulf of Corinth (C. Greece), a strong motion network, the so called RASMON (Makropoulos et al., 1993) was installed in July 1991.
During the two years period of the network's operation, a total of 550 3-component accelerograms were recorded from the 8 SSA-1 digital

V. Schenk (ed.), Earthquake Hazard and Risk, 241–252.

stations that correspond to 130 seismic events having local Richter magnitudes (M) ranging from 2.5 to 5.3 R.

One of the main objectives of this experiment was the collection of acceleration data from earthquakes with different magnitudes in a geotectonically complex area in order to study the importance of possible effects of the local site conditions on the amplification of strong ground motion due to the complex geology. This target is of great importance since the number of strong motion records available from earthquakes in the area of Greece is limited, especially for the case of near field and small events, thus, allowing the comparison of the site effects on the records of a wide range of magnitude earthquakes. The latter was attempted by setting the trigger level to the lowest permitted value.

In the area of gulf of Corinth there are two characteristic types of geological formations, Plio-quaternary sediments along the south coast and mesozoic, limestones and flysch on the north coast. Therefore, considerable differences due to site are expected to be observed in the records of the instruments located on the south (3 units) and north (5 units) coasts respectively.

It is evident that this objective is of especially great importance to engineering seismology since the results can assist seismic hazard assessment studies, improve the capability of calculating synthetic strong ground motion at specified sites of major constructions and enrich our knowledge about the effects of local site conditions on strong ground motion.

DATA USED

The scope of this paper is to present and evaluate the results obtained from the processing of the 18 selected events (figure 1) during the first two years operational period of the network with their focal parameters listed in table 1.

For the period July-August 1991 the location parameters of the events recorded by RASMON network were adopted from the results of the detailed large microseismicity experiment in the area (Rigo et al, 1993) during the same time period. In order to determine the location of the recorded events after the end of August 1991, the records of the National Observatory of Athens (NOA) were initially examined. Using the velocity model obtained from the microseismicity campaign, the NOA phase data in combination with the S-P arrivals of the accelerograms the recorded events were re-located. The epicentre map of figure 1 shows the location of the 18 larger events that are used in this paper.

TABLE 1. Focal parameters of the 18 events used

DATE	H/M	SEC	LAT	LON	DEPTH	M
911020	5:35	16.22	38-22.07	21-48.92	12.78	3.90
911025	21:24	21.25	38-20.51	22-12.45	13.16	4.50
911212	3:37	8.29	38-23.71	22-8.55	7.36	3.50
911218	2:57	2.38	38-21.93	22- 3.97	8.13	3.60
920421	21:12	19.80	38-23.71	22-12.25	12.11	3.40
920530	18:55	53.31	37-59.07	21-30.38	15.06	4.70
920606	15:41	15.34	38-25.76	22- 8.42	11.16	3.20
920707	13:59	32.09	38-27.30	21-56.55	9.50	3.00
920801	12:27	21.17	38-26.30	22- 3.75	10.79	3.00
920915	2:59	32.54	38-19.52	21-46.17	8.88	3.40
930318	15:47	4.13	38-33.34	21-58.82	46.25	4.90
930512	21:15	53.03	38-22.07	22-3.08	8.03	3.50
930513	21:33	4.70	38-22.70	21-59.65	5.70	3.40
930607	12:25	33.85	38-24.92	21-59.52	8.15	3.20
930714	12:31	49.35	38-13.01	21-45.52	21.82	5.00
930714	12:33	27.00	38-19.93	21-54.78	19.94	4.40
930714	12:54	7.95	38-14.87	21-42.75	21.68	3.50
931104	5:18	35.69	38-22.95	21-58.15	10.00	4.90

The digital data files recorded at 200 samples per second and collected in the field were scaled and converted to acceleration values using the sensitivity and damping of each instrument. Next, the scaled accelerograms were corrected for instrument frequency response, and baseline adjustment was performed. Initially, a low pass Ormsby (1961) filter with a cut-off frequency of 47 Hz and a roll-off frequency of 50 Hz was applied. Next, instrument correction was performed using the instrument specific characteristics. For the baseline correction a high-pass Ormsby filter was used with a cut-off and roll-off frequency of 0.25 and 0.5 Hz respectively. From the obtained corrected accelerograms the response spectra were calculated for different periods and damping values, using the approach based on the exact analytical solution of the Duhamel integral for successive linear segments of excitation.

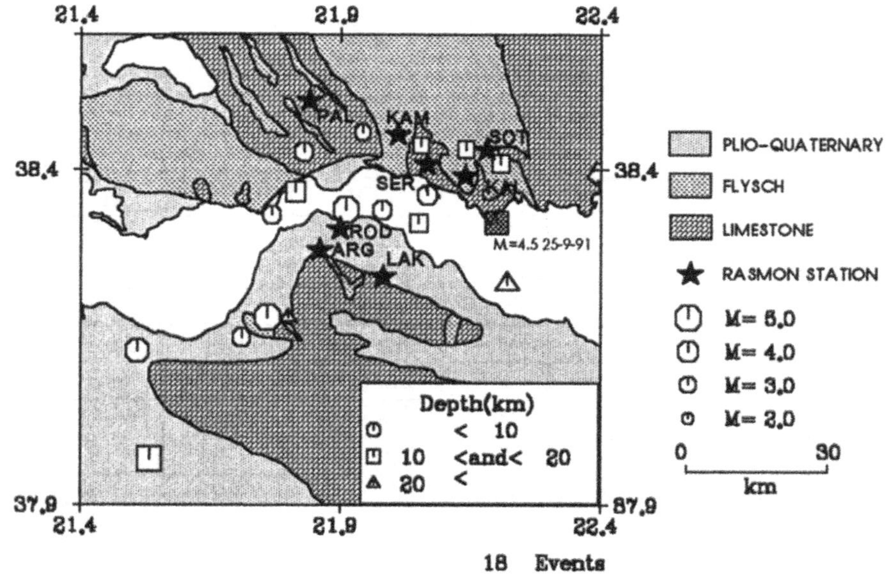

Figure 1. Location map of the RASMON network along with the epicentre locations of the 18 selected events. Epicentre location of the October 25, 1991 with local Richter magnitude M=4.5 event.

ANALYSIS - DISCUSSION

The earthquake of October 25, 1991, with local Richter magnitude M=4.5 which occurred almost in the centre of the network (figure 1), was selected as a characteristic event in order to present the variations in time and frequency domain at each site. Figure 2 shows the corrected N-S acceleration components of all stations that recorded this earthquake.

As it was shown by various authors (Scherbaum 1987a,b and Hutching and Wu, 1990), the records obtained from small and average magnitude near field events are not significantly influenced by source characteristics and hence they can provide useful information for site effect studies. Within this concept, the data recorded by RASMON network allows the examination of variations in frequency content, amplification and energy distribution, in relation to the different geological formations upon which the network was installed, a fact which is verified by the analysis of the October 25, 1991 event.

By examining the corrected acceleration waveforms of the north-south components obtained at different sites (figure 2), it can be seen that there are distinct differences between the first three records of the south stations located on Plio-quaternary sediments and the remaining four records of the north stations installed on alpine formations. The larger peak acceleration values in the recordings of the north stations are evident, while additional frequency variations can also be observed in the records of the south stations.

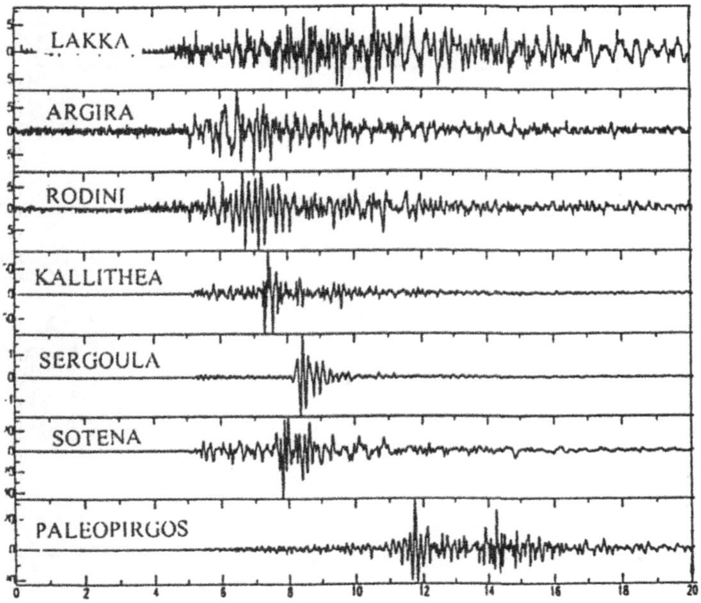

Figure 2. Corrected acceleration of the October 25, 1991 earthquake with local Richter magnitude M=4.5, recorded on the N-S components of 7 stations.

Furthermore as it can be seen in figure 3, where an example of the Fourier spectra for the characteristic stations of Lakka (south) and Sergoula (north) is shown, in the south site the released seismic energy is dispersed across a wider frequency band, whereas the spectra of the north station show an energy concentration in a limited and better defined frequency range, resulting to larger peak acceleration values. Similar features can be observed in the response spectra of the N-S components of all stations presented in figure 4. Thus, in the case of the response spectra of the south stations (Rodini (ROD), Argira (ARG),

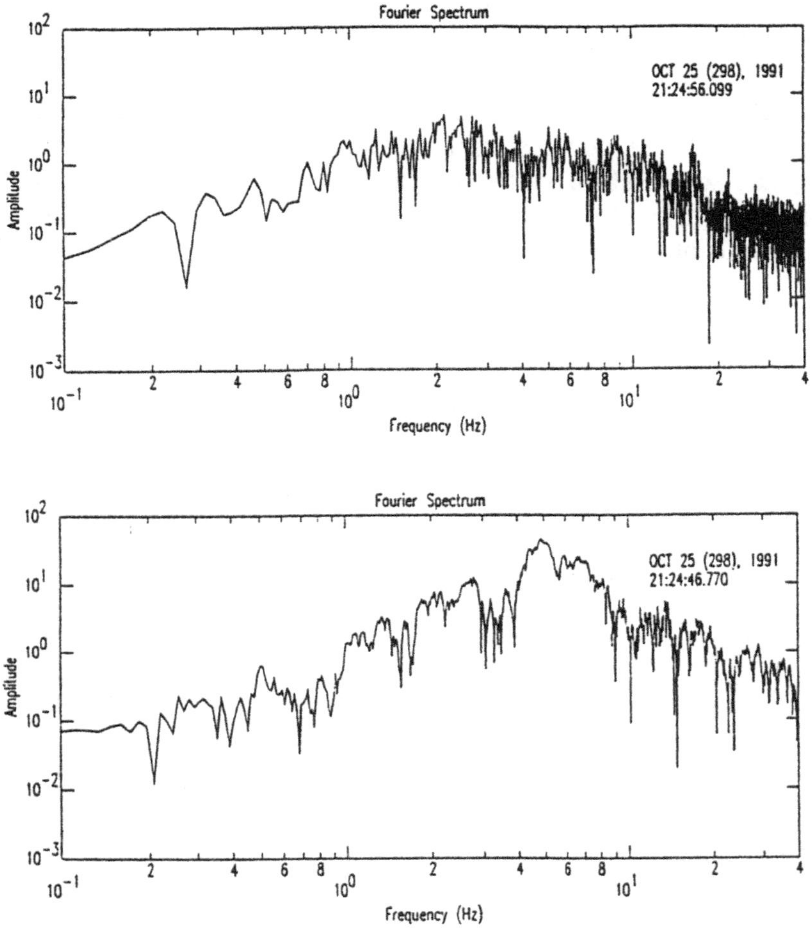

Figure 3. Fourier amplitude spectra derived from the N-S components of the October 25, 1991 earthquake M_L=4.5 recorded by Lakka (a) and Sergoula (b) stations.

Lakka (LAK), figure 4a,b,c) we can once again notice that the seismic energy is distributed across a wider period band with a variety of peaks, especially in the lower limit of the band, most likely caused by the energy scattering in the relatively poorly consolidated Plio-quaternary sediments. On the other hand, in the response spectra of the north stations (Kallithea (KAL), Sotena (SOT), Sergoula (SER), Paleopirgos (PAL), figure 4d,e,f,g) located on the older better consolidated alpine formations, the seismic energy release in low periods (less than 0.1 sec) is limited and the energy appears to be concentrated in a narrower band

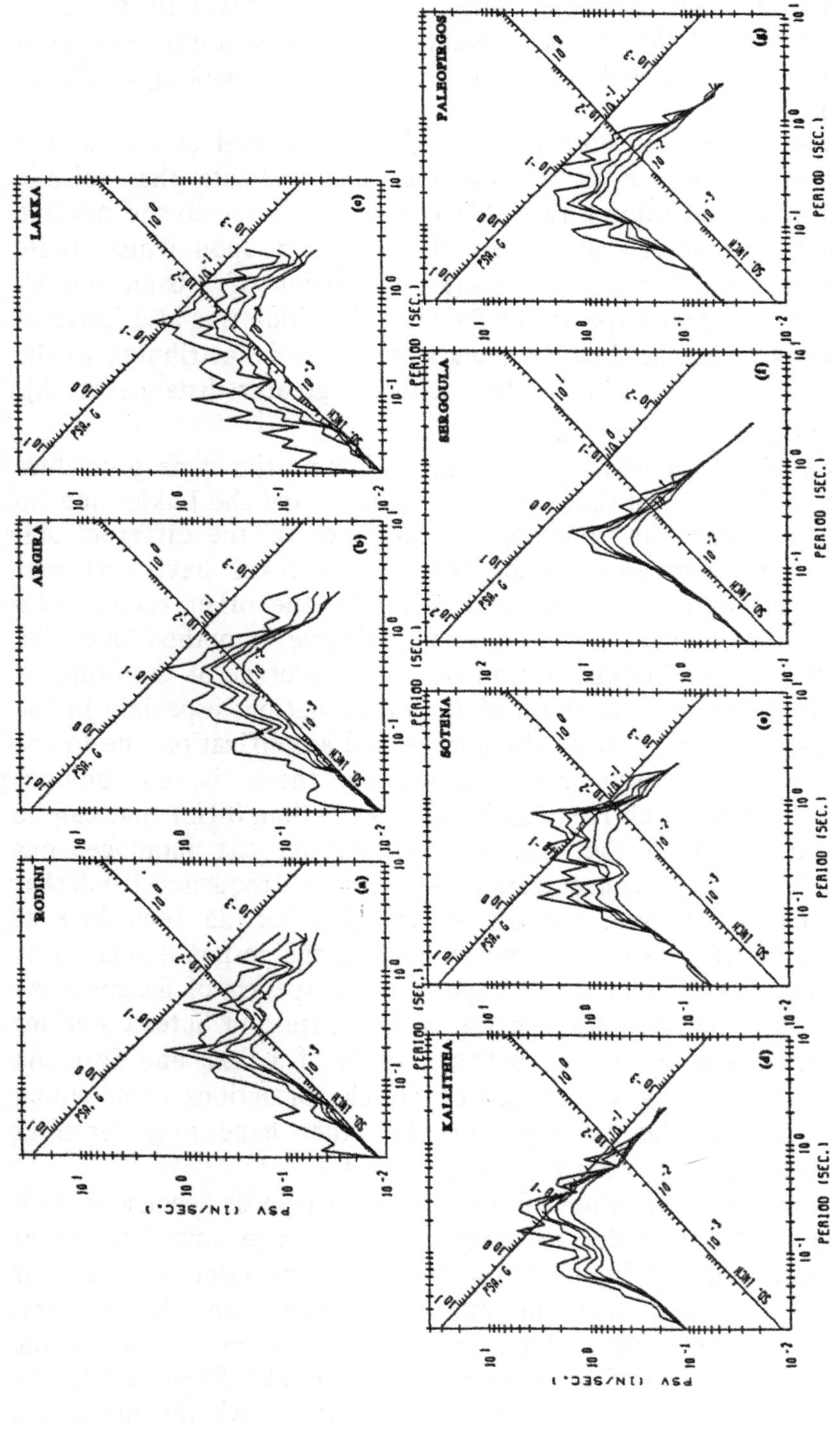

Figure 4. Response spectra of the October 25, 1991 earthquake derived from the longitudinal components recorded at Rodini (a) Argira (b) Lakka (c) Kallithea (d) Sotena (e) Sergoula (f) and Paleopirgos (g) stations for 0%, 2%, 5%, 10% and 20% damping.

(0.1-0.5 sec). This is more pronounced in the case of the Sergoula station located on Cretaceous limestone which yielded the maximum observed acceleration value of 181.5 cm/sec^2 where the released seismic energy is concentrated in the limited period band of 0.1-0.3 sec peaking at 0.2 sec (figure 4f).

The response spectra of Sotena and Kallithea, located at almost the same epicentral distance as the previous one, indicate that seismic energy is less concentrated a fact that can be attributed to the relative heterogeneity characterising the flysch formation upon which these stations were sited. Finally, the greater dispersion of seismic energy observed in the response spectra of Paleopirgos which was also installed on Cretaceous limestone, like Sergoula station, can be attributed to the effects of the ray path due to the relatively greater distance of this station from the epicentre (≈45km).

Considering the 18 records together and following the same procedure for zero damping for the three northern stations and the Lakka one (to the south), presented in figure 5, the influence of the different soil conditions is now apparent. Since these earthquakes have different magnitudes, epicentral distances and azimuths, the other factors like source and path effects are to a certain degree smoothed out. The spectral amplitudes of Lakka station are on the average of the order of one magnitude higher than those of the other stations especialy in the lower frequency band showing the long period amplification due to the sediments of the south coast. Moreover there is an obvious amplification on the high frequency band (more than 8 Hz) that can be probably attributed to the energy scattering on the inomogeneous sedimends. The energy concentration in a narrow frequency band that appeared in the response spectra of the October 25 recording at Sergoula station (figure 4f), is now obvious in the response spectra of all the events as well as in the average response spectra of Sergoula. As it was expected, the response spectra of Kallithea and Sotena stations display average spectral values between those of Lakka and Sergoula due to the relatively less consolidated flysch formations upon which they are installed. The envelopes on the other hand may serve as indicators when trying to design response spectra.

In figure 6 the average values of the calculated respose spectra at each station are plotted in order to compare the average amplification at each frequency. In this figure, the average spectral values of the four characteristic stations, confirm our observations on the spectral amplification in low and in high frequency bands at Lakka station, observed in the analysis of the October 25 earthquake. Even though the response spectra in this paper are not normalised with the maximum

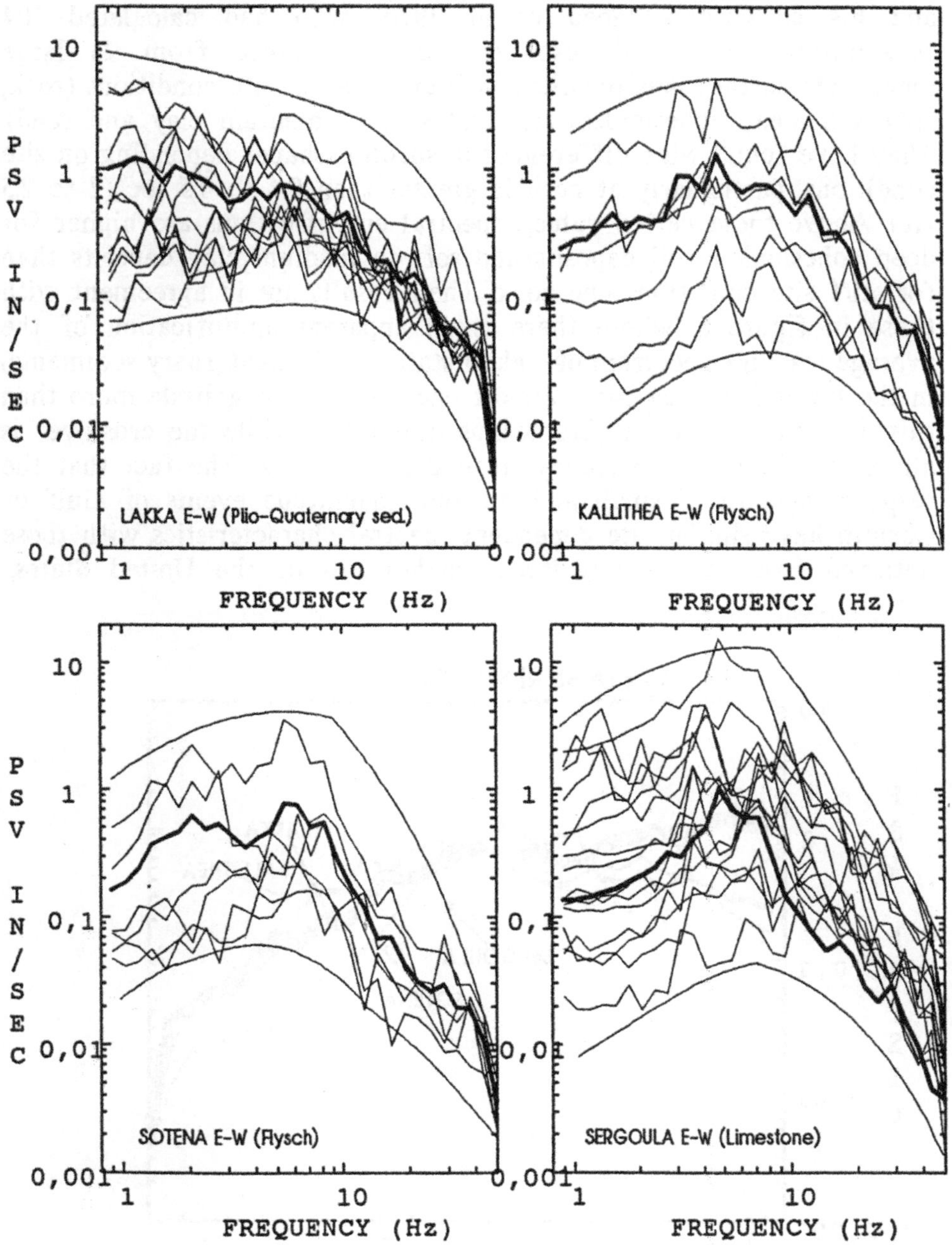

Figure 5. Response spectra of the E-W components of the 18 events recorded at Lakka, Kallithea, Sotena and Sergoula stations. Thick lines are the average response spectra.

acceleration, figure 6 is however in agreement with the work of Seed and his co-workers (Seed et al. 1976) who had calculated 104 acceleration normalized response spectra obtained from 23 large magnitude earthquakes recorded in four different site conditions (rock, stiff soil, deep cohesionless soil and soft to medium clay and sand). They have found wide differences in spectral shapes depending on site conditions, particularly at periods greater than 0.4 to 0.5 sec (2 to 2.5 Hz). Above these period values, spectral amplifications are higher for deep cohesionless soil deposits and soft to medium clay deposits than for stiff site conditions and rock. Their results are in agreement with those in figure 6, where there is an apparent amplification of the average response spectrum of Lakka station (Plio-quaternary sediments) in the lower frequncy band, almost one order of magnitude more than this of Sergoula station (Cretaceous limestone), while the crossover is almost in the same frequency, around 2 to 2.5 Hz. The fact that the response spectra of small and medium magnitude events of Gulf of Corinth have similar site dependent spectral characteristics with those obtained from large magnitude earthquakes of the United States,

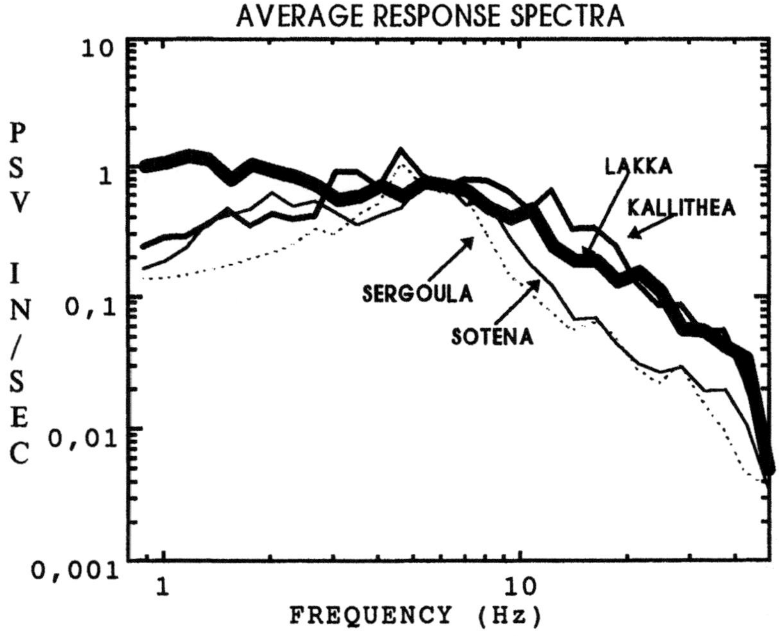

Figure 6. Average response spectra of the four characteristic stations of Sergoula, Sotena Kallithea (north) and Lakka (south).

permits the prediction of the spectral characteristics of large events, by combining the spectral observations of smaller events with the theoretical spectral scaling laws.

CONCLUSIONS

The obtained data set satisfies our expectations while designing RASMON network and proves the necessity of such local strong motion networks for more detailed earthquake engineering and seismic hazard assessment studies. This is even more important when dealing with a seismotectonically complex area like the one covered by RASMON network. Furthermore the spectral analysis of the October 25, 1991 event shows relatively large variations in peak acceleration values as well as in frequency content at different sites with comparable epicentral distances, underlining the importance of considering the role of site conditions in aseismic design studies. This is more pronounced in the case of the comparison of the response spectra of all the 18 events. However, it is well understood that realistic aseismic design requires a multidisciplinary approach involving not only the above mentioned site effect but also local and regional attenuation laws, high quality local event recordings to be used as empirical Green's functions in order to synthesise expected strong ground motion etc. The above data set can be used as a basis for such detailed studies towards which further work is oriented.

REFERENCES

Diagourtas. D., Makropoulos. K., Gariel J.-C., Wajeman N., Hatzfeld D. and Bard P.-Y., 1993. Simulation of strong ground motion using empirical Green's function method. Preliminary results. 2nd Intern. Geophysical Congress, Florina, v1, 222-235.

Hudson, D.E., 1979. Reading and interpreting strong motion accelerograms. ERRI, Pasadena, California, 1-112.

Hutching L. and Wu F., 1990. Empirical Green's Functions from small earthquakes: A waveform study of locally recorded aftershocks of the 1971 San Fernando earthquake. Journ. Geoph. Research, Vol. 95, No B2, 1187-1214.

Makropoulos, K.C., Diagourtas, D., Voulgaris, N., 1993. The Western part of Gulf of Corinth (Greece) strong motion network (RASMON). Preliminary results. . 2nd Intern. Geophysical Congress, Florina, v1, 460-471.

Ormsby, J.F.A., 1961. Design of numerical filters with application to missile data processing. Assoc. Comp. Machinery J. 8:440-460.

Rigo, A., Lyon-Caen, H., Armijo,R., Bernard,P., Makropoulos, K.C., Papadimitriou, P. and Deschamps,A., 1993. Fault plane solutions of microearthquakes and tectonics analysis in the western part of the gulf of Corinth (Greece). 2nd Intern. Geophysical Congress, Florina, v1, 275-285.

Scherbaum F., 1987a. Seismic energy imagining of the site-response using microearthquake recordings, I, Method. Bull. Seismol. Soc. Am., 77, 1905-1923.

Seed, H.B., Ugas, C. and Lysmer, J., 1976. Site-dependent spectra for earthquake-resistant design, Bull. Seism.Soc.Am., 66, 221-243.

Scherbaum F., 1987b. Seismic energy imagining of the site-response using microearthquake recordings, II, Application to the Swabian Juras, southwest Germany, seismic network. Bull. Seismol. Soc. Am., 77, 1924-1944.

Trifunac, M.D. and Lee, V.M. 1979. Automatic digitisation and processing of strong-motion accelerograms. Part 1, Univ. Southern California, Tech. Rep. CE, 79-151.

STRONG MOTION MODELING AT A NEAR-SOURCE SITE IN AVEZZANO, ITALY

A. MENDEZ[(1)], F. PACOR[(1)], R. BERARDI[(2)] and C. PETRUNGARO[(2)]
(1) ISMES S.p.A., Via Pastrengo 9, 24068 Seriate (BG), Italy
(2) ENEL S.p.A., Rome, Italy

This paper describes the results of a seismic risk study at a near-source site in Avezzano, Italy. The goal was to quantify seismic hazard through the calculation of site-dependent response spectra and ground motion time histories. The methodology used consists in combining deterministic and stochastic modeling of radiated seismic energy from an extended fault. The use of an extended source is crucial as the site is in the near-source region of the Serrone fault, which is thought to have ruptured during the 1915 Avezzano earthquake (Ms = 6.9). In a first phase of the study, we determined plausible ranges for the parameters used in characterizing possible rupture scenarios on the Serrone fault. These parameters range from the purely geometrical (i.e., fault orientation and dimensions) to kinematic (i.e., plausible hypocenters, rupture velocity, seismic moment); and were subsequently used to generate a representative family of low-frequency synthetics. The simulation of high-frequency ground motion was based on the stochastic technique of Boore (1983). Since this method implicitly assumes a point source, we developed a hybrid technique which combines Boore's method with elements of the isochron formulation of Spudich and Frazer (1984) and Bernard and Madariaga (1984), used to generate high-frequency synthetic waveforms. The isochron formulation, as used here, provides the physical basis for the generation of realistic waveform envelopes resulting from the rupture of an extended fault. In a final step, broadband synthetics were constructed by merging deterministic low-frequency and stochastic high-frequency waveforms. Response spectra for the site of interest were then calculated, along with bounds reflecting the different rupture scenarios considered.

1. Introduction and Methodology

The problem at hand was to obtain a representative family of acceleration time histories and corresponding response spectra for a near-source site assuming that an earthquake with similar characteristics as the 1915 earthquake could occur on the Serrone fault (Figure 1). The requirement was to furnish engineers with pertinent parameters such as peak acceleration values, strong motion duration, and spectral content of the ground motion, to help evaluate the seismic hazard as part of a site selection study.

V. Schenk (ed.), Earthquake Hazard and Risk, 253–266.

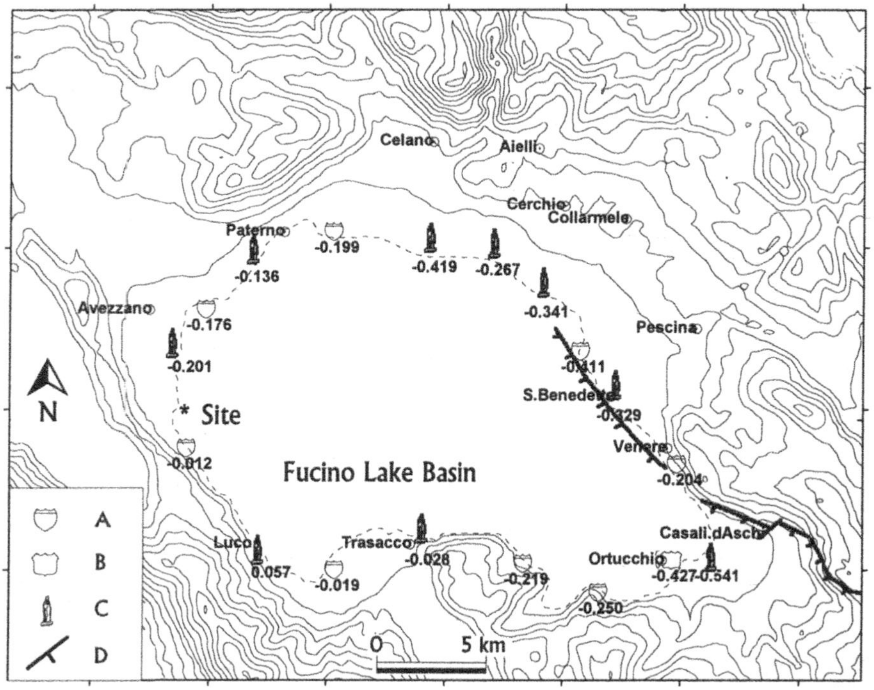

Figure 1. Epicentral region of the 1915 Avezzano earthquake (modified from Galadini et al., 1995). The fault scarp of the 1915 earthquake is represented with the symbol labelled D. As proposed by Ward and Valensise (1989), this earthquake is thought to have activated the Serrone fault, rupturing an area approximately 24 km in length and 16 km wide underlying the Conca del Fucino. The asterisk represents the near-source site under study. The symbols labelled A, B, and C are the benchmarks of two leveling campaigns performed in a time interval spanning the occurrence of the earthquake. The geodetic data was the starting point for a parametric study to define plausible rupture scenarios for a future earthquake.

The methodology used consists in the combination of deterministic and stochastic modeling of radiated seismic energy from an extended fault plane. Both aspects of the forward modeling require the selection of physical parameters to characterize a possible rupture of the Serrone fault. However, the number of ground motion time histories that can be generated from an exhaustive list of parameters describing fault, propagation medium, and site characteristics can quickly become unwieldy. Starting at the low frequency limit of 0 Hz, we conducted a study to define plausible ranges of parameters such as hypocentral location, spatial slip distribution and seismic moment, which would serve as the basis for defining plausible rupture scenarios. To complete the definition of a rupture scenario, the range of variation of other parameters such as rupture velocity and slip rise time was selected so as to encompass those values which produced significant variations (in amplitude, duration, and/or frequency content) in the low-frequency synthetic acceleration time histories.

Having defined 6 plausible rupture scenarios of the Serrone fault, low-frequency synthetics were calculated using the complete response of the medium (Figure 2). For this purpose, the Serrone fault was divided into a number of subfaults and the response at the site of interest from a slip dislocation placed on each subfault was calculated using PROSE (Luco and Apsel, 1983; Apsel and Luco, 1983). The total motion at the site was calculated by summing over the contributions from subfaults, with appropriate time delays corresponding to the passage of the rupture front through each subfault. A representative family of high-frequency acceleration time histories was generated using a hybrid method which combines the stochastic modeling approach of Boore (1983) with elements of the isochron formulation of Bernard and Madariaga (1984) and Spudich and Frazer (1984). In this approach, the high-frequency oscillations observed in actual ground motion records are generated stochastically, while the envelopes of these oscillations are calculated in a deterministic fashion from plausible rupture scenarios. The broadband synthesis of representative acceleration time histories was the final step to define the input for the calculation of response spectra.

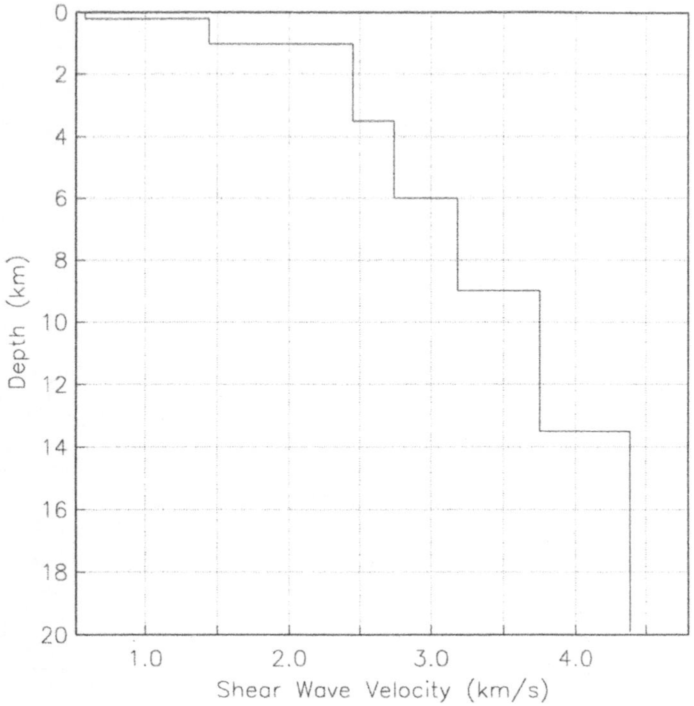

Figure 2. Profile of shear wave velocity vs. depth for the Conca del Fucino. The first two layers, characterizing the medium to a depth of 1.0 km, are intended to represent the sediment filled bed of an old lake that once partially filled the Fucino depression. The amplification factor derived from the impedance contrast between the velocity structure representative of the lake bed and that at depth is 1.7. The velocity structure is used in the calculation of complete Green's functions for the inversion of the geodetic data (0 Hz) and for the deterministic modeling of synthetic ground motion (0.-1.5 Hz). On the high frequency side, it is used to calculate travel times from the extended fault plane to the near-source site. This travel time information is employed in the construction of high-frequency envelopes from a series of proposed rupture scenarios for the Serrone fault.

2. Definition of Rupture Scenarios and Generation of Low-Frequency Synthetics

The parametric study, briefly outlined above, was carried out starting at the low frequency limit of 0 Hz and proceeding toward higher frequencies. At low frequencies, one of the most important parameters is the resulting spatial slip distribution from the rupture of an extended fault. Plausible spatial slip distributions were obtained by repeating the inversion of geodetic data used by Ward and Valensise (1989) in their study of the 1915 earthquake. Figure 3 shows an idealized Serrone fault and its geometric relation to the near-source site and to the 18 benchmarks of 2 surveying campaigns, one conducted in 1862 and the other in 1916 after the earthquake (Loperfido, 1919).The faulting mechanism is assumed to be normal. The fault dips towards the SW at an angle of 63° and is buried between 1.0 and 15.25 km in depth. The average medium parameters in this depth range are: P-wave velocity 5.31 km/s, S-wave velocity 3.07 km/s, density 2.65 gm/cm^3 and shear modulus 3.0×10^{11} dyne/cm^2.

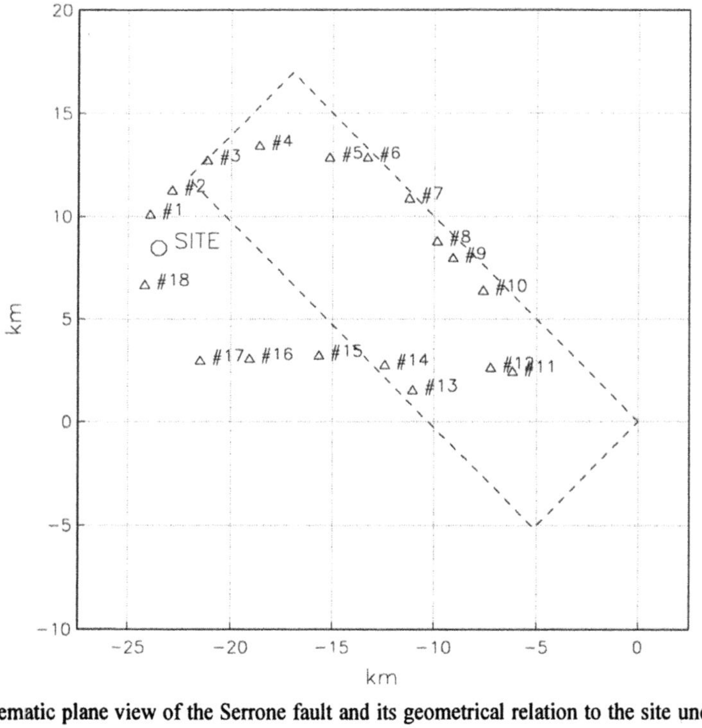

Figure 3. Schematic plane view of the Serrone fault and its geometrical relation to the site under study and to the 18 benchmarks where vertical surface deformation is available. This data was inverted to obtain plausible slip distributions, and derived parameters such as seismic moment, to be used in the definition of plausible rupture scenarios.

The goal of the inversion was not so much a study of the exact source characteristics of the 1915 earthquake but a study to obtain plausible faulting scenarios for the case of a future earthquake on the Serrone fault. Using a method described in

Mendez and Pacor (1994), the inversion was performed to obtain not one, but a family of slip distributions compatible with the geodetic measurements. In this context, compatible means that all inferred slip distributions have seismic moments within certain bounds and produce a predefined tolerable amount of misfit to the actual data. The important characteristics of this method that concern us here are: 1) it provides a quantitative way to bound the parameter space of plausible slip models as well as, to a certain extent, rupture models, and, 2) it provides spatial slip distributions, and derived parameters such as seismic moment distribution, which are input for the calculation of both low-frequency and high-frequency synthetic ground motion time histories.

Figure 4 displays the geodetic data (triangles), and the best fit (asterisks) with error bounds obtained from the inversion. The best fit is produced by the spatial slip distribution shown in Figure 5 which has a seismic moment of 9.7×10^{25} dyne-cm and 84 cm of average dip-slip motion. The inversion process also yields two extremal slip models (not shown here) which encompass slip models producing data fits lying within the error bounds of Figure 4. The bounds on seismic moment derived from this study are 8.8×10^{25} - 10.6×10^{25} dyne-cm.

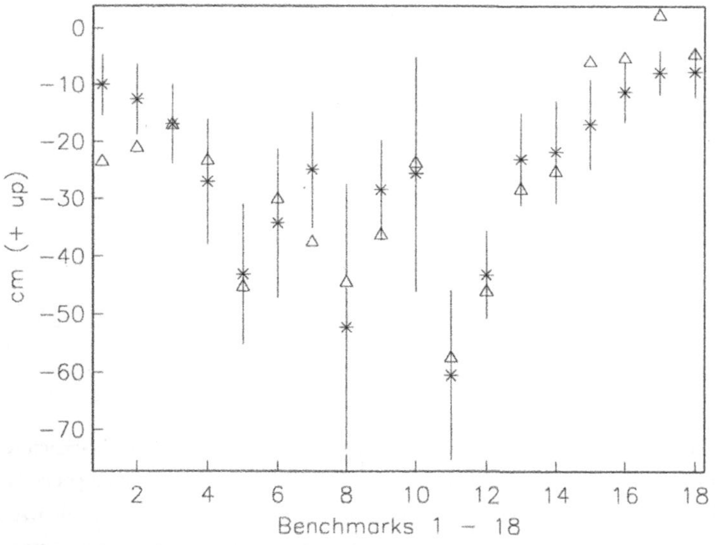

Figure 4. Vertical surface deformation data (triangles), data fit (asterisks) from the inferred slip distribution shown in Figure 5, and error bounds (vertical lines).

The results of the geodetic inversion can be used as the basis for proposing plausible faulting scenarios of the Serrone fault. From Figure 5, three hypocenters can be selected. Two would be centered near the two areas of maximum slip derived from the geodetic inversion (H1 and H3). This selection was also intended to be representative of ruptures initiating at opposite ends of the fault, and therefore representative of the different directivity effects that could arise during a future earthquake. Without further analysis, it is difficult to determine if the central area of the fault experienced further slip after the 1915 earthquake. A conservative assumption would be that this area is locked and is therefore a candidate for the nucleation of a future rupture. The third hypocenter would be placed here (H2). For the scenarios H1 and H3, a plausible spatial slip distribution would be that found from the inversion. For hypocenter H2 centered in the area of low slip, a complementary slip distribution would be used. It would present large values of slip where the geodetic inversion inferred low values and vice versa.

The results of the geodetic inversion serve to partially define the space of parameters used to generate low-frequency synthetics: hypocenter locations and spatial slip distributions. A further parametric study was carried out to gauge the influence of other variables needed to complete the description of a possible rupture on the Serrone fault: total seismic moment, slip rise time, rupture velocity, rupture front geometry, and rake of the slip vector. The goal was to identify the most relevant rupture parameters to be used for the generation of a representative family of low-frequency synthetics.

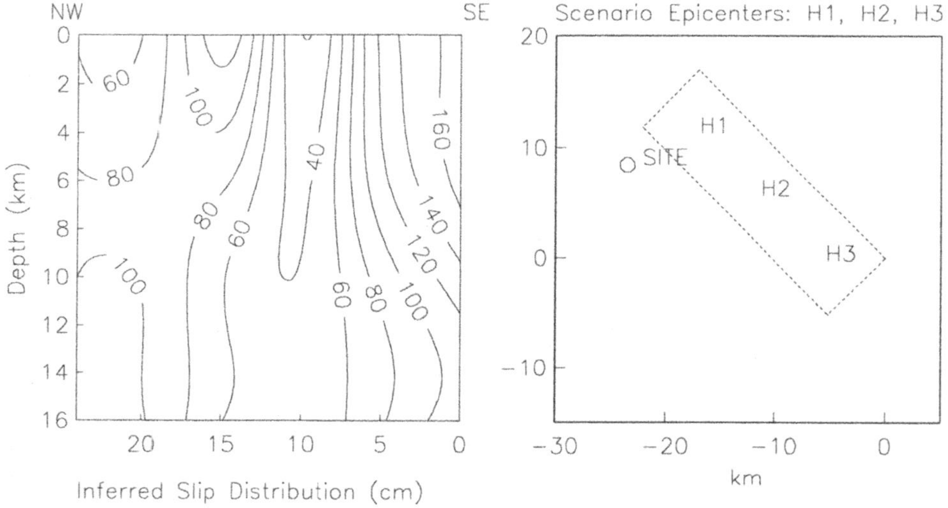

Figure 5. (Left) Spatial slip distribution on the Serrone fault inferred from the inversion (contour values in cm). *(Right)* The two areas of high slip values were selected as possible nucleation points for a future rupture (H1 and H3). H1 is the hypocenter closest to the site under study. The area of low slip was also proposed as a possible nucleation point (H2). In this case, a complementary slip distribution was used to generate low-frequency synthetics.

In some cases, the influence of a parameter was determined by its effect in the response spectra generated from the three rupture scenarios H1, H2 and H3. For example, the geodetic inversion was performed assuming that the vertical surface deformation could be attributed completely to dip-slip motion. In order to judge if this assumption was too restrictive, response spectra were also generated introducing strike-slip motion on the fault. In other cases, the influence of a parameter was determined by its effect on the low-frequency synthetics. For example, it was seen that using rupture velocities in the range between 70% and 85% of the mean shear wave velocity of the medium had a greater influence (in terms of peak amplitudes, duration, and/or frequency content of synthetics) than the variation of seismic moment between the bounds inferred from the geodetic inversion.

In summary, a total of six plausible rupture scenarios were selected for the generation of synthetic ground motion. The kinematic description of a rupture scenario consists in specifying a hypocenter, from which the rupture propagates in a radial fashion with a velocity equal to some percentage of the mean shear wave velocity of the medium, and with a constant slip rise time. The circular rupture front is perturbed in shape by the addition of a small stochastic component to the actual time of rupture of each point on the fault, as determined by the average rupture propagation value. For each of the three selected hypocenters, H1, H2, and H3, two rupture velocities were used. A 'slow' event was characterized by an average rupture velocity (V1) equal to 70% of the mean shear wave velocity of the medium: 3.07 km/s. A 'fast' event had a mean rupture velocity (V2) of 85% of the mean shear wave velocity. The classification of ruptures in slow and fast is purely kinematic and is not meant to imply any underlying physical difference between these two types of rupture.

For the purpose of calculating low-frequency synthetics the Serrone fault was divided into subfaults of area 0.5×0.5 km^2. The triggering time of each subfault was defined as the time of rupture arrival at the center of each subfault. The response at the site of interest due to a purely normal dislocation of these subfaults was calculated using PROSE (Luco and Apsel, 1983; Apsel and Luco, 1983) assuming a vertically stratified medium consisting of six layers overlying a halfspace and a Poisson's ratio of 0.25 (see Figure 2). The total response at the site was calculated by summing over the time delayed subfault contributions. Figure 6 illustrates the response spectra calculated at the site of interest from the low-frequency horizontal component synthetics generated for these six scenarios. The range of variability of response spectra below 1 Hz is a direct reflection of the different scenarios considered. As will be discussed in Section 4, a broadband synthesis of low- and high-frequency synthetics was performed in order to obtain response spectra in the frequency range 0.1-50 Hz. Thus the spectral values above 1 Hz in Figure 6, calculated using as input only the low-frequency synthetics, should not be considered significant.

3. High-Frequency Synthetics

For the generation of high-frequency synthetics, we have developed a hybrid technique based on the stochastic simulation method of Boore (1983) and the isochron formulation of Bernard and Madariaga (1984) and Spudich and Frazer (1984). In a typical application of stochastic modeling, a large number of synthetic time histories are generated such that on the average, the properties of the time histories mimic those of earthquake ground motion. In Boore's method, this is accomplished by windowing white Gaussian noise with an envelope function which generally has some simple mathematical representation. In order to replicate the spectral characteristics of earthquake ground motion, these time limited noise waveforms are filtered with a reference spectra such as an ω^2 model with anelastic attenuation.

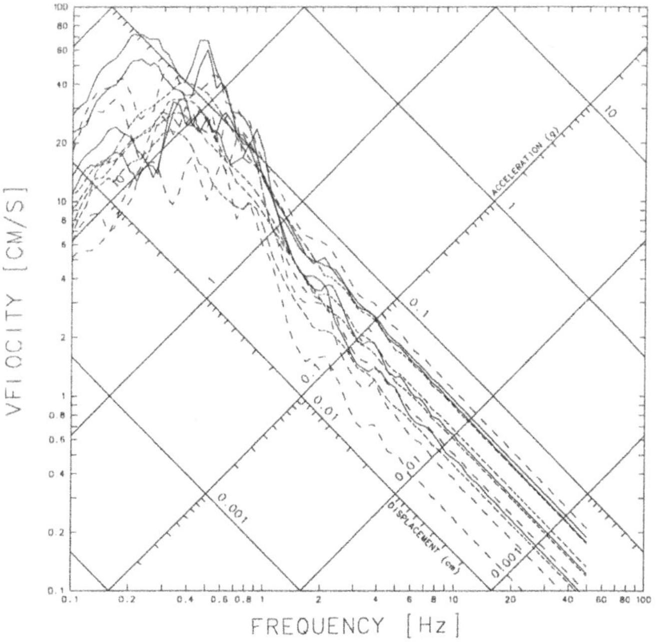

Figure 6. Response spectra for the site of interest calculated from horizontal component low-frequency synthetics generated for the case of six rupture scenarios described in text: hypocenter H1 and rupture velocities V1 and V2 (dashed traces), hypocenter H2 and rupture velocities V1 and V2 (fine dashed traces); and hypocenter H3 and rupture velocities V1 and V2 (solid traces).

This technique is capable of efficiently generating realistic earthquake ground motion time histories and requires the specification of only a limited number of parameters to characterize the source, propagation medium, and site. One major limitation is that this method is based on a point source description of an earthquake. For the case of near-source modeling, it would be desirable to take into account the finite dimensions of the fault so as to incorporate effects such as source directivity. We have approached this problem by using the isochron formulation for the calculation of radiated seismic energy from an extended fault. In our application, the isochron formulation is used to generate ground motion envelopes based on a simple kinematic description of a rupture. Following Boore (1983), these envelopes are subsequently filled in with Gaussian noise with spectral characteristics equivalent to those of actual earthquake ground motion.

This idea is illustrated with an example showing the calculation of envelopes at the near-source site under study and at benchmark #11 (see Figure 3). Figure 7a represents a hypothetical rupture of the Serrone fault. The contours represent time in seconds when the rupture arrives at different portions of the fault. In Figure 7b, the contour plots represent the time of arrival of shear waves from the Serrone fault to the two sites. In the isochron formulation, ground motion at a given time is calculated as the contribution from all points on the fault where emitted seismic radiation has the same arrival time at the site of interest. This locus of points defines a curve on the fault called

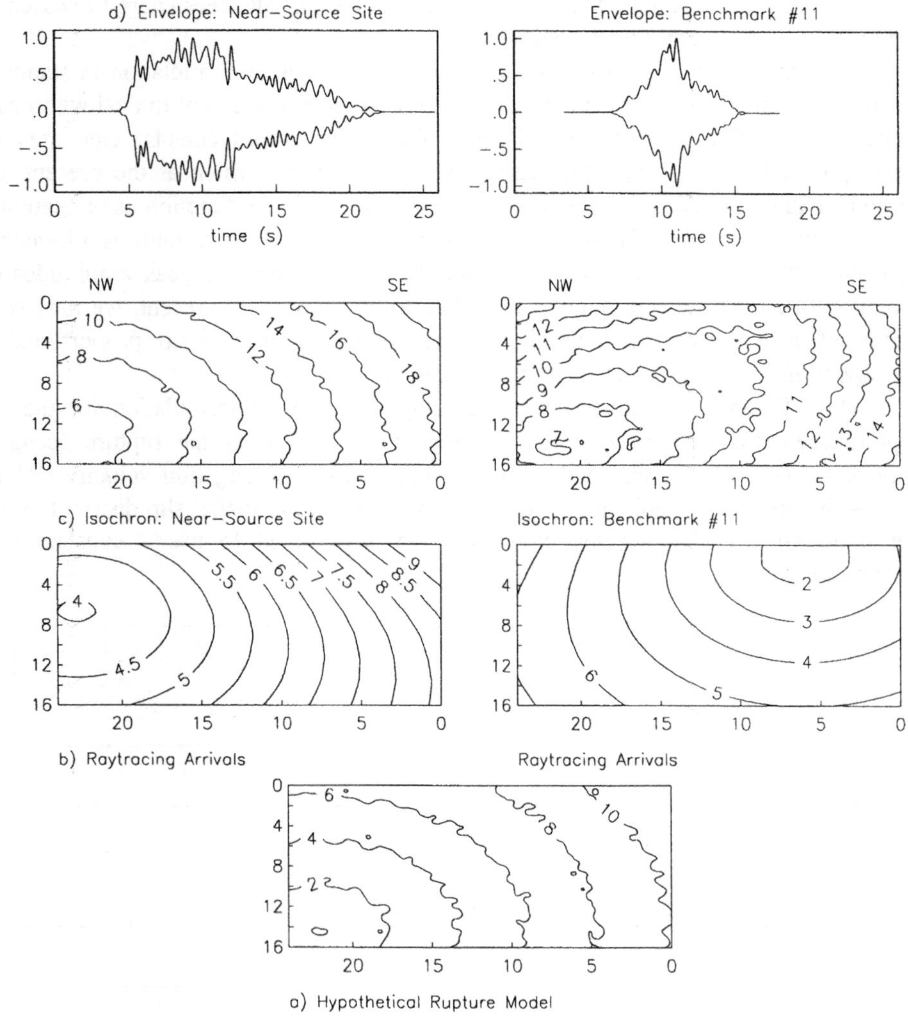

Figure 7. ***a)*** **Hypothetical rupture of the Serrone fault. Contours represent time in seconds at which each point on the fault ruptures. The contours are slightly irregular because a small stochastic component has been added to the rupture velocity.** ***b)*** **Shear wave raytracing and** ***c)*** **isochron arrivals for the near-source site and benchmark #11. The rupture model times of** ***a)*****, added to the raytracing times** ***b)*** **for each site, produce the corresponding isochron times shown in** ***c).d)*** **Isochron generated envelopes normalized to unit peak amplitudes, for the near-source site and benchmark #11, for the rupture scenario depicted in** ***a)*****.**

an isochron. The isochrons for the two sites are shown in Figure 7c, and clearly show that the energy release on the fault plane is perceived in different fashions depending on the position of the observer. One would therefore expect that the envelopes of high-frequency radiation at the two sites should reflect the different way that the two sites receive information about the rupture (Figure 7d). In this figure, the shorter duration of the envelope for benchmark #11 is due to the fact that this site sees energy released at the fault coming in at an accelerated rate.

While the use of envelopes calculated using the isochron formulation in standard stochastic modeling methods appears to be a promising enhancement to deal with near-source site conditions, there is still need for further improvements. One area of development has to do with the scaling of the envelopes, which at the present are smoothed and normalized to unit area. Since the envelope function is a mapping function which describes how seismic moment release along a fault is ultimately perceived at a site, it contains information about the power and peak amplitudes of ground motion. This information is lost after normalization. At present, we deal with this problem by scaling the stochastic waveforms with a constant factor proportional to the power of the envelope before its normalization.

Figure 8 shows envelopes generated using the isochron formulation for the six rupture scenarios considered. Each envelope is identified by the rupture scenario hypocenter (H1, H2, or H3) and the mean rupture front propagation velocity (V1 or V2). Although it is possible to include information about the spatial slip distribution on the fault when calculating the envelopes, we have opted to use a constant slip distribution.

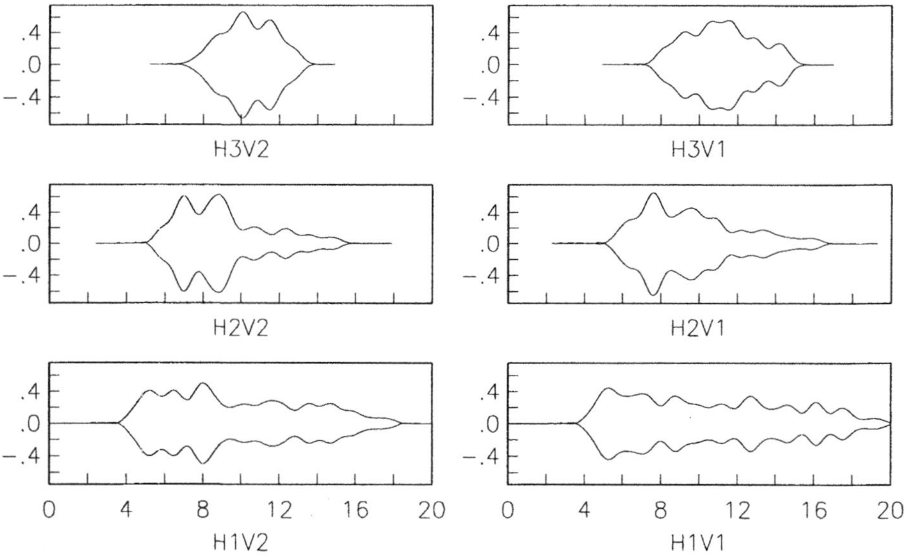

Figure 8. Envelope functions generated using the isochron formulation for the six faulting scenarios described in the text. These envelope functions replace the standard envelope used in Boore's method for the purpose of generating high-frequency synthetics.

Figure 9 displays representative response spectra for the site of interest obtained from the stochastic modeling of acceleration time histories incorporating the envelopes generated using the isochron formulation. This example corresponds to the rupture scenarios H3V1 and H3V2, and for each scenario five realizations were performed. It can be seen that the high frequency asymptote of the spectra for the case of H3V2 is greater than that for H3V1, reflecting the fact that the average rupture velocity of the former scenario is larger than that of the latter.

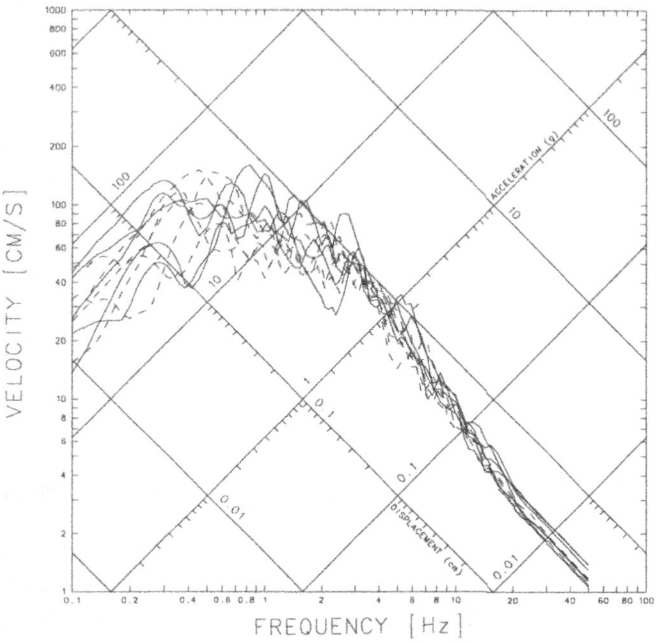

Figure 9. Response spectra (5% damping) for rupture scenarios H3V1 (dashed traces) and H3V2 (solid traces). For each rupture scenario, 5 realizations of Boore's method were performed using the corresponding envelopes shown in Figure 8. These 10 acceleration time histories were the input for the calculation of the response spectra shown here.

4. Broadband Synthesis and Response Spectra

The low- and high-frequency synthetics were combined using a broadband synthesis technique. The six faulting scenarios were used to generate 12 low-frequency horizontal acceleration time histories. For each low-frequency horizontal component, two realizations of stochastic modeling were performed to obtain high-frequency counterparts. The broadband synthesis was performed in the frequency domain; below 0.6 Hz, the contribution to the broadband trace is exclusively from the low-frequency trace, while above 1.0 Hz, the contribution is exclusively from the high-frequency trace. In the frequency interval 0.6 - 1.0 Hz, a weighted sum of low- and high-frequency spectral components is used to synthesize the broadband signal. The total number of broadband synthetics obtained from this process was 24. Figure 10 shows 6 example acceleration time histories, one for each rupture scenario considered.

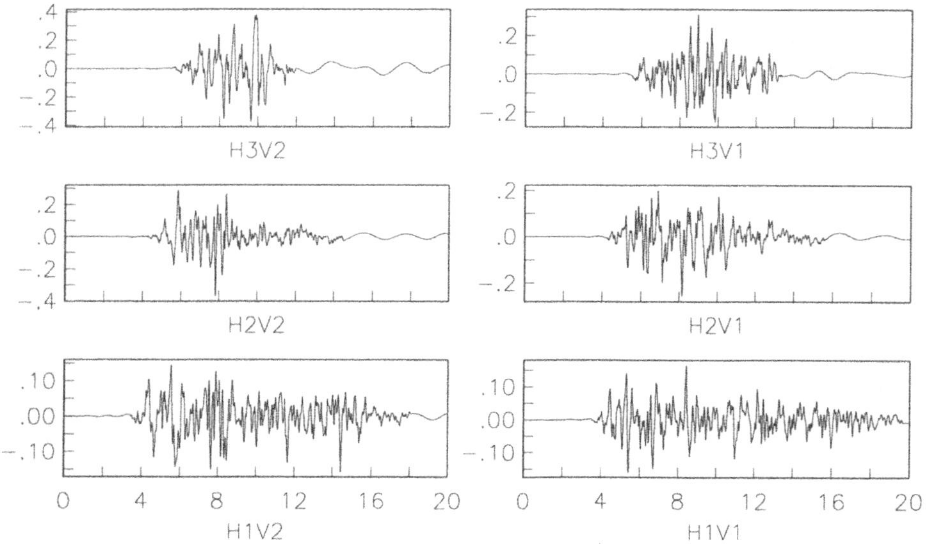

Figure 10. Example broadband acceleration traces generated for each of the six rupture scenarios. Units are in g. Only the first 20 seconds are shown.

The 24 acceleration traces were used as input to calculate response spectra (5% damping) at the site of interest. Figure 11 displays the resulting mean and 84.1 percentile response spectra. It is important to note that the variability associated with the mean response spectral values is representative of the different rupture hypothesis considered. This figure also shows the mean and 84.1 percentile response spectra calculated using the attenuation relation of Pugliese and Sabetta (1989) for pseudo-velocity spectral values as a funtion of magnitude, distance, and site conditions. This attenuation relation was developed using an accelerogram database consisting of 95 strong-motion recordings from 17 Italian earthquakes of magnitudes between 4.6 and 6.8 (Sabetta and Pugliese, 1987). The values of the parameters used to calculate the displayed spectra are: M_s=6.6, epicentral distance equal to 12.7 km (equivalent to the distance between the site and the surface projection of the fault baricenter), and rigid site conditions. It should be noted that Pugliese and Sabetta's attenuation relation for pseudo-velocity spectral values was developed for frequencies greater than 0.25 Hz, reflecting the fact that the employed acceleration traces have been highpass filtered to remove low-frequency noise such as that introduced by the analog recording of ground acceleration on film and its subsequent digitization. Figure 11 also displays for comparison the US Atomic Energy Commission RG1.60 (1973) response spectrum scaled to 0.25 g. The major difference between the response spectra lies in the estimate of spectral velocities in the low-frequency interval below 0.9 Hz. On the whole, however, the similarity in response curves is encouraging as it serves to validate the methodology used here and to qualify it as a method which can complement existing techniques used in the estimate of seismic input parameters for risk assessment.

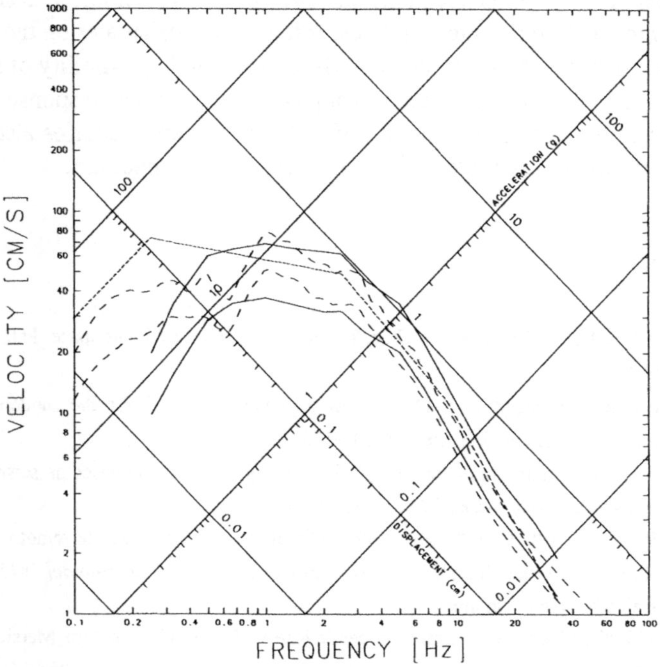

Figure 11. Comparison between mean and 84.1 percentile response spectra (5% damping) obtained using the method described in the text (dashed traces), those obtained using Pugliese and Sabetta's regression results (solid traces), and the RG1.60 response spectrum (fine dashed trace) scaled to the mean PGA of the previous two mean response spectra: 0.25g.

5. CONCLUSIONS

This paper has presented a methodology to generate broadband ground motion records, and corresponding response spectra, based on a combination of deterministic and stochastic modeling of radiated seismic energy. An important element of this method is the extension of standard stochastic modeling techniques, such as that of Boore (1983), to take into account effects due to extended fault planes. This was accomplished by calculating strong motion envelopes in a deterministic fashion based on a physical representation of the rupture process on a finite fault. The method was illustrated here for a near-source site study in Avezzano, Italy.

In a first step, the study concentrated on defining plausible rupture scenarios for the Serrone fault, thought to have ruptured during the 1915 Avezzano earthquake. A parametric study was performed to evaluate the importance on ground motion of the various parameters used to characterize a possible rupture scenario. In particular, the variation of rupture velocity and hypocenter location had a greater influence on ground motion (in terms of peak amplitudes, duration, and/or frequency content) than the variation of seismic moment between the bounds inferred from a geodetic inversion.

The parametric study permitted to suggest six plausible rupture scenarios for a hypothetical earthquake on the Serrone fault. These rupture scenarios are characterized by three different hypocenter locations and two rupture velocities. The scenario which presents the most risk to the site of interest is that which originates on the SE section of the fault and propagates towards the NW. Because of the impossibility of specifying the manner in which a possible future earthquake could occur, response spectra were calculated for all six scenarios. The use of different rupture scenarios also permitted to quantify the variability associated with the mean response spectrum.

6. REFERENCES

APSEL R. J. AND J. E. LUCO (1983) On the Green's functions for a layered half-space. Part II, *Bull. Seismol. Soc. Am.*, **73**, 931-951.

BERNARD P. AND R. MADARIAGA (1984). A new asymptotic method for the modeling of near-field accelerograms, *Bull. Seismol. Soc. Am.*, **74**, 539-558.

BOORE D. M. (1983) Stochastic simulation of high-frequency ground motion based on seismological models of the radiated spectra, *Bull. Seismol. Soc. Am.*, **73**, 1865-1894.

GALADINI F., P. GALLI AND C. GIRAUDI (1995) Gli effetti geologici del terremoto del 1915, in S. CASTENETTO, F. GALADINI (editors): *Il terremoto di Avezzano del 13 gennaio del 1915*. Monografia del Servizio Sismico Nazionale, in press.

LOPERFIDO A. (1919) Indagini asronomiche e geodetiche nell'area del terremoto Marsicano, *Atti Lavori Pubblici*, pp 95.

LUCO J. E. AND R. J. APSEL (1983) On the Green's functions for a layered half-space. Part I, *Bull. Seismol. Soc. Am.*, **73**, 909-929.

MENDEZ A. AND F. PACOR (1994) A study of the rupture characteristics of the 40s subevent of the 1980 Irpinia earthquake, *Annali di Geofisica*, **37**, 1601-1620.

PUGLIESE A. AND F. SABETTA (1989) Stima di spettri di risposta da registrazioni di forti terremoti italiani, *Ingenieria Sismica*, **6**, 3-14.

SERVA L., A. M. BLUMETTI AND A. M. MICHETTI (1986) Gli effetti sul Terremoto del Fucino (13 Gennaio 1915); Tentativo di interpretazione della evoluzione tettonica recente di alcune strutture, *Mem. Soc. Geol. It.*, **35**, 893-907.

SPUDICH P. AND L. N. FRAZER (1984) Use of ray theory to calculate high-frequency radiation from earthquake sources having spatially variable rupture velocity and stress drop, *Bull. Seismol. Soc. Am.*, **74**, 2061-2082.

SABETTA F. AND A. PUGLIESE (1987) Attenuation of peak horizontal acceleration and velocity from Italian strong-motion records, *Bull. Seismol. Soc. Am.*, **77**, 1491-1513.

UNITED STATES NUCLEAR REGULATORY COMMISSION (1973) Design response spectra for seismic design of nuclear power plants, *Regulatory Guide*, No 1.60, pp. 7.

WARD S. N. AND G. R. VALENSISE (1989) Fault parameters and slip distribution of the 1915 Avezzano, Italy earthquake derived from geodetic observations, *Bull. Seismol. Soc. Am.*, **79**, 690-710.

APPLICATION OF MICROSEISMS TO MICROZONING OF AN URBAN AREA IN A SEDIMENTARY BASIN

JUNPEI AKAMATSU[1], HIROSHI YASUI[2],
KEIICHI NISHIMURA[3] and MASAO KOMAZAWA[4]

1 Disaster Prevention Research Institute, Kyoto University
Gokasho, Uji, Kyoto 611, Japan
2 School of Civil Engineering, Kyoto University
Gokasho, Uji, Kyoto 611, Japan
3 Geophysical Institute, Faculty of Science, Kyoto University
Kitashirakawa-Oiwakecho, Sakyo, Kyoto 606-01, Japan
4 Geological Survey of Japan
1-1-3, Higashi, Tsukuba 305, Japan

Abstract

Taking account of the necessity for extending the frequency range to lower frequency in the seismic microzoning in an urban area, a comparative analysis of microseisms (long-period microtremors) and gravity data was carried out in the Kyoto basin, southwest Japan. The stability of spectral ratios of microseisms (soil site/rock site) was first examined through repeated simultaneous observation at soil sites and rock sites. Then spatial variation of the spectral ratios in the whole basin was mapped and compared with the characteristic feature of Bouguer gravity anomaly which was bandpass filtered using the upward-continuation technique. Main results obtained are as follows: (1) spectral features of microseisms observed both at soil sites and rock sites are remarkably affected by weather condition, but when averaged for different weather conditions, the spectral ratios (soil site/rock site) exhibit similar features, which enable us to take them as time-invariant and site-specific amplification factors; (2) the amplification factors for both horizontal and vertical components become larger stepwise from north to south in the basin, and correspondingly the frequency range having large amplification extends to lower frequency; (3) the amplification factors for horizontal components are larger than those for vertical component in the low frequency range; (4) frequency characteristics of the amplification factors for vertical component exhibit a similar shape to those for horizontal components with one-octave shift in frequency; (5) horizontal component of vibration oriented parallel to the long axis of the basin is amplified more than that oriented normal to the axis; (6) the spatial distribution of amplification factors has a close correlation with the general trend of the bandpass-filtered Bouguer gravity anomaly. These features of amplifications are interpreted from the viewpoint of Rayleigh surface waves incident to the basin with 3-D structure of bedrock, though it is suggested that some of them may be explained approximately by 1-D resonance of S and P waves in the soil sediments. It is pointed that comparative analysis of microseisms and gravity data are very useful for microzoning of an urban area with irregular configuration of bedrock, especially in the case where the information of subsurface structure is not available.

V. Schenk (ed.), Earthquake Hazard and Risk, 267–279.

1. Introduction

In the study of predicting earthquake ground motion in an urban area, long-period vibrational characteristics of a soil ground in a sedimentary basin are of particular importance. This comes from the fact that many urban areas are situated on thick soil sediments with long characteristic period, and that the need for various urban facilities with long natural period has been increasing rapidly. For example, it is known that the natural period of high rise buildings is considerably longer than 1 second, sometimes reaching 3 seconds or more. This requires that the frequency range for microzoning in an urban area should be extended toward lower frequency. Thus in this study we will focus on, so to say, the low-frequency microzoning.

As an empirical approach to the ground vibration characteristics, the applicability of long-period microtremors or microseisms has become recognized (e.g. Lermo and Chavez-Garcia, 1994). However, this is presently limited to areas with relatively simple geology. For areas with complicated geology, such as a sedimentary basin with irregular configuration of bedrock, more systematic researches on the natures of microseisms are necessary to clarify the effectiveness and limitation of their application (Seo, 1992).

It is well known that microseisms (i.e. microtremors with period longer than about 1 s) are generated mainly by sea-waves (Darbyshire, 1990), and accordingly their spectral contents are affected to a considerable extent by weather condition. Moreover, microseisms are considered as a non-stationary stochastic process (Morikawa et al., 1994). Therefore, prior to utilizing them for the study of ground vibration characteristics, it is an important task to examine whether the time-invariant site-specific characteristics can be obtained in spite of temporal variation of incident microseisms. And, if the site-specific characteristics are obtained, the next task is to examine in what manner they reflect the subsurface structure under the study area. However, a difficulty often faced with at this point is that the sufficient information about the structure is not always available. In such cases, an effective, if not complete, way of overcoming the difficulty is to use other geophysical data, such as Bouguer gravity anomaly, which are helpful to discuss the subsurface structure.

From these standpoints, we conducted two-step study of microseisms in the Kyoto Basin, southwest Japan, which is one of the typical urban areas in Japan: the first step is to examine the natures of microseisms in and around the basin, and the second is to examine the spatial variation of ground vibration characteristics of microseisms in the whole basin. The results of the second step were compared with the characteristic features of the bandpass-filtered Bouguer gravity anomaly, focusing on their relationship with the shape of the basin and the configuration of bedrock.

The present work is the first attempt of comparative analysis of microseisms and gravity data for the purpose of seismic microzoning in an urban area in Japan.

2. Geological Setting and General Features of Microseisms in the Kyoto Basin

The Kyoto basin is situated in an inland area. Figure 1 shows its location and the geological map. The basin is surrounded by the Paleo-Mesozoic complex consisting mainly of sandstone, slate and chart. The surface velocities of the complex are Vp = 4.66 km/s and Vs = 2.58 km/s to a depth of about 1 km (Akamatsu et al., 1975). The shape of basin is long in the north-south direction (20 km), and the east-west width increases stepwise from north to south, reaching about 10 km in the southern part.

The soil deposits filling the basin are the Plio-Pleistocene groups covered by thin Holocene deposits. The velocity structure of the basin is represented by the following two layer model which was obtained by seismic refraction survey carried out along an east-west line in the southern part: Vp = 2.0 - 2.1 km/s for soil layer and Vp = 4.4 - 5.2 km /s for bedrock; the configuration of bedrock is shaped like a ship bottom; the depth to bedrock is approximately 700 m in the center of the southern part (Kitsunezaki et al., 1971). As for the northern part of the basin, there is no available information about the depth and configuration of bedrock.

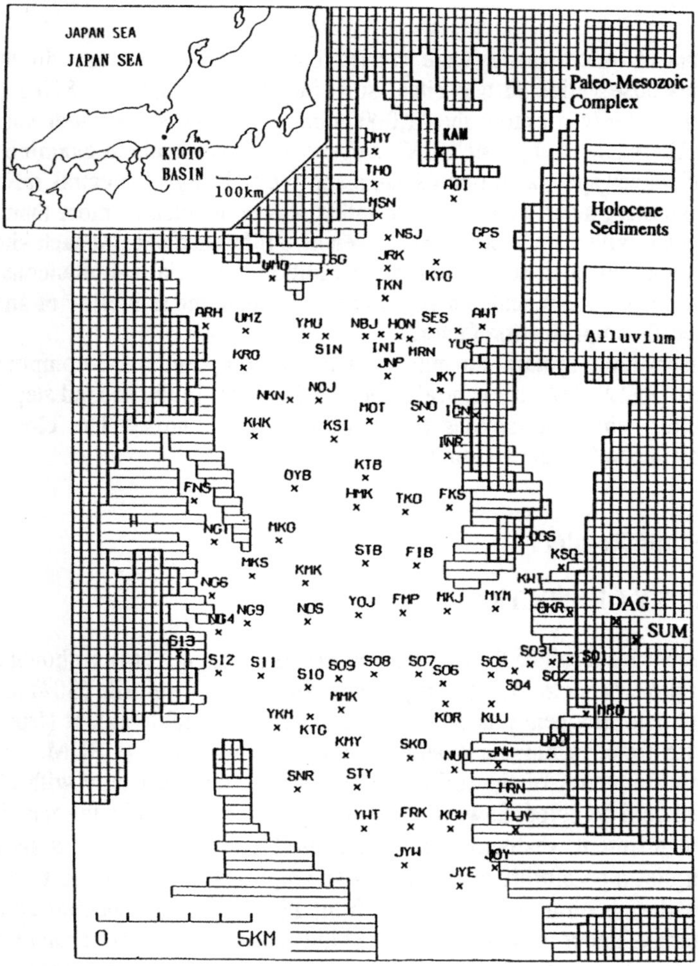

Figure 1. Geological map and site location in the Kyoto basin. SUM and DAG denote rock sites. DAG was used for a reference site during the second step observation to obtain spatial distribution of amplifications.

The nature of microseisms observed in and around the Kyoto basin was studied in detail by Okano (1961a, b, 1963). His important result is that microseisms are composed mainly of Rayleigh waves generated by sea-waves in the coastal region of the Japan Sea under a condition of winter monsoon and by swells in the Pacific coastal region due to cyclones or typhoons in summer season. Although the Kyoto basin is situated far (around 100 km) from the coastal region of the Japan Sea, a good positive correlation between the amplitude of microseisms and the height of sea-waves of the Japan Sea was observed (Akamatsu,1984). Simultaneous observations at rock sites around the basin revealed that microseisms with the same characteristics are incident to a wide area including the Kyoto basin (Akamatsu et al., 1992b).

3. Observation

Observations of microseisms were carried out through two steps. In the first step, simultaneous observation at two sites, soil site S04 and rock site SUM (Fig. 1), was performed repeatedly for more than 1,000 hours over three years under various weather condition (Akamatsu et al., 1992b). This observation was made to examine the temporal variation of power spectra of microseisms and the stability of spectral ratio, S04/SUM. In the second step, a survey of spectral ratios was conducted at more than 90 soil sites inside the basin with a rock site, DAG, taken as a reference site. At each site 15 minutes observation was repeated several times. In addition, a special simultaneous observation at KAM and DAG was repeated many times to evaluate the reliability of amplitude ratio as a site-specific amplification factor.

The observation system for microseisms is composed of 3-component velocity seismometer (PELS-73) with natural period of 10 s and, in the second step, analog 1 Hz low-pass filter which suppresses the artificial ground noise. The clock signal was calibrated by Japan Standard Time (JJY).

4. Analyses and Results

4.1 STABILITY OF SPECTRAL RATIO

Figure 2 shows the result of the first step observation. In Fig. 2 (a) is shown the temporal variation of power spectra of NS components observed at a soil site (S04) and a rock site (SUM). The peak frequency at S04 changes to a considerable extent (from 0.25 Hz to 0.6 Hz) with time, while the change is not so remarkable at SUM. Therefore, the characteristic frequency of the soil site hardly can be obtained only with observation at that site. It should be noted, however, that the spectral ratios, soil site/rock site, averaged for the three different weather conditions, exhibit a similar feature, as shown in Fig. 2 (b). This implies a possibility of estimating an amplification factor due to soil sediments with use of the spectral ratio. As seen from Fig. 2 (a), the spectral contents in the frequency range higher than about 1 Hz exhibit the typical diurnal change, showing the effect of artificial ground noise. This led us to focus on the frequency range lower than 1 Hz in the second step observations.

4.2 AMPLIFICATION FACTOR AND THEIR SPATIAL DISTRIBUTION

Figure 3 shows examples of spectral ratios for three components and their variation with site location from north to south. It should be mentioned here that the spectral ratio between KAM and DAG, estimated from the special observation repeated 120 times, is reliable to the extent that more than 85% of observed values fall within +-20% range of

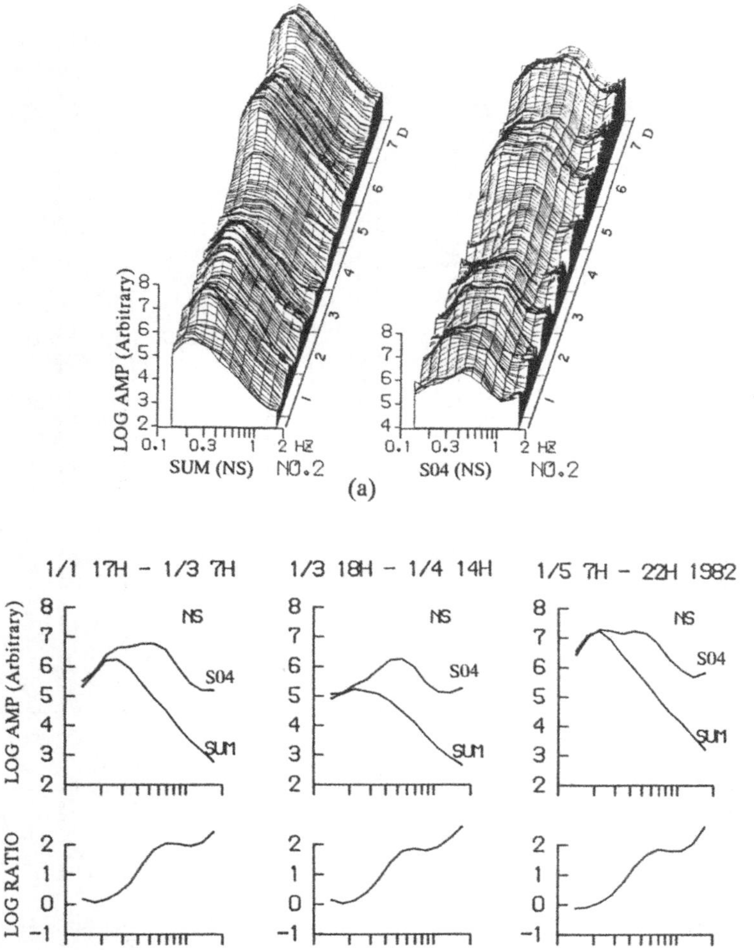

Figure 2. An example of temporal variation of power spectra observed at rock site, SUM, and soil site, S04, under various weather condition for more than a week (a), and averaged spectra under typical weather conditions and their ratios (b). Note the time-invariant characteristics of the ratios.

their mean value. We assume that the spectral ratio between other sites and DAG has the same reliability as KAM/DAG.

As can be seen from Fig. 3, for both horizontal and vertical components, the amplification factors become larger from north to south, and correspondingly the frequency ranges with large amplification extend to lower frequency; a closer look at the horizontal components shows that , once the peak of the ratios appears in the frequency range under consideration, its value does not increase significantly. For a given site, the amplification factor for horizontal components is larger than that for vertical one in the low frequency range. It is very interesting that, for the site, the frequency characteristics

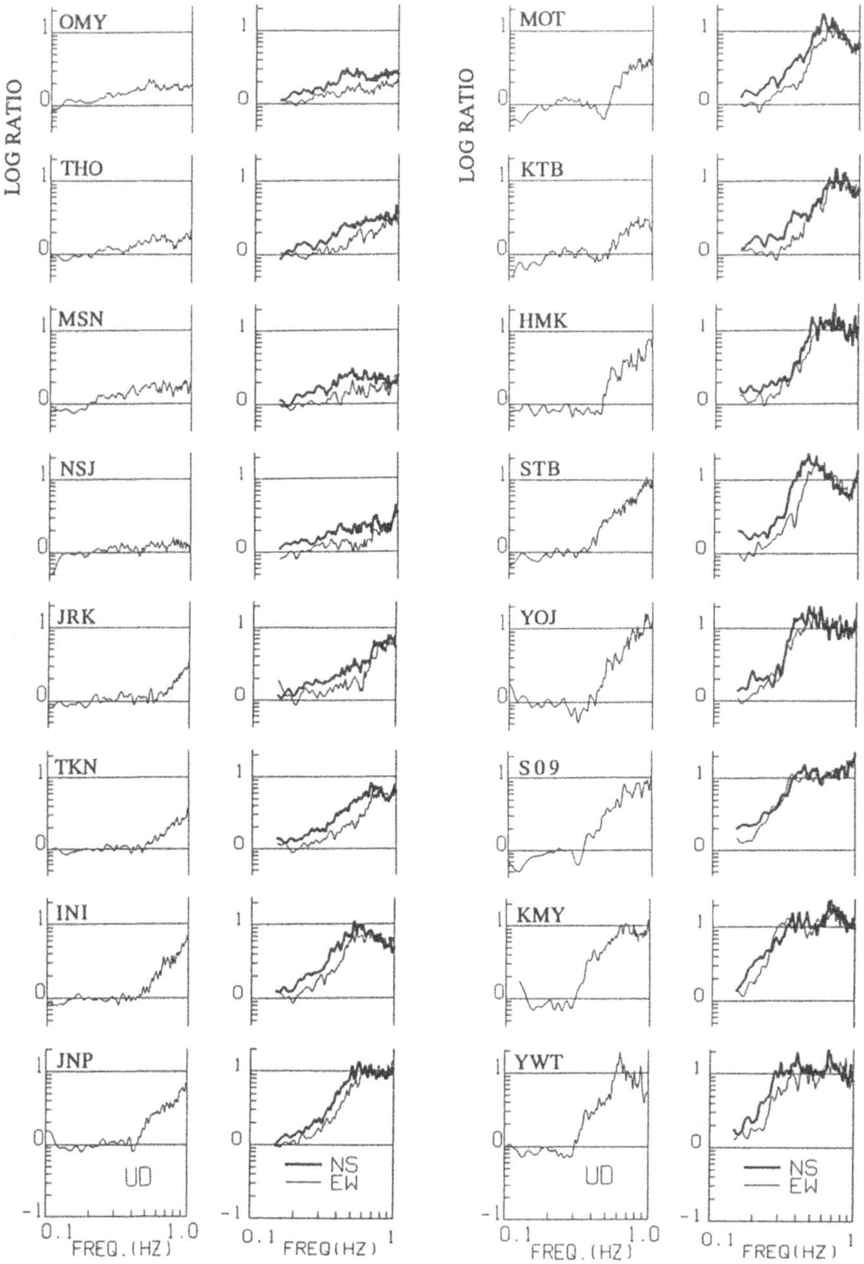

Figure 3. **Spatial variation of amplitude-spectrum ratio (soil/rock) along a NS line in the Kyoto basin, the top left panel being for the northernmost site and the bottom right for the southernmost site. Note the systematic increase in amplitude of the ratios and the continuous shift of frequency range having large amplification toward low frequency with the change of site location from north to south.**

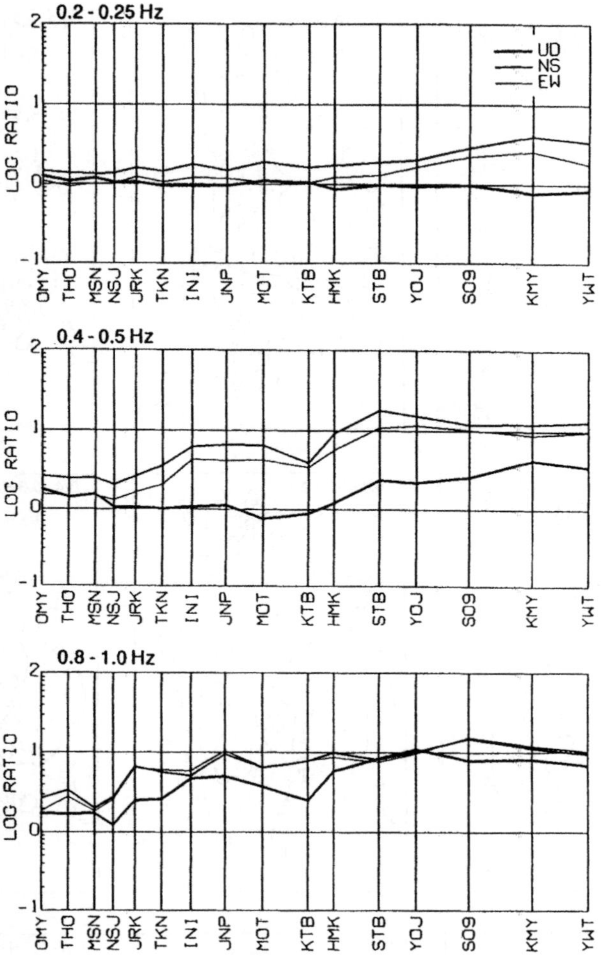

Figure 4. Spatial variation of amplification factors in the various frequency bands separated by one octave, along the same NS line as Fig. 3, showing step-like increase in amplification for both vertical and horizontal components. Note the similarity in the site-dependency of amplification between the 0.4 - 0.5 Hz vertical component and the 0.2 - 0.25 Hz horizontal component, and between the 0.8 - 1.0 Hz vertical component and the 0.4 - 0.5 Hz horizontal component.

of spectral ratios for vertical component resemble those for horizontal ones, if we shift the curves of the horizontal components toward higher frequency by about one octave. It is also noteworthy that amplifications for the horizontal component parallel to the long axis of the basin (NS) are larger than those normal to the axis (EW) and that the frequency range in which the difference in amplification occurs depends on the site location.

Figure 4 shows the spatial distribution of amplification factors along a north-south line in the various frequency bands separated by one octave. As seen from this figure, the amplifications increase stepwise from north to south (from left to right in the figure).

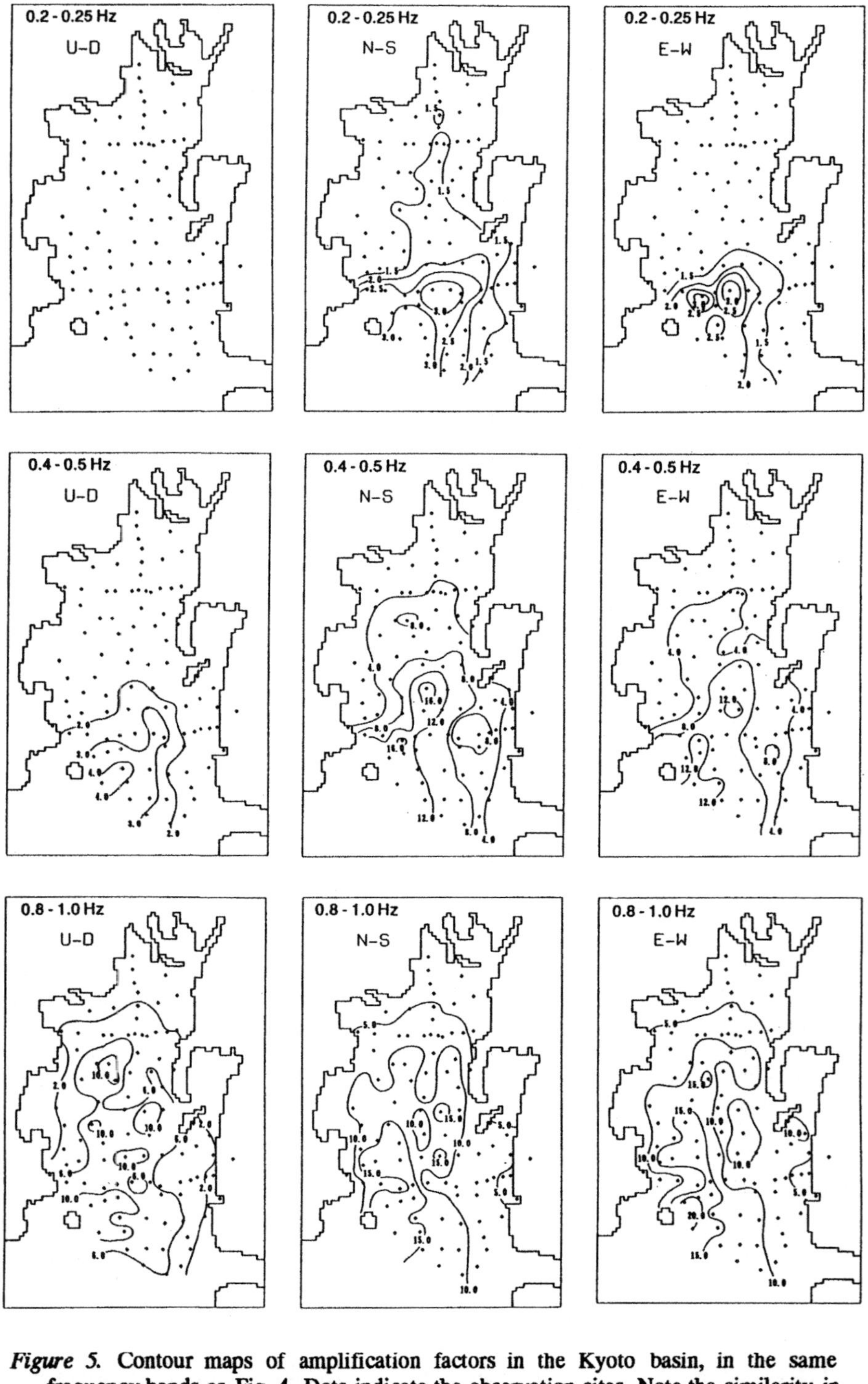

Figure 5. Contour maps of amplification factors in the Kyoto basin, in the same frequency bands as Fig. 4. Dots indicate the observation sites. Note the similarity in the pattern of spatial distribution of amplification factors between vertical and horizontal components with one-octave shift in frequency.

The amplification for horizontal components is larger than that for vertical one, except in 0.8-1.0 Hz band in the southern part of the basin. The one-octave shift of frequency characteristics between the vertical and horizontal components observed in Fig. 3 appears in such a way that the spatial distribution of vertical amplifications in a certain frequency band is similar to that of horizontal component in the frequency band lower by one octave (i.e. UD in 0.4-0.5 Hz to NS and EW in 0.2-0.25 Hz and UD in 0.8-1.0 Hz to NS and EW in 0.4-0.5 Hz). It is also noted that the amplification of NS component is larger than that of EW component in the frequency range lower than 0.5 Hz.

Figure 5 shows the contour maps of amplification factors in the various frequency bands. Here again, the systematic spatial variation in amplification is observed very clearly: namely, increase of amplifications from north to south and from the marginal parts to the center of the basin; the larger amplifications of horizontal components than those of vertical component; the similarity of contours for the vertical component in a certain frequency band to those for horizontal components in the one-octave lower band; and the larger amplification of NS component than EW one.

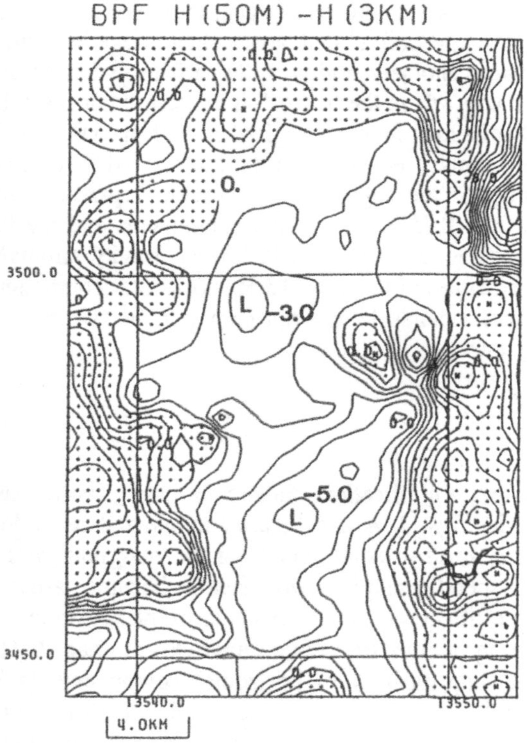

Figure 6. Band-pass filtered Bouguer gravity anomaly in the Kyoto basin, the same area as Fig.5. Contour interval is 1 m Gal. White and dotted regions show negative and positive anomaly, respectively. The lowest anomaly (-5.0 m Gal) is located in the southern part of the basin, and the relatively low anomaly (-3.0 m Gal), in the northern part. Note that general feature of the gravity anomaly has a close correlation with the spatial distribution of amplification factors shown in Fig. 5, especially those for the 0.4 - 0.5 Hz horizontal components.

5. Comparison with Bandpass-Filtered Bouguer Gravity Anomaly

To discuss the spatial distribution of amplification factors in relation to the subsurface structure, we analyzed the gravity data. The data set was provided by Japan Geological Survey. The analyzing procedure is as follows: (1) calculation of the Bouguer gravity anomaly as mesh data with 500 m grid interval; (2) separation of the residual component from the regional one with combined use of upward-continuation filters (Gupta and Ramani, 1980). The original Bouguer gravity anomaly was calculated with a reduction density of 2.3 gr/cm^3, focusing on the shallow structure. The residual component of gravity anomaly was obtained as the difference between the anomalies continued to the heights of 50 m and 3,000 m. This means that the gravity anomaly was bandpass filtered so as to suppress both the regional component with the horizontal wavelength longer than about 10 km and the so-called noise with extremely short wavelength. According to Komazawa(1984), the bandpass-filtered anomaly thus obtained are due to the density distribution down to the depth of 2 to 3 km. In this depth range the difference in density is largest between the soil sediments and the bedrock, therefore it is highly probable that the bandpass-filtered anomaly reflects the configuration of bedrock.

Figure 6 shows the bandpass-filtered Bouguer gravity anomaly. The contour intervals are 1.0 m Gal. The white and dotted areas show negative and positive anomaly, respectively. The lowest anomaly region up to -5.0 m Gal is located in the center of southern part of the basin, where the bedrock is considered deepest. In the northern part, there is also a relatively low anomaly region of -3.0 m Gal. It is interesting that the general trend of the anomaly is similar to the spatial distribution of the amplification factors shown in Fig. 5: the region in the southern part of the basin with large amplification of the 0.2 - 0.25 Hz horizontal components and the 0.4 - 0.5 Hz vertical component corresponds to the region with lowest gravity anomaly; the region with relatively large amplification of the 0.4 - 0.5 Hz horizontal components and the 0.8 - 1.0 Hz vertical component extends to the northern part of the basin where the gravity anomaly is relatively low.

6. Discussion

The characteristics of amplification factors in the Kyoto basin inferred from microseisms, shown in Figs. 3, 4 and 5, are summarized as follows: (1) as the site location changes from north to south and from the marginal parts of the basin to its central part, the site-specific amplification factors become larger and the frequency range having large amplification extends to lower frequency, though the values of peak amplification appearing in the frequency range lower than 1 Hz do not increase significantly; (2) the amplification of horizontal components is larger than that of vertical component in the low frequency range; (3) the amplification factors for vertical component resemble those for horizontal components, if the horizontal amplification factors are shifted toward higher frequency by about one octave; and (4) the amplification of horizontal component parallel to the long axis of the basin (NS) is larger than that normal to the axis (EW).

The bandpass filtered Bouguer gravity anomaly (Fig. 6) shows the irregular distribution of low anomaly region and its contour map resembles the contour maps of amplification factors for microseisms. As the anomaly is considered to reflect the configuration of bedrock under the basin, we will discuss the characteristics of (1), (2),

(3) and (4) in relation to the configuration of bedrock and the shape of the basin.

The characteristic frequency of soil ground over bedrock is often interpreted from the view point of multiple reflections of S wave. At first glance, such an interpretation seems not to be applicable to the vibrational characteristics inferred from microseisms which are considered to be composed mainly of Rayleigh surface waves. However, it is known that in the case of large impedance ratio between soil sediments and bedrock, the excitation of surface wave can be approximated by 1-D resonance of S wave (Allam and Shima, 1967). Therefore, as a first approximation, we might relate the spatial distribution of amplification factors, (1), to the depth of bedrock: the bedrock seems to become deeper from north to south in the basin, not gradually but stepwise, as supposed from the distributions of amplification factors shown in Fig. 4.

The step-like increase in the depth to bedrock from north to south is supported by the distribution of the gravity anomaly shown in Fig. 6. The relatively low anomaly in the northern part of the basin corresponds to the observation sites from TKN to KTB (see Fig. 4), where the amplification is relatively large in the 0.4 - 0.5 Hz band, and the still larger amplification is found from STB to YWT, where the gravity anomaly becomes lowest in the Kyoto basin. Geographically, the step-like increase of the depth to bedrock in the basin corresponds with the step-like increase of the east-west width of the basin.

The characteristic, (2), is in general accord with the calculated result of Drake (1980), who demonstrated that, for long-period surface-waves transmitted from rock site to alluvial valley, the horizontal surface amplifications are considerably larger than the vertical one. Large horizontal amplifications of microseisms are also observed on the bay-mud in the San Francisco Bay area (Akamatsu et al.,1991, 1992a).

The characteristics, (3), is very interesting but difficult to interpret. If the argument of Allam and Shima (1967) can be generalized, one possibility is that the amplification of horizontal components is governed by the resonance of SH waves, while that of vertical component by the resonance of P waves. This interpretation, of course, is to be discussed theoretically, but it is remarkable that Vp/Vs in the soil sediments is about two on average, which bring the one-octave difference in resonant frequency between P wave and S wave.

It is obvious that the directional characteristics, (4), can hardly be explained by a horizontally layered structural model, because microseisms observed in the Kyoto basin are generated in wide coastal regions and are regarded as superposition of Rayleigh waves propagating in various directions (see, for example, Okano, 1961a, b): the directional characteristics cannot be attributed to any particular direction of propagation. One of possible reasons for such characteristics is azimuthal anisotropy of sediments, that is, the difference in physical properties between NS and EW directions. Although the differences in velocity between the vertical and horizontal directions was found by Kitsunezaki (1971) in the southern part of the basin, we have no evidence of azimuthal anisotropy of sediments which could explain the directional characteristics.

Another likely reason for the directional characteristics, (4), is the effect of irregular configuration of bedrock on wave propagation. As demonstrated by a numerical simulation (Khair et al., 1991), Rayleigh waves obliquely incident to a cylindrical valley exhibit selective amplification: the vibration in the longitudinal direction is amplified much more than that in the transverse direction. This may be taken as a basis for the explanation of the directional characteristics, (4), because the shape of the Kyoto basin is long in the north-south direction. In this connection, it should be noted that the directional characteristics almost disappear in the frequency range higher than about 0.4 Hz for the sites, such as YOJ, S09 and KMY, located around the center of the southern part of the basin where the east-west width of the basin increase remarkably (see Figs. 1 and 3). This may suggest that the basin here is wide enough to suppress the selective amplification.

7. Conclusive Remarks

Taking account of the necessity for the low-frequency microzoning in urban areas, a comparative analysis of microseisms and gravity data was carried out in the Kyoto basin, southwest Japan. The stability of spectral ratios (soil site/rock site) of microseisms was examined through repeated simultaneous observation at soil sites and rock sites. Then spatial variation of the spectral ratios in the whole basin was mapped and compared with the characteristic feature of bandpass-filtered Bouguer gravity anomaly. The main results are summarized as follows:

(1) Spectral features of microseisms observed both at soil sites and at rock sites are remarkably affected by weather condition, but averaged spectral ratios, soil site/rock site, exhibit time-invariant and site-specific features, which can be used as amplification factors at the soil sites.

(2) Reliable amplification factors can be estimated through repeated observations with a suitable recording length.

(3) Amplification factors for horizontal components are larger than those for vertical component in the low frequency range.

(4) Frequency characteristics of the amplification factors for vertical component exhibit a similar shape to those for horizontal components with one-octave shift in frequency.

(5) Horizontal component of vibration oriented parallel to the long axis of the basin is amplified more than that oriented normal to the axis, though this difference is smaller than the difference between the vertical and horizontal components.

(6) Spatial distributions of amplification factors are similar to the general trend of the bandpass-filtered Bouguer gravity anomaly.

These features of amplifications are interpreted from the view point of Rayleigh surface waves incident to the basin with 3-D structure of bedrock, though it is suggested that some of them may be explained approximately by 1-D resonance of S and P waves in the soil sediments.

In conclusion, it is emphasized that comparative analysis of microseisms and gravity data are very useful for microzoning of an urban area in the case where detailed information of subsurface structure is not available; especially when bandpass filtered, the Bouguer gravity anomaly provides some images of bedrock configuration responsible for ground vibration characteristics inferred from microseisms.

Acknowledgement

The authors express their sincere thanks to Dr. J. Zahradnik, Charles University, Praha and an anonymous reviewer for reviewing the manuscript and the valuable comments.

References

Akamatsu, J. (1984) Seismic amplification by soil deposits inferred from vibrational characteristics of microseisms, Bull. Disas. Prev. Res. Inst., Kyoto Univ. 34, 105-127.

Akamatsu, J., Fujita, M. and Kameda, H. (1991) Long-period (1-10 s) microtremor measurements in the area affected by the 1989 Loma Prieta earthquake, Proc. 4th Int'l Conf..Seismic Zonation, Stanford, Calif. II, 393-400.

Akamatsu, J., Fujita, M., Kameda,H., Celebi, M. and Borcherdt, R. D. (1992a) Assessment of the effects of local geology using long-period microtremors and the 1989 Loma Prieta Earthquake motions, U. S. Geological Survey Open-File Report 92-214.

Akamatsu, J., Fujita, M. and Nishimura, K. (1992b) Vibrational characteristics of microseisms and their applicability to microzoning in a sedimentary basin, J. Phys. Earth 40,137-150.

Akamatsu, J., Furuzawa, T. and Irikura, K. (1975) On natures of S waves from local small earthquakes observed at Amagase Crustal Movement Observatory, Annuals Disas. Prev. Res. Inst., Kyoto Univ. 18B, 11-21, (in Japanese).

Allam, A. and Shima, E. (1967) An investigation into the nature of microtremors, Bull. Earthq. Res. Inst., Univ. of Tokyo 45, 43-60.

Darbyshire, J. (1990) Analysis of twenty microseism storms during the winter of 1987-1988 and comparison with wave hindcasts, Phys. Earth Planet. Inter. 63, 181-195.

Drake, L. A. (1980) Love and Rayleigh waves in an irregular soil layer, Bull. Seism. Soc. Am. 70, 571-582.

Gupta, V. K. and Ramani, N. (1980) Some aspects of regional-residual separation of gravity anomalies in a Precambrian terrain, Geophysics 45, 1412-1426.

Khair, K. R., S. K. Datta, S. K. and Shah, A. H. (1991) Amplification of obliquely incident seismic waves by cylindrical valleys Part II. Incident SH and Rayleigh waves, Bull. Seism. Soc. Am. 81, 343-357.

Kitsunezaki, C., Goto, N. and Iwasaki, Y. (1971) Underground structure of the southern part of the Kyoto Basin obtained from seismic explosion and some related problems of earthquake engineering, Annuals Disas. Prev. Res. Inst., Kyoto Univ. 14-B, 203-215 (In Japanese).

Komazawa, M. (1984) On the quantitative gravimetric analysis in the Hokuroku district, Geophy. Explor. 37, 123-134 (in Japanese with Eng. abstract).

Lermo, J. and Chavez-Garcia, F. J. (1994) Are microtremors useful in site response evaluation?, Bull. Seism. Soc. Am. 84, 1350-1364.

Morikawa, H., Akamatsu, J., Nishimura, K., Onoue, K. and Kameda, H. (1994) Conditional stochastic simulation for an interpretation of microseisms, Bull. Seism. Soc. Am. (submitted).

Okano, K. (1961a) Observational study on microseisms (Part 1), Bull. Disas. Prev. Res. Inst., Kyoto Univ. 44, 2-22.

Okano, K.(1961b) Observational study on microseisms (Part 2), Bull. Disas. Prev. Res. Inst., Kyoto Univ. 47, 2-15.

Okano, K. (1963) Simultaneous observation of microseismic waves at two stations, Geophys. Pap. dedicated to Prof. K. Sassa, 387-391.

Seo, K. (1992) A joint work for measurements of microtremors in the Ashigara valley, Int'l Symp. on the Effect of Surface Geology on Seismic Motion, ESG1992, II, 43-52.

PECULIARITIES OF GROUND SEISMIC REACTION IN KISHINEV, MOLDOVA

A.ROMAN, V.ALCAZ
Institute of Geophysics and Geology, Moldavian Academy of Science
Academiei str.3, Kishinev, 277028, Moldova

1. Introduction

Systematic instrumental observations on the grounds with various engineering - geological properties were started in Moldova in the 1970-s. Three subsequent strong intermediate-depth Carpathian earthquakes: 4.03.1977 (M=7.2), 30.08.1986 (M=7.0), 30.05.1990 (M=6.9) have also provided rich macroseismic material.

Joint analysis of instrumental and macroseismic data has revealed some peculiarities in the manifestation of different local geologico-engineering factors during the above mentioned earthquakes. Hence, it has been proposed to re-consider the existing practice of evaluation the seismic hazard in respect to this territory.

2. Brief characteristics of Kishinew's engineering-geological structure

Kishinev is situated at 200 km from the Vrancea (the Carpathians) epicentral zone. The surface deposits are represented by alternating loams, sand loams, sands, clays and silts. These sandy-clayish loose deposits overlay on rocks - Sarmathian limestones (bedrocks for Kishinev zone).

A special distinctive feature of the grounds structure on this territory is the stratification and substantial variations of the thickness of sandy-clayish deposits within a small area: from 0 in the flood area of the city to 200 m on the watersheds.

The ground waters have been located at different depths on the territory of the city. On watersheds and high parts of slopes their depth is 10-15 m or more, while in the low sites of the city is 0-5 m.

Propagation velocity of S-waves in covering loose deposits is 200-300 m/s and hard rock (limestones) it is 1000-1300 m/s.

3. Instrumental date. Working methods and results

78 recordings of Carpathian earthquakes, whose magnitude was within the M = 2.8-7.2 interval and the focal depth H = 100-150 km, were used as factual material. The above mentioned earthquakes were registered as follows: the strong ones (M > 6.0) - by 6

V. Schenk (ed.), Earthquake Hazard and Risk, 281–287.

seismometric channels operating in the waiting regime, and the weak ones- by 6 portable seismic stations in permanent operation. The range of frequencies, registered without distortions, was in the 0.6 - 30 Hz interval.

The stations were arranged as follows: one control station was placed on bedrocks, the others on sites that differed from the geologico-engineering point of view. Thus, observation was exercised over sites with different so called seismic category of grounds (depending of physico-mechanical properties of the upper 10m grounds; SNiP II-7-81, 1982), with different level of ground waters as well as with different thickness of surface (quaternary) deposits and loose over-limestone deposits on the whole.

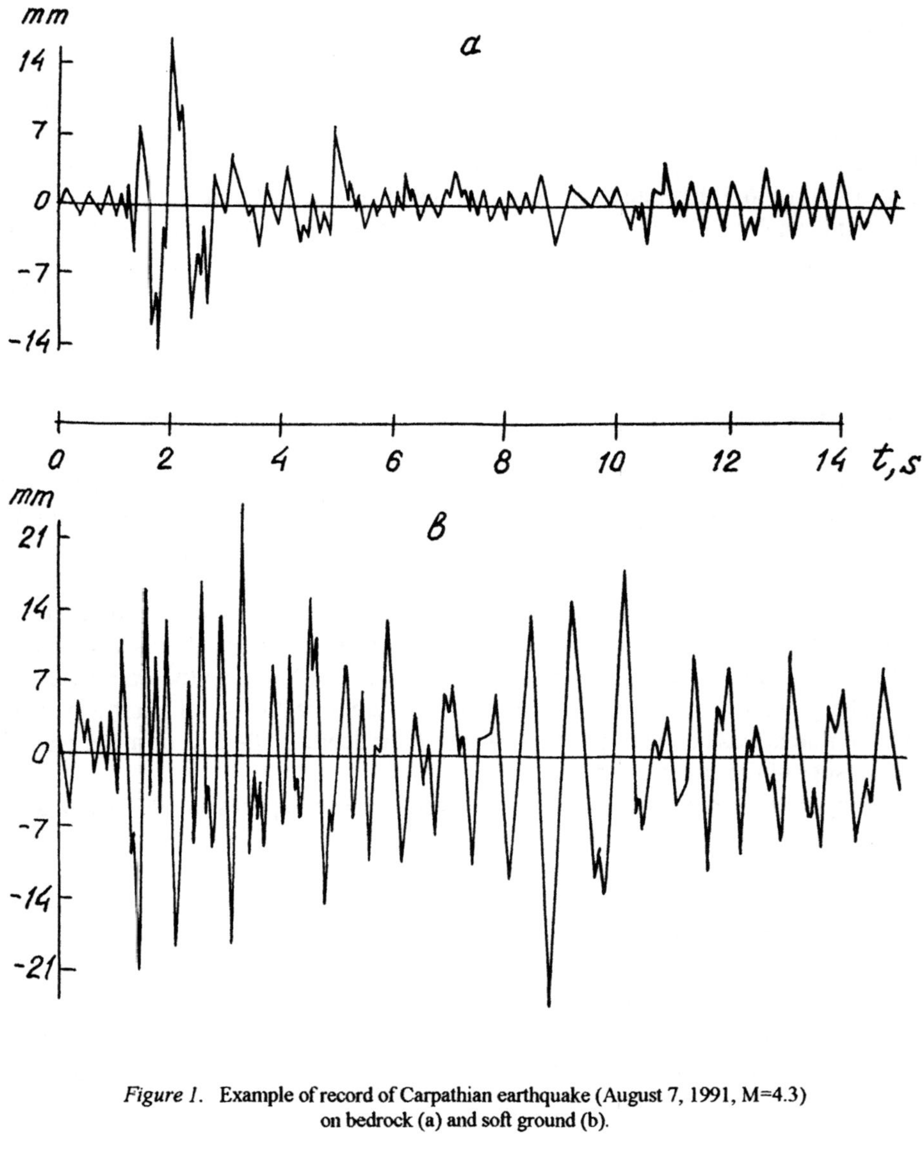

Figure 1. Example of record of Carpathian earthquake (August 7, 1991, M=4.3) on bedrock (a) and soft ground (b).

The example of an earthquake registered on hard and soft grounds is shown in Fig.1.

The obtained data processing consisted in the estimation of points increment (correction of seismic intensity) basing upon the calculation of ratio of average amplitudes and Fourier amplitude spectra (for the range 0.6 to 10 Hz) at a site to the average amplitudes (Fourier spectra) of the reference station (Djurik *et al.*, 1988).

Having the values of P and S seismic waves velocities which had been measured by seismic well logging, the amplitude-frequency characteristics on the sites of seismic stations' location (the ground amplification factors) were calculated simultaneously to facilitate the interpretation.

The analysis of obtained points increment values in comparison with the specific section structure in each registration point made it possible to establish that the intensity and spectral composition of oscillations on the surface is determined mainly by the thickness of cover deposits. On the sites of loose deposits with 80 m or more thickness the dominating factor, which determines the spectral composition of oscillations, is the capability of the ground to considerably intensify the seismic waves. Thus, the comparison of fixed Fourier spectra (displacement) of 4.03.1977 and 30.08.1986 strong earthquake motions on the site where the bedrocks are at about 180 m depth shows, that in spite of the difference of initial parameters of these two earthquakes, the spectral composition of oscillations registered in this place is rather close (Fig. 2).

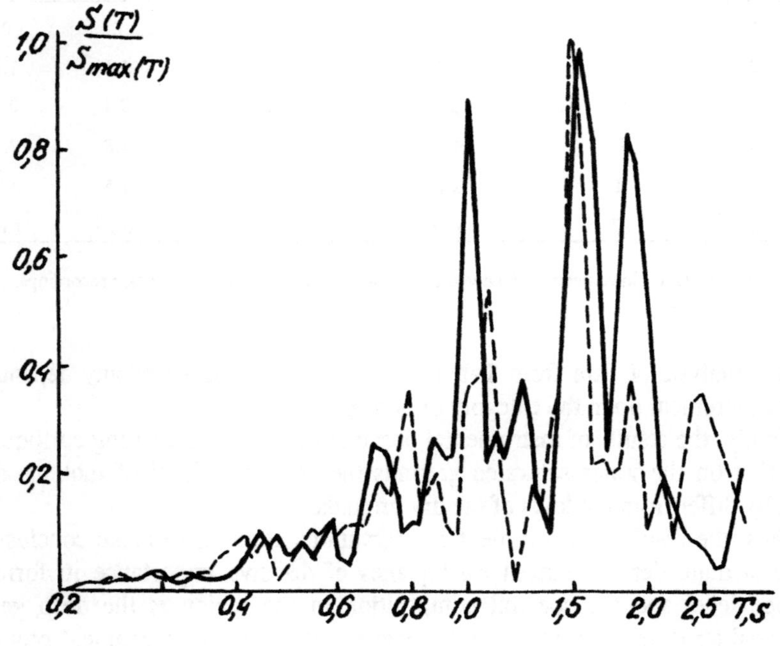

Figure 2. Comparison of Fourier spectra of strong Carpathian earthquakes: 4.03.1977 (dashed line) and 30.08.1986 (solid line).

The increment of intensity on the sites composed of such kind of grounds (of great thickness) in respect to hard rocks range from 1.5 to 2.5 points MSK-64 scale. The amplification factor range 2 to 14.

The results of obtained recordings processing has also revealed that both during weak and strong earthquakes, the intensity increment is absent or negligibly small on sites with ground water close to the surface; the maximum increment not exceeding 0.5 points.

Along with the water content we have also analyzed thoroughly the values of oscillation intensity on grounds of various seismic categories. The forecasted (according to SNiP II-7-81, 1982) increased intensity on III category grounds has not been confirmed experimentally.

An example for confirming the above mentioned could serve the data processing results of May 30, 1990 strong earthquake (Tab. 1). The station N=1, located on bedrock, has been taken as the basic station in respect to which the point increment was calculated. One can see, that the difference in the intensity values at various sites of the city reach 2 points and more, first of all due to essential amplification of the oscillations as a result of resonance effects in the grounds considerable of thickness.

TABLE 1. Values of intensity increment according to instrumental data during the May 30, 1990 earthquake

Station N	Ground categ. according to SNiP II-7-81	Ground water lev. m	Depth to bedrock m	Points incr. ΔI, points $\Delta I_{N\text{-}S}$	$\Delta I_{E\text{-}W}$
1	I	-	0	0	0
2	III	0-2	20	1.1	1.2
3	II	> 5	60	0.4	0.5
4	III	> 15	120	1.8	2.4
5	II	> 10	170	1.5	1.7
6	III	> 10	60	0.7*	1.0*

*) The values have been calculated on the basis of May 31, 1990 aftershock's recordings.

The analysis of data from Table 1 confirms the absence of any dependence of intensity increment upon the category of non-rocky grounds.

Finally, the results of instrumental observations of the last strong earthquake also shows, that on the water saturated grounds the amplitude level of motions does not practically differ from the level of the dry grounds.

Thus, the results of instrumental observations allow us to make conclusion, that for intermediate depth Vrancea earthquakes of decisive importance in forming the amplitude level and the spectral composition in the region is the deep geologico-geophysical structure, and to a smaller degree - the physico-mechanical properties of the upper grounds. We suppose that the peculiarities of the region are the most probable reasons of this phenomenon: deep earthquake focus, radiating intensive long-

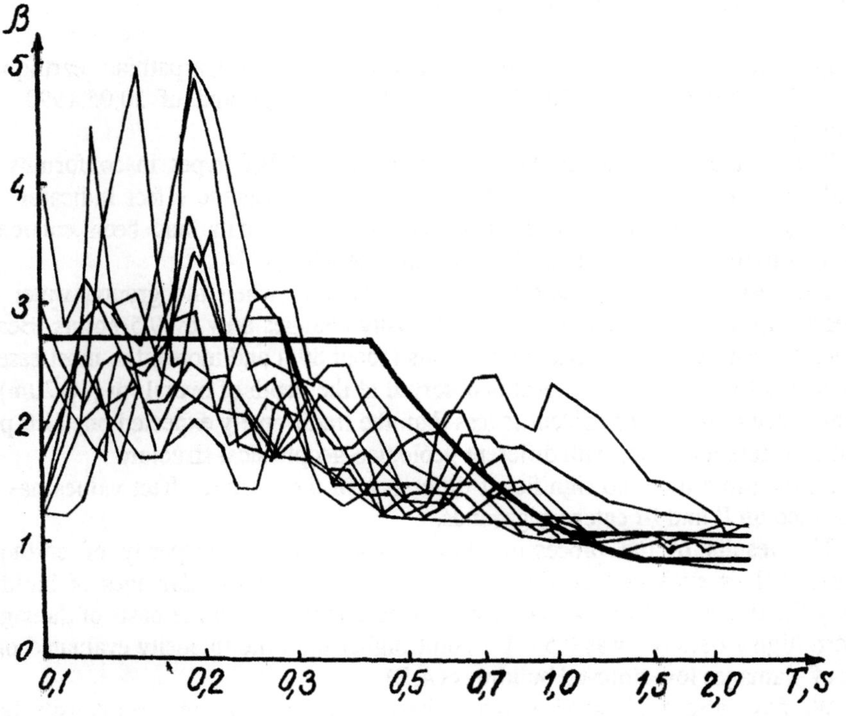

Figure 3. Comparison of normative spectral curve (thick line) and response spectra of the 4.03.1977, 30.08.1986 earthquakes.

period oscillations, for which the upper layers are enough transparent; big thicknesses of loose grounds, when the seismic effect on the surface is to a great extent formed by all the loose thickness on the whole; little difference of physico-mechanical characteristics of the surface grounds.

It is also interesting to compare the real seismic impacts occurring during earthquakes with the existing normative ones (SNiP II-7-81, 1982). For this purpose the response spectra of ground expressed in absolute values of spectral curve of the dynamic coefficient β(T), have been compared with the standard curve (Fig. 3). It should be noted that the real impacts on the ground exceed 1.5-1.8 times those given in the normative document (SNiP II-7-81, 1982) in the high frequency range (0.1-0.3 s). In the 0.3-0.6 s interval the normative curve b overlaps the real impact on the ground. In the range of long periods (over 0.7 s) the real seismic loads again exceed the rated ones.

4. Macroseismic data. Method and results

Use has been made of macroseismic data of 3 strong Carpathian earthquakes: 4.03.1977 - 700 buildings, 30.08.1986 - 2100 buildings and of 30.05.1990 - 660 buildings.

The degrees of buildings damage classified by A,B,C types in conformity with MMSK scale (Djurik *et al.*, 1988) have been used as seismic effect indicator. Joint processing of macroseismic and geologico-engineering data has been carried out according to the "average damage" method of Shebalin (1974).

The results of this processing have shown, that when the ground water level decreases up to 10-12m the earthquake intensity changes only by 0.5 points. Besides, on sites consisting of thick sedimentary soils (about 80m and more), the most essential influence of the ground water level is observed at the depth of over 10m (10-20m). So, the influence of the water content is less than the traditionally expected one (1.0 point) and it is different on sites with different geologico-geophysical structure.

At the same time, no significant difference in the seismic effect values has been established on II and III category grounds.

The results of the processing have shown also the disparity of earthquake intensity values evaluated on the basis of types A, B and C damages of buildings, namely the intensity of Carpathian earthquakes determined on the basis of damages of modern high - rises (C) was 0.5 - 1.0 point higher than the intensity evaluated on the basis of damages low-storeyed buildings (A,B).

We have also established a stable disparity in determining the intensity basing upon instrumental (maximum accelerations) and macroseimic data: the values of registered accelerations produce the intensity which is excessive by 1.0 - 1.5 points MSK-64 scale.

5. Conclusions

It has been established that the availability of rather thick sedimentary soils (80-200m) on the territory of the city leads to the appearance, in the oscillations spectra on the free surface, of sharp peaks of resonance character. The seismic oscillations intensity in these sites is 1.5 - 2.5 higher compared to those on hard rocks. As a result of closeness of grounds resonance periods to the periods of proper oscillations of high-rises, one can also observe the increased level of modern structure damages. Taking into consideration the existing tendency of building high-rises, one should expect that the role of the resonance as one of the factor, determining the seismic hazard of this territory, will increase in the future.

Since it has been established that the decisive significance in forming the seismic effect on the surface rests with the deep geologico-geophysical structure and, to a smaller degree, with the geologico-engineering characteristics of upper grounds, it is expedient, as a first step, to re-consider the role of ground water content and category of ground in forming the seismic effect with account for the obtained real estimations of the influence of these factors in the region (0.5 point).

It is also necessary to correct the spectral curve β(T), first of all in the range of long periods (>0.6 s) and to elaborate a regional instrumental scale for estimating the intensity of the past earthquake.

6. References

Djurik, V.I., Sevastianov, V.V. and Potapov, V.A. (1988) *Soils Conditions and Seismic Danger Estimation*, Nauka, Moscow (in Russian).

Shebalin, N.V. (1974) Process of cataloging, in *Catalogue of Balkan Earthquakes*, Survey of Seismicity of the Balkan Region, UNDP-UNESCO, Skopje.

SNiP II-7-81 (1982) *Building Norms and Rules. Construction in Seismic Areas*, Stroyizdat, Moscow (in Russian).

Advances in Natural and Technological Hazards Research

Series Editor: Prof. Dr. Mohammed I. El-Sabh, *Département d'Océanographie, Université du Québec à Rimouski, 310 Allée des Ursulines, Rimouski, Québec, Canada G5L 3A1*

Publications

1. S. Tinti (ed.): *Tsunamis in the World.* Fifteenth International Tsunami Symposium (1991). 1993 ISBN 0-7923-2316-5
2. J. Nemec, J.M. Nigg and F. Siccardi (eds.): *Prediction and Perception of Natural Hazards.* Symposium Perugia, Italy (1990). 1993 ISBN 0-7923-2355-6
3. M.I. El-Sabh, T.S. Murthy, S. Venkatesh, F. Siccardi and K. Andah (eds.): *Recent Studies in Geophysical Hazards.* 1994 ISBN 0-7923-2972-4
4. Y. Tsuchiya and N. Shuto (eds.): *Tsunami: Progress in Prediction, Disaster Prevention and Warning.* 1995 ISBN 0-7923-3483-3
5. A. Carrara and F. Guzzetti (eds.): *Geographical Information Systems in Assessing Natural Hazards.* 1995 ISBN 0-7923-3502-3
6. V. Schenk (ed.): *Earthquake Hazard and Risk.* 1996 ISBN 0-7923-4008-6

Kluwer Academic Publishers – Dordrecht / Boston / London

GPSR Compliance

The European Union's (EU) General Product Safety Regulation (GPSR) is a set of rules that requires consumer products to be safe and our obligations to ensure this.

If you have any concerns about our products, you can contact us on ProductSafety@springernature.com

In case Publisher is established outside the EU, the EU authorized representative is:

Springer Nature Customer Service Center GmbH
Europaplatz 3
69115 Heidelberg, Germany

Batch number: 10755292

Printed by Printforce, the Netherlands